ELECTRODYNAMICS OF HIGH-TEMPERATURE SUPERCONDUCTORS

LECTURE NOTES IN PHYSICS - Vol. 48

ELECTRODYNAMICS OF HIGH-TEMPERATURE SUPERCONDUCTORS

Alan M Portis

Department of Physics
University of California at Berkeley

World Scientific
Singapore • New Jersey • London • Hong Kong

Published by

World Scientific Publishing Co. Pte. Ltd.
P O Box 128, Farrer Road, Singapore 912805
USA office: Suite 1B, 1060 Main Street, River Edge, NJ 07661
UK office: 57 Shelton Street, Covent Garden, London WC2H 9HE

Library of Congress Cataloging-in-Publication Data
Portis, Alan M., 1926–
Electrodynamics of high temperature superconductors / Alan M. Portis
p. cm.
Includes index.
ISBN 9810212151. -- ISBN 9810212488 (pbk.)
1. High temperature superconductors. 2. Electrodynamics.
I. Title.
QC611.98.H54P67 1992
537.6'23--dc20 92-43560
CIP

First published 1996
First reprint 1998

Printed in Singapore by JCS Office Services & Supplies Pte Ltd

PREFACE

These lecture notes were first prepared for a series of six seminars that were held during Winter 1987-88 at the Institute for Intermediate Energy Physics/ETH Zürich, located at the Paul Scherrer Institute near Villigen, Switzerland. The notes were revised for a one-week inter-term course at the Physics Institute of the University of Zürich in March 1988.

The subject matter of these seminars was suggested by my research interests at the time—muon and microwave studies of high temperature superconductivity. Although much of the background material is not new, going back twenty years or more, what is new is emphasis on the modern high-temperature superconductors discovered by K. A. Müller and J. G. Bednorz, whom it was my pleasure to know while working at IBM Rüschlikon.

I was on sabbatical leave from the Department of Physics of the University of California at Berkeley during this early period and wish to acknowledge my appreciation to those colleagues who arranged for my stay in Switzerland and provided such a supportive and stimulating environment for research, teaching and learning: Dr. Alex Schenck at the Institute for Intermediate Energy Physics/ETH Zürich; Dr. Claude Petitjean at the Paul Scherrer Institute; Professors Walter Kundig and Franz Waldner and Dr. Bruce Patterson at the University of Zürich; Professor Alex Müller and Drs. Georg Bednorz and Keith Blazey at the IBM Zürich Research Laboratory, Rüschlikon.

The notes were revised and considerably expanded for a special topics graduate seminar, offered in the Physics Department of the University of California at Berkeley during Fall 1990. I am grateful to those graduate research students who participated in the seminar and who gave me stimulation and encouragement. I want particularly to mention Chandu Karadi, Harry Lam, Andrew Miklich, Beth Parks and Tim Shaw.

Drs. Paul Berdahl and Nathan Newman and visiting research physicist Antonello Andreone made many valuable suggestions during the course of a workshop offered at the Lawrence Berkeley Laboratory in Spring 1992.

These chapters took their present form for a colloquium series in Summer 1992 at the University of Wuppertal where, through the Minister for Science and Research of the State of Nordhein-Westfalen, I held an appointment as University Professor. I thank Professor Helmut Piel for making the necessary arrangments and for his unbounded personal support and hospitality. I want particularly to mention Susannah Orbach, Matthias Hein and Martin Strupp, who participated actively in the colloquium series and made a number of very helpful suggestions.

My colleague and collaborator D. W. Cooke has generously allowed me to include material on which we worked together. I am also grateful to J. R. Clem, John Clarke, Vladimir Kresin, Jürgen Halbritter, P. L. Richards and D. J. Scalapino for instructive discussions of the microwave properties of superconductors.

The preparation of this work in its final form was supported in part by the Director, Office of Energy Research, Office of Basic Energy Sciences, Materials Sciences Division, of the U.S. Department of Energy under Contract No. DE-AC03-76SF00098.

Alan M. Portis

Berkeley, California

September 1992

CONTENTS

1
HIGH TEMPERATURE SUPERCONDUCTORS

1.1. Background

The solid state physics and materials science communities were entirely unprepared for the Sept. 1986 publication by J. G. Bednorz and K. A. Müller[1] of the discovery of percolative superconductivity near 30 K in a polycrystalline sample in the La-Ba-Cu-O system.[2]

Following the 1911 discovery of superconductivity in mercury at 4.1 K by H. Kamerlingh Onnes[3] there had been steady progress in finding materials with increasing transition temperatures up to Nb_3Ge with $T_c = 23.3$ K, reported in 1973 and 1974 by Gavaler[4] and Testardi *et al.*[5] Although Bernd Matthias and his colleagues[6,7] had achieved early success with additional compounds in the cubic A-15 structure, nothing had been found to surpass Nb_3Ge.

1.1.1. Search and discovery

Sleight *et al.*[8] had found superconductivity at 13 K in the mixed valence oxide $Ba(Pb_{1-x}Bi_x)O_3$, which crystallizes in the perovskite structure. This work undoubtedly caught the attention of Bednorz and Müller, who had devoted much of their scientific careers to research on structural and ferroelectric phase transitions in oxide insulators, many of which

---

[1] J. G. Bednorz and K. A. Müller, "Possible high T_c superconductivity in the Ba-La-Cu-O system," *Z. Phys. B* **64**, 189 (1986).

[2] T. H. Geballe and J. K. Hulm, *Science* **239**, 367 (1988).

[3] H. K. Onnes, *Commun. Phys. Lab. Univ. Leiden* **120b**, 3 (1911).

[4] J. R. Gavaler, *Appl. Phys. Lett.* **23**, 480 (1973).

[5] L. R. Testardi, J. H. Wernick and W. A. Royer, *Solid. State Commun.* **15**, 1 (1974).

[6] B. T. Matthias, J. K. Hulm and E. J. Kunzler, "The road to superconducting materials," *Physics Today* **34**, 34 (January 1981).

[7] A. M. Clogston, T. H. Geballe and J. K. Hulm, "Bernd T. Matthias," *Physics Today* **34**, 84 (January 1981).

[8] A. W. Sleight, J. L. Gillson and P. E. Bierstedt, *Solid State Commun.* **17**, 27 (1975).

were perovskites.[9] Although Bednorz had participated in a study of Nb-doped $SrTiO_3$[10] and Müller had worked on percolative superconductivity in granular aluminum,[11,12] neither was known to be deeply engaged in research on superconductivity. What had stimulated Müller's interest was the suggestion[13] by H. Thomas at the Erice International School in Summer 1983 that the Jahn-Teller effect[14] might lead to polaron-induced high-temperature superconductivity in transition-metal oxides.[15]

In late Summer 1983 Müller engaged Bednorz[16] in a study of mixed perovskites of composition $La(Ni_{1-x}Al_x)O_3$. Although these materials gave some evidence of a Jahn-Teller effect, there was no indication of a drop in resistivity that might signal a superconducting transition. After two years of part-time effort with no indication of superconductivity, it was decided to shift to copper oxides of mixed valence. Raveau's group at the University of Caen had been actively studying the crystal chemistry of such materials for possible use in catalysis.[17] The mixed perovskite $La_4BaCu_5O_{5(3-y)}$ appeared promising and Bednorz initiated a program of preparation of the system $(La_{5-x}Ba_x)Cu_5O_{5(3-y)}$. Whereas Michel *et al.*[12] had sintered mixed oxides at 1000 °C to achieve solid state reaction, Bednorz was accustomed to precipitate the more finely divided oxalates from solution[18] and was able to sinter at the lower temperature of 900 °C.

A sharp drop in the resistivity of the compound with nominal Ba fraction $x = 0.75$ was observed below 30 K, suggesting the presence of a superconducting phase transition and leading Bednorz and Müller to submit a manuscript[1] to *Zeitschrift für Physik* with the cautious title "Possible High T_c Superconductivity in the Ba-La-Cu-O System."

1.1.2. Confirmation

Bednorz and Müller did not have long to wait for confirmation following publication of their results in *Zeitschrift für Physik* . On November 28, 1986 the Japanese newspaper *Asahi Shinbun*, whose *International Satellite Edition* is received in Zürich, reported that Professor S. Tanaka at the University of Tokyo had observed bulk diamagnetism around 30 K in the Ba-La-Cu-O system, confirming the presence of superconductivity in this system. Further confirmation came at the Fall 1986 meeting of the Materials Research Society where C. W. Chu announced that his group at the University of Houston had reproduced the results of Bednorz and Müller. At the same meeting Dr. K. Kitazawa presented an informal report of the activities of Tanaka's group, of which he was a member.[19,20]

1.1.3. Extension

---

[9] W. H. H. Gränicher, "K. Alex Müller and J. Georg Bednorz as graduate students at ETH Zürich," *Ferroelectrics* **89**, iii (1989).

[10] G. Binnig, A. Baratoff, H. E. Hoenig and J. G. Bednorz, *Phys. Rev. Lett.* **45**, 1352 (1980).

[11] K. A. Müller, M. Pomerantz, C. M. Knoedler and D. Abraham, *Phys. Rev. Lett.* **45**, 832 (1980).

[12] M. Pomerantz and K. A. Müller, *Physica* **107B**, 325 (1981).

[13] K.-H.. Höck, H. Nickisch and H. Thomas, *Helv. Phys. Acta* **56**, 237 (1983).

[14] C. Kittel, *Introduction to Solid State Physics, Sixth Edition* (Wiley, New York, 1986) p. 409.

[15] K. A. Müller and J. G. Bednorz, *Science* **237**, 1133 (1987).

[16] J. G. Bednorz and K. A. Müller, Nobel Lecture, *Rev. Mod. Phys.* **60**, 585 (1988).

[17] C. Michel, L. Er-Rakho and B. Raveau, *Mater. Res. Bull.* **20**, 667 (1985).

[18] J. G. Bednorz, K. A. Müller, H. Arend and H. Gränicher, *Mater. Res. Bull.* **18**, 181 (1983).

[19] H. S. Takagi, S. Uchida, K. Kitazawa and S. Tanaka, *Jpn. J. Appl. Phys.* **26**, L123 (1987).

[20] S. Uchida, H. Takagi, K. Kitazawa and S. Tanaka, *Jpn. J. Appl. Phys.* **26**, L151 (1987).

By placing their sample of La-Ba-Cu-O under hydrostatic pressure, C. W. Chu *et al.*[21,22] were able to increase the onset temperature from 35 K to over 50 K. Bednorz, Müller and Takashige performed studies[23] in which La was partially replaced by Sr or Ca rather than by Ba. With Sr substitution the transition could be raised to around 40 K at ambient pressure while with Ca the transition temperature could not be raised above 22 K.

Researchers were startled to find that replacing La by the smaller Y raises the onset temperature above the temperature of liquid N_2 to 92 K.[24,25] Superconductivity above 100 K was obtained in the Bi-Sr-Ca-Cu-O system.[26] More recently, Sheng and Herman[27,28] discovered superconductivity near 120 K in the Tl-Ba-Ca-Cu-O system and Parkin *et al.* in a closely related system have obtained a transition to superconductivity at the current record temperature of 125 K.[29]

J. Georg Bednorz and K. Alex Müller were awarded the 1987 Nobel Prize in Physics on 8 December 1987, less than two years after their discovery of high-temperature superconductivity.

1.2. Structure

Much of the systematic work of Bednorz and Müller and their colleagues at the IBM Rüschlikon Laboratory anticipated approaches that would continue to be essential to the analysis of the high-temperature superconductors and to the study of their physical properties. In this section we discuss three approaches that have been found particularly useful: structure analysis, crystal chemistry and the observation of microstructure.

1.2.1. Systems

Structure determination has played a central role in characterizing the high-temperature superconductors. Materials prepared by solid state reaction are commonly a mixture of phases, only some of which become superconducting. Crystallographic studies are crucial to understanding the mechanisms of superconductivity and to the development of improved superconducting materials.[30]

Bednorz and Müller[1] obtained x-ray powder diffractograms of their sintered material and were able to distinguish three phases:

i. The dominant phase was a layer-type perovskite that appeared to have the K_2NiF_4 structure with lattice constants a = 3.79 Å and c = 13.21 Å. This phase turned out to be the superconductor $La_{2-x}Ba_xCuO_4$ and its structure was confirmed by Tyagi *et al.*[16]

ii. A second phase appeared to be the oxygen-deficient perovskite $(La_{1-x}Ba_x)CuO_{3-y}$ with the $LaNiO_3$ structure. This phase is not superconducting.

---

[21] C. W. Chu, P. H. Hor, R. L. Meng, L. Gao and Z. J. Huang, *Science* **235**, 567 (1987).

[22] C. W. Chu, P. H. Hor, R. L. Meng, L. Gao, Z. J. Huang and Y. Q. Wang, *Phys. Rev. Lett.* **58**, 405 (1987).

[23] J. G. Bednorz, K. A. Müller and M. Takashige, *Science* **236**, 73 (1987).

[24] M. K. Wu, J. R. Ashburn, C. J. Torng, P. H. Hor, R. L. Meng, L. Goa, Z. J. Huang, Y. Q. Wang and C. W. Chu, *Phys. Rev. Lett.* **58**, 908 (1987).

[25] P. H. Hor, L. Gao, R. L. Meng, Z. J. Huang, Y. Q. Wang, K. Forster, J. Vassilious, C. W. Chu, M. K. Wu, J. R. Ashburn and C. J. Torng, *Phys. Rev. Lett.* **58**, 911 (1987).

[26] H. Maeda, Y. Tanaka, M. Fukutomi, and T. Asano, *Jpn. J. Appl. Phys.* **27**, L209 (1988).

[27] Z. Z. Sheng and A. M. Hermann, *Nature* **332**, 55 (1988).

[28] Z. Z. Sheng and A. M. Hermann, *Nature* **332**, 138 (1988).

[29] S. S. Parkin, V. Y. Lee, A. I. Nazzal, R. Savoy, R. Beyers and S. La Placa, *Phys. Rev. Lett.* **61**, 750 (1988).

[30] R. J. Cava, "Superconductors beyond 1-2-3," *Scientific American* **263**, 42 (August 1990).

iii. A third phase was stable up to sintering temperatures of 1000 °C with a volume fraction in excess of 30 % . At higher temperatures this phase reacted to form the oxygen-deficient perovskite obtained by Michel *et al.*[14] Chemical analysis suggested the identification of this third phase with unreacted common CuO. Later preparations halved the amount of Cu, eliminating this phase but retaining the other two phases. Single-crystal precession measurements[31] later confirmed the identification of these phases.

Hazen[32,33] has collected and tabulated data on the atomic structures of all known high-temperature copper oxide superconductors. The twenty-nine variants of twenty topologically distinct structures are given in Table 1. Values of T_c are taken in part from the compilation of Junod.[34] The most obvious common structural theme that unites the known layered copper oxide high-temperature superconductors are the corner-linked square-plane coordinated coppers together with oxygen nonstoichiometry in layers that interleave the CuO_2 sheets. In structures with more than one consecutive CuO_2 sheet, divalent or trivalent cations in eight coordination are always found to act as spacers.

The structures of the high-temperature copper oxide superconductors are closely related to that of the mineral perovskite $CaTiO_3$ shown in Fig. 1. At the center of the cubic cell is the ion Ti^{4+} surrounded by an octahedron of O^{2-} ions. At the corners are Ca^{2+} ions. This and subsequent figures are scaled to the covalent radii given in Table 2.

Alternatively, perovskite may be regarded as a layered structure in which TiO_2 planes alternate with CaO planes as shown in Fig. 2. Designating these planar structures as 1A and 2A, the structure of perovskite may be represented by the stacking sequence

[2A (CaO), 1A (TiO_2)], [2A (CaO), 1A (TiO_2)], [2A (CaO), 1A (TiO_2)], *etc.*

La-Cu-O and related structures

The structure of La_2CuO_4 is shown in Fig. 3. In each of the three 2-1-4 structures listed in Table 1, layers of copper in square-plane coordination with oxygen are separated by layers of cations in eight or nine coordination. All three structures have the same cation arrangement, but they differ in the positions of oxygen atoms above and below the Cu-O planes. The structures 1b, 1c and 1d are topologically identical to the 1a structure but have lower symmetry.

---

[31] J. G. Bednorz, M. Takashige and K. A. Müller, *Mater., Res. Bull.* **22**, 819 (1987).

[32] R. M. Hazen, "Crystal structures of high-temperature superconductors," in *Physical Properties of High Temperature Superconductors II*, ed. D. M. Ginsberg (World Scientific, Singapore, 1990) ch. 3.

[33] R. M. Hazen, *The Breakthrough: The Race for the Superconductor*, (Summit Books, New York, 1988) is a personal account of the determination of the 123 structure together with an account of the discovery of this compound.

[34] A. Junod, "Specific heat of high temperature superconductors: a review," in *Physical Properties of High Temperature Superconductors II*, ed. D. M. Ginsberg (World Scientific, Singapore, 1990) ch. 2.

Table 1. Copper oxide high-temperature superconducting structures.

Number	Composition	Space Group	Abbreviation	T_c
1a	La_2CuO_4	I4/mmm	214-T	22-37 K
1b		$P4_2/ncm$		
1c		Bmab		
1d		Fmmm		
2	Nd_2CuO_4	I4/mmm	214-T'	25 K
3	$(Nd, Ce, Sr)_2CuO_4$	P4/mmm	214-T*	
4	$(La, Sr)_2CaCu_2O_6$		2126	60 K
5a	$YBa_2Cu_3O_6$	P4/mmm	123-T	
5b	$YBa_2Cu_3O_7$	Pmmm	123-O	92 K
6	$YBa_2Cu_4O_8$	Ammm	124	80 K
7	$Y_2Ba_4Cu_7O_{15}$	Ammm	247	40 K
8	$(Ba, Nd)_2(Nd, Ce)_2Cu_3O_8$	I4/mmm	223	≈ 40 K
9a	$Pb_2YSr_2Cu_3O_8$	P4/mmm	2123	≈ 70 K
9b		Cmmm		
10a	$Bi_2Sr_2CuO_6$	Amaa	Bi-2201	12 K
10b		A2/a		
10c		C2		
11a	$Bi_2Sr_2CaCu_2O_8$	Fmmm	Bi-2212	90 K
11b		Amaa		
12	$Bi_2Sr_2Ca_2Cu_3O_{10}$	I4/mmm	Bi-2223	110 K
13a	$Tl_2Ba_2CuO_6$	I4/mmm	Tl-2201	11 K
13b		Fmmm		
14	$Tl_2Ba_2CaCu_2O_8$	I4/mmm	Tl-2212	110 K
15	$Tl_2Ba_2Ca_2Cu_3O_{10}$	I4/mmm	Tl-2223	121 K
16	$Tl_2Ba_2Ca_3Cu_4O_{12}$	I4/mmm	Tl-2234	
17	$TlBa_2CuO_5$	P4/mmm	Tl-1201	
18	$TlBa_2CaCu_2O_7$	P4/mmm	Tl-1212	90 K
19	$TlBa_2Ca_2Cu_3O_9$	P4/mmm	Tl-1223	110 K
20	$TlBa_2Ca_3Cu_4O_{11}$	P4/mmm	Tl-1234	122 K

The compound Nd_2CuO_4 is the first electron high-temperature superconductor with excess negative charge per unit cell. The substitution of Ce and Sr for Nd leads to a new 2-1-4 structure that incorporates aspects of the T and T' topologies.

The compound $(La, Sr)_2CaCu_2O_6$ shown in Fig. 4 is the least complex of all the structures with double layers of copper oxide pyramids common to the compounds with highest T_c. Superconductivity in this structure, first reported[35] for $La_2SrCu_2O_6$ and $La_2CaCu_2O_6$ has been achieved by Cava *et al.*[36] The highest transition temperature observed is 60 K at the composition $La_{1.6}Sr_{0.4}CaCu_2O_6$.

---

[35] N. Nguyen, L. Er-Rakho, C. Michel, J. Choisnet and B. Raveau, *Mater. Res. Bull.* **15**, 891 (1980).

[36] R. J. Cava, B. Batlogg, R. B. van Dover, J. J. Krajewski, J. V. Waszczak, R. M. Fleming, W. F. Peck Jr., L. W. Rupp Jr., P. Marsh, A. C. W. P. James and L. F. Schneemeyer, *Nature* **345**, 602 (1990).

Table 2. Covalent radii.

atom	radius, nm.	atom	radius, nm.	atom	radius, nm
O	0.073	Y	0.162	Ba	0.198
Cu	0.117	La	0.169	Ln	0.156-0.169
Ti	0.132	Ca	0.174		

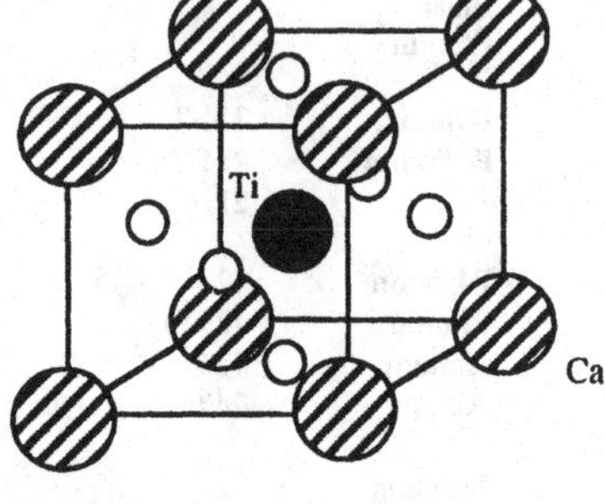

Figure 1. The mineral perovskite $CaTiO_3$. Ti is at the body center, O at the face centers and Ca at the corners.

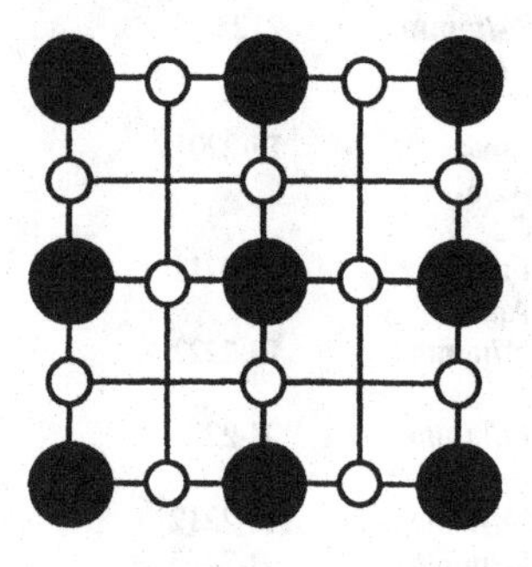

1A (TiO_2) plane

2A (CaO) plane

Figure 2. Planes of the mineral perovskite $CaTiO_3$

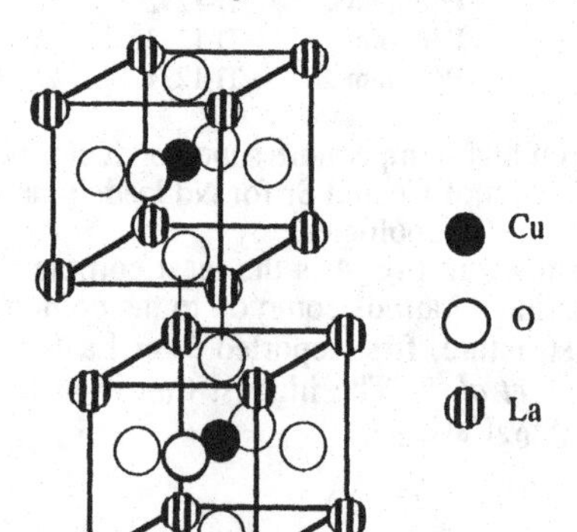

Figure 3. The crystal structure of La_2CuO_4.

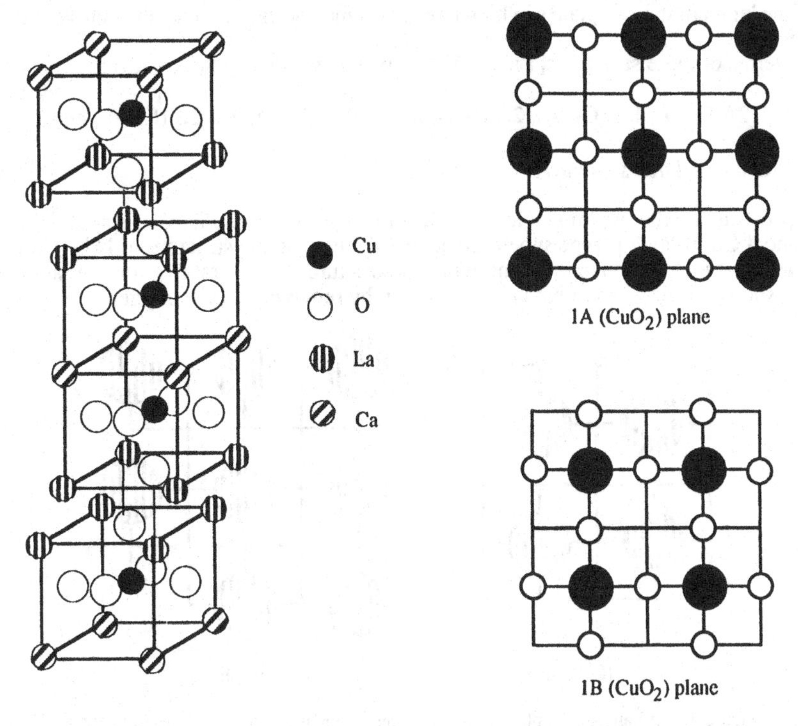

Figure 4. The crystal structure of $La_2CaCu_2O_6$. This structure is related to that of La_2CuO_4 shown in Fig. 3 by replacing the layer of octahedra by a double layer of pyramids.

Figure 5. The two CuO_2 planes, displaced along a face diagonal

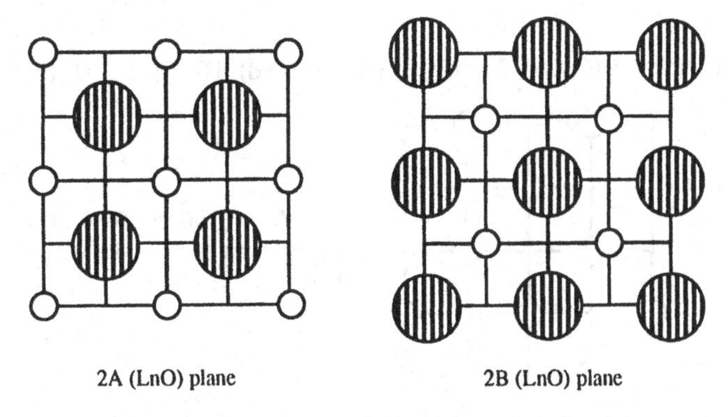

Figure 6. The two lanthanide LnO planes, displaced along a face diagonal.

The 2-1-4 and related structures may be developed from the perovskite planes shown in Fig. 2 with the addition of planes shifted by half a face diagonal as shown in Fig. 5 for the CuO_2 planes, which have the same configuration as the perovskite TiO_2 planes and in Fig.

6 for the lanthanide LnO planes, which have the same configuration as the perovskite CaO planes.

In terms of these structures, La_2CuO_4 may be represented by the six layers

[2A (LaO), 1A (CuO_2), 2A (LaO)|2B (LaO), 1B (CuO_2), 2B (LaO)], *etc.*

Nd-Cu-O and related structures

$Nd_{2-x}Ce_xCuO_4$ is the model single CuO_2-layer n-type superconductor as $La_{2-x}Sr_xCuO_4$ is the model CuO_2-layer p-type superconductor. To indicate the structure of $Nd_{2-x}Ce_xCuO_4$ it is necessary to introduce three more planar structures. The first two are the lanthanide planes shown in Fig. 7 and obtained from Fig. 6 by removing all the oxygen.

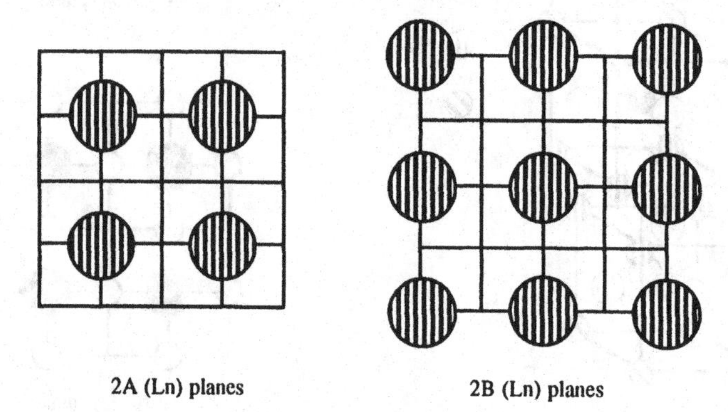

Figure 7. Lanthanide (Ln) planes formed by removing all the oxygen from LnO planes

The third structure is that of the oxygens in the CuO_2 planes shown in Fig. 8 and obtained from Fig. 5 by removing all the copper.

With these additional layers, the structure of $Nd_{2-x}Ce_xCuO_4$ is represented by the eight layers

[2A (Nd), 1A (CuO_2), 2A (Nd), 1 (O_2)| 2B (Nd), 1B (CuO_2), 2B (Nd), 1 (O_2)], *etc.*

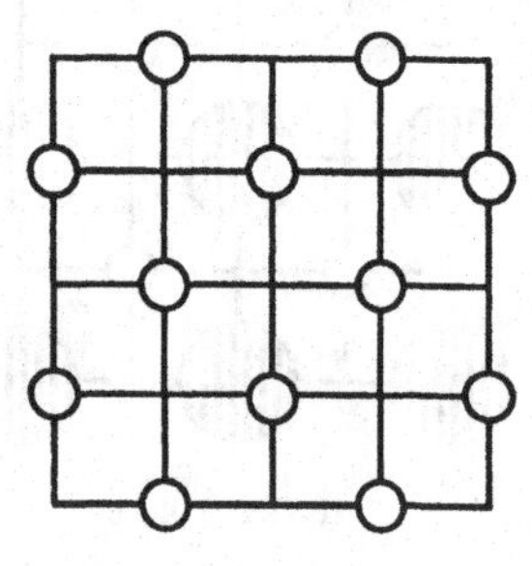

Figure 8. 1(O_2) planes. The oxygens in this plane are in the same position as in the CuO_2 planes.

Y-Ba-Cu-O and related structures

The discovery by Wu *et al.*[21] of the superconductor $YBa_2Cu_3O_{7-d}$ with a transition temperature of 92 K has stimulated a massive amount of crystallographic research. The structure is based on a simple unit cell composed of a stack of three perovskite-like cubes shown in Fig. 9. It was not, however, until the first neutron powder diffraction studies that the details of the oxygen positions were resolved with certainty. The 123 structure crystallizes as tetragonal at high temperature, but converts by oxygen ordering to an orthorhombic form on cooling. Orthorhombic 123 with small domain size may appear tetragonal in single-crystal x-ray studies. The most precise structure determinations have relied on Rietveld refinement of neutron powder diffraction data. Compositional variants of the 123 structure have been summarized by Beyers and Shaw.[37]

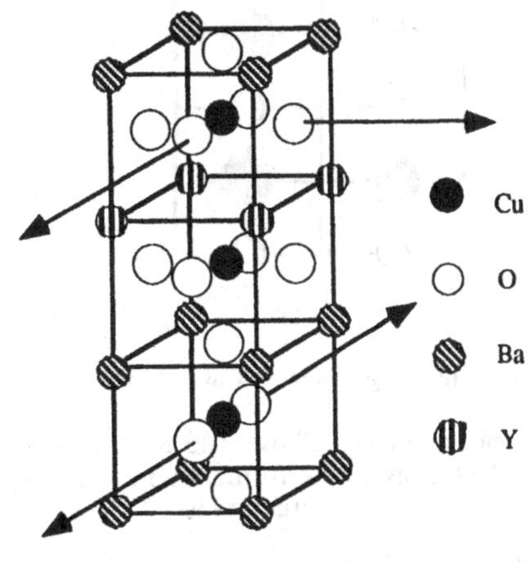

Figure 9. The crystal structure of $YBa_2Cu_3O_7$ showing the one-dimensional Cu-O chains and the two-dimensional pyramidal planes. This structure is derived from the structure of $La_2CaCu_2O_6$ by adding the Cu-O chains and shifting the pyramidal planes to align their apices.

Morris *et al.*[38] have produced eight members with the 1-2-4 structure, and six variants of the 247 structure,[39] with both groups topologically equivalent to 123.

Cava *et al.*[40] have described a new family of near-70K superconductors with a 2-1-2-3 structure that bears a close relationship to the tetragonal 1-2-3 structure.

In order to represent the structure of $YBa_2Cu_3O_7$, two additional planar structures are required. In Fig. 10 are represented the CuO chain structures obtained from Fig. 5 by removing half the oxygens.

The structure of $YBa_2Cu_3O_7$ is represented by the six layers

$$[2A\ (BaO),\ 1A\ (CuO_2),\ 2A\ (Y),\ 1A\ (CuO_2),\ 2A\ (BaO),\ 1A\ (CuO)],\ etc.$$

Bi-Sr-Ca-Cu-O structures

Michel et al. reported the first of a series of modular layer structures in which the copper and oxygen form in sheets typical of all the high-temperature superconductors spaced by

---

[37] R. Beyers and T. M. Shaw, "The Structure of $YBa_2Cu_3O_{7-\delta}$ and its Derivatives," in *Solid State Physics*, Volume 42, eds. H. Ehrenreich and D. Trunbull (Academic, Boston, 1989) p 135.

[38] D. E. Morris, J. H. Nickel, J. Wei, N. G. Asmar, J. S. Scott, U. M. Scheven, C. T. Hultgren, A. G. Markelz, J. E. Post, P. J. Heaney, D. R. Veblen and R. M. Hazen, *Phys. Rev. B* **39**, 7347 (1989).

[39] D. E., Morris, N. G. Asmar, J. Wei, J. H. Nickel, R. L. Sid, J. S. Scott and J. E. Post, *Phys. Rev. B* **40**, 11406 (1989).

[40] R. J. Cava, B. Batlogg, J. J. Krajewski, L. W. Rupp, L. F. Schneemeyer, T. Siegrist, R. B. van Dover, P. Marsh, W. F. Peck Jr., P. K. Gallagher, S. H. Glarum, J. H. Marshall, R. C. Farrow, J. V. Waszczak, R. Hull and P. Trevor, *Nature* **336**, 211 (1988).

alkaline-earth cations and interleaved with BiO layers.[41] The simplest structure is that of $Bi_2Sr_2CuO_6$

[2A (BiO), 2B (SrO), 1B (CuO_2), 2B (SrO), 2A (BiO)|

|2B (BiO), 2A (SrO), 1A (CuO_2), 2A (SrO), 2B (BiO)], *etc.*

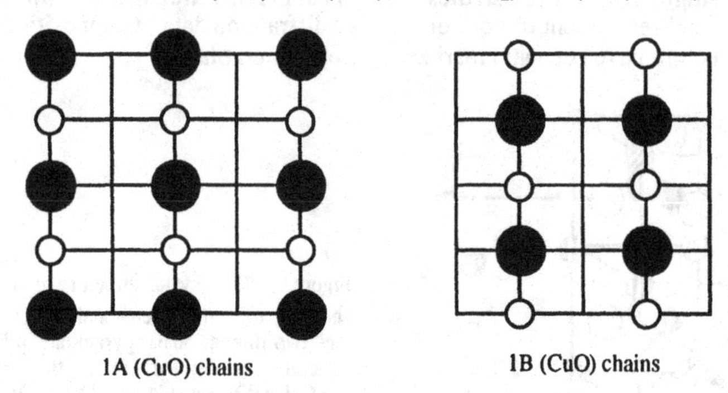

Figure 10. CuO chains formed by the removal of half the oxygen from the CuO_2 planes

The addition of Ca by Maeda *et al.*[23] raised T_c for this system above 100 K. Ca forms multiple layers of CuO_2 separated by Ca. For n CuO_2 layers the formula unit is $Bi_2Sr_2Ca_{n-1}(CuO_2)_nO_4$. For 2 CuO_2 layers (n = 2) per formula unit the structure is

[2A (BiO), 2B (SrO), 1B (CuO_2), 2B (Ca), 1B (CuO_2), 2B (SrO), 2A (BiO)|

|2B (BiO), 2A (SrO), 1A (CuO_2), 2A (Ca), 1A (CuO_2), 2A (SrO), 2B (BiO)], *etc.*

Tl-Ba-Ca-Cu-O structures

Sheng and Hermann[24,25] discovered superconductivity near 120 K in the Tl-Ba-Ca-Cu-O system. The structures of thallium with three new superconducting phases are strikingly similar to the bismuth superconductors. The compound $Tl_2Ba_2CuO_6$ with the structure

[2B (TlO), 2A (BaO), 1A (CuO_2), 2A (BaO), 2B (TlO)|

|2A (TlO), 2B (BaO), 1B (CuO_2), 2B (BaO), 2A (TlO)], *etc.*

has a transition temperature of 11 K. As with the Bi-compounds, the addition of Ca forms multiple layers of CuO_2 separated by Ca. For n CuO_2 layers the formula unit is $Tl_2Ba_2Ca_{n-1}(CuO_2)_nO_4$. For 2 CuO_2 layers (n = 2) per formula unit, the structure is

[2B (TlO), 2A (BaO), 1A (CuO_2), 2A (Ca), 1A (CuO_2), 2A (BaO), 2B (TlO)|

[2A (Tl(O), 2B (BaO), 1B (CuO_2), 2B (Ca), 1B (CuO_2), 2B (BaO), 2A (TlO)], *etc.*

---

[41] C. Michel, M. Hervieu, M. M. Borel, A. Grandin, F. Deslandes, J. Provost and B. Raveau, *Z. Phys. B* **68**, 421 (1987).

Shortly after Sheng and Hermann's work on the Tl-Ba-Ca-Cu-O system, Parkin *et al.*[42,43] reported yet another homologous series of thallium superconductors, which differ from the earlier series in containing only one TlO layer in the unit cell. The compound without Ca

[2A (BaO), 1A (CuO_2), 2A (BaO), 2B (TlO)], *etc.*

is not superconducting. Again, as with the Bi-compounds, the addition of Ca forms multiple layers of CuO_2 separated by Ca. For n CuO_2 layers the formula is $TlBa_2Ca_{n-1}(CuO_2)_nO_3$. For 2 CuO_2 layers (n = 2) the structure is

[2A (BaO), 1A (CuO_2), 2A (Ca), 1A (CuO_2), 2A (BaO), 2B (TlO)], *etc.*

1.2.2. Block layers

Tokura and Arima[44] have introduced the idea of block layers to characterize the functional units that separate CuO_2 planes. For example, $La_{2-x}Sr_xCuO_4$ may be regarded as CuO_2 sheets separated by La_2O_2 layers with the rocksalt structure (the **L**-layer) and written as

L-[CuO_2]-L-[CuO_2]-L-[CuO_2]-L-[CuO_2]-

with **L** = 2A (LaO), 2B (LaO) or 2B (LaO), 2A (LaO). On the other hand, $Nd_{2-x}Ce_xCuO_4$ may be regarded as CuO_2 sheets separated by N_2O_2 layers with the fluorite structure (the G-layer) and written as

G-[CuO_2]-G-[CuO_2]-G-[CuO_2]-G-[CuO_2]-

with **G** = 2A (Nd), 1 (O_2), 2B (Nd), or 2B (Nd), 1 (O_2), 2A (Nd).

The block-layers may or may not provide apical oxygens for the CuO_2 planes. For example, the **L**-layers do supply apical oxygens while the **G**-layers do not. The oxygen structure that surrounds the copper (squares, pyramids or octahedra) is determined by the block layers. A second function of the block-layers, doping of the CuO_2 planes, is discussed below.

1.2.3. Structural chemistry

Sleight has discussed the chemistry of the oxide superconductors with reference to the oxidation states of copper.[45] One of the simplest methods of connecting CuO_4 units together forms the infinite sheets that are present in all the copper oxide superconductors with the corner oxygens shared by adjacent copper atoms. High-temperature superconductivity occurs only in oxides where there is high covalency, which is necessary for the formation of bands that support metallic behavior. All the oxide superconductors contain in addition mixed-valent cations, a requirement of the dispropor-tionation mechanisms for superconductivity.

---

[42] S. S. P. Parkin, V. Y. Lee, A. I. Nazzal, R. Savoy, R. Beyers and S. La Placa, *Phys. Rev. Lett.* **61**, 750 (1988).

[43] R. Beyers, S. S. P. Parkin, V. Y. Lee, SA. I. Nazzal, R. Savoy, G. Gorman, T. C. Huang and S. La Placa, *Appl. Phys. Lett.* **53**, 432 (1988).

[44] Y. Tokura and T. Arima, "New classification method for layered copper oxide compounds and its application to design of new high T_c superconductors," *Jpn. J. Appl. Phys.* **29**, 2388 (1990).

[45] A. W. Sleight, *Science* **242**, 1519 (1988).

The structural chemistry of $YBa_2Cu_3O_7$ is best understood[39] by first examining $YBa_2Cu_3O_6$, which has the planar structure

[2A (BaO), 1A (CuO_2), 2A (Y), 1A (CuO_2), 2A (BaO), 1A (Cu)], *etc.*

where the 1A (Cu) planes are the 1A (CuO_2) planes with all the oxygen removed. This compound is an antiferromagnetic insulator. When $YBa_2Cu_3O_6$ is annealed in an oxygen-containing atmosphere, it acquires oxygen to form $YBa_2Cu_3O_7$, which is metallic and superconducting with the planar structure

[2A (BaO), 1A (CuO_2), 2A (Y), 1A (CuO_2), 2A (BaO), 1A (CuO)], *etc.*

Raveau *et al.*[46] and Greene and Bagley[47] have reviewed studies of oxygen nonstoichiometry in connection with the mixed valence of copper. Cava[48] has reviewed the known copper oxide superconductors with particular emphasis on the way in which they fall into structural families.

Tokura and Arima[44] have described the crystal chemistry of the CuO_2-layer compounds in terms of the block-layers that they have identified. An important role of the block-layers is to supply the CuO_2 planes with carriers. As they explain from the viewpoint of ionic crystals, an insulating CuO_2 sheet consists of Cu^{2+} and O^{2-} and hence is charged -2 per formula unit. To maintain charge neutrality, the average charge of the block-layer must be +2 in the insulating compound. The charge of the L- and G- layers described above is just +2. The substitution of Sr^{2+} for La^{3+} or Ce^{4+} for Nd^{3+} reduces or increases the charge per formula unit in the CuO_2 planes. In $La_{2-x}Sr_xCuO_4$ the charge of the L-layer is 2 - x and the effective charge per CuO_2 unit is - 2 + x. In $Nd_{2-x}Ce_xCuO_4$ the charge of the G-layer is 2 + x and the charge per formula unit in the CuO_2 planes is - 2 - x. The former process corresponds to hole-doping and the latter to electron-doping. In this way the block-layers control the carrier concentration in the CuO_2 planes.

1.2.4. Microstructure

Microstructure—twin and grain boundaries, second phases, incommensurate modulation, intergrowth defects, *etc.* have profound effects on the physical properties of the high-temperature superconductors. Much of what has been learned through transmission electron microscopy (TEM) on well characterized materials is reviewed by Chen.[49]

Powders and ceramics

Oxygen stoichiometry significantly affects the properties of the high-temperature superconductors. Cava found in the Y-Ba-Cu-O system[50] a plateau in T_c and a minimum

[46] B. Raveau, C. Michel, M. Hervieu, J. Provost and F. Studer, "Oxygen nonstoichiometry and valence states in superconductive cuprates," in *Earlier and Recent Aspects of Superconductivity*, eds. J. G. Bednorz and K. A. Müller (Springer-Verlag, Berlin, 1990) pp. 66-95.

[47] L. H. Greene and B. G. Bagley, "Oxygen stoichiometric effects and related atomic substitutions in the high-T_c cuprates," in *Physical Properties of High Temperature Superconductors II*, ed. D. M. Ginsberg (World Scientific, Singapore, 1990) ch. 8.

[48] R. J. Cava, *Science* **247**, 656 (1990).

[49] C. H. Chen, "The microstructure of high-temperature oxide superconductors," in *Physical Properties of High Temperature Superconductors II*, ed. D. M. Ginsberg (World Scientific, Singapore, 1990) ch. 4.

[50] R. J. Cava, B. Batlogg, C. H. Chen, E. A. Rietman, S. M. Zahurak and D. J. Werder, *Nature* **329**, 423 (1987) and *Phys. Rev. B* **36**, 5719 (1987).

in the room temperature resistivity as a function of oxygen concentration. This observation suggested the possibility of oxygen-vacancy ordering, which has since been confirmed by both electron and x-ray diffraction studies.

Defects and second phases concentrate at grain boundaries in sintered ceramic samples, leading to critical current densities as low as 400 A/cm^2. Grain boundaries are predominantly parallel to the (001) basal planes in at least one of the adjacent grains, suggesting that the basal plane may be the fastest growing face.

Crystals

A number of the high-temperature superconductiors undergo a transition from tetragonal to orthorhombic as they are cooled. This transition can be observed in TEM with the nucleation and growth of twin domains as the temperature is reduced.

Studies of the surface layers of oriented single crystals indicate that the orthorhombicity and defect density at the surface may differ significantly from the bulk of the crystal. Single crystals also contain intergrowth defects that result from reduced oxygenation.

Epitaxial films

Significant advances have been made in the preparation of crystalline films, grown epitaxially on suitably oriented insulating crystals. Although these films are not entirely defect free, those defects that are present appear to be well localized.

1.3. Physical Properties

1.3.1. Resistivity

Bednorz and Müller[1] measured dc resistivity with a four-point method that had the advantage of reducing the sensitivity to contact resistance. Gold electrodes were attached to rectangular samples that had been cut from the sintered pellets. Contact to the electrodes was made with indium wires.

Samples with fractional Ba concentration $x < 1$ exhibited a metallic-like resistivity that above 100 K dropped linearly with temperature. With further reduction in temperature the resistivity of samples annealed in air increased somewhat, followed by a precipitous drop in the 30 K region. Annealing in a reducing atmosphere enhanced the resistivity maximum observed below 100 K. Current densities above 0.5 A/cm^2 led to partial suppression of the resistivity maximum.

1.3.2. Magnetic susceptibility

In a second paper,[51] submitted for publication in October 1986, Bednorz, Müller and Takashige measured the magnetic susceptibility of their ceramic samples as a function of temperature. The very considerable importance of magnetic susceptibility is that the development of macroscopic diamagnetism, more than a resistivity drop alone, is required to establish the presence of superconductivity.

Samples were measured in a commercial variable-temperature susceptometer at temperatures down to 4.2 K and in fields up to 5 T. The presence of a transition to superconductivity in $(La_{1-x}Ba_x)CuO_4$ was confirmed. For samples cooled in fields below 0.1 T, bulk diamagnetism appeared at a somewhat lower temperature than did the drop in

[51] J. G. Bednorz, M. Takashige and K. A. Müller, *Europhys. Lett.* **3**, 379 (1987).

resistivity. Cooling in fields above 1 T suppressed the development of bulk diamagnetism. If, however, the samples were cooled below T_c in zero field and the field was then applied, diamagnetism could be observed in fields above 1 T.[52] The origin of this difference in magnetic behavior has received considerable attention and is considered in Ch. 6.

1.3.3. Surface impedance

Because alternating electric and magnetic fields are largely excluded from conductors, it is useful to characterize the response of conductors in terms of what is called a surface impedance rather than in terms of bulk quantities like resistivity and inductivity. The theoretical background to this approach is discussed in Ch. 3. The approach is particularly useful at radio and microwave frequencies but also of use for magnetic susceptibility measurements at low alternating frequencies.

Blazey *et al.*[53] measured the modulated microwave surface resistance using a commercial electron spin resonance spectrometer and obtained large low-field signals, which were ascribed to the granular structure of their superconducting samples in the La-Cu-O system. This technique has proved useful in the study of intergranular coupling[54,55] further discussed in Ch. 13.

---

[52] K. A. Müller, M. Takashige and J. G. Bednorz, *Phys. Rev. Lett.* **58**, 1143 (1987).

[53] K. W. Blazey, K. A. Müller, J. G. Bednorz, W. Berlinger, G. Amoretti, E. Buluggiu, A. Vera and F. Mattacotta, *Phys. Rev.* **36**, 7241 (1987).

[54] K. W. Blazey, "Microwave absorption in granular guperconductors," in *Earlier and Recent Aspects of Superconductivity,* eds. J. G. Bednorz and K. A. Müller (Springer-Verlag, Berlin, 1990) pp. 262-277.

[55] A. M. Portis, "Microwaves and superconductivity: processes in the intergranular medium," in *Earlier and Recent Aspects of Superconductivity,* eds. J. G. Bednorz and K. A. Müller (Springer-Verlag, Berlin, 1990) pp. 278-303.

2
THEORIES OF SUPERCONDUCTIVITY

2.1. Introduction

This chapter reviews the history of what is now called Type I superconductivity with emphasis on the electrodynamics of superconductivity. As noted in Ch. 1, the discovery of superconductivity by H. K. Onnes at Leiden in 1911 was made possible by the availability of liquid helium. During a period of activity following this discovery it was found in 1914 by Onnes that the superconducting state was stable only below a critical magnetic field H_c or alternatively below a corresponding critical current density J_c.

A second period of activity began with the discovery of the specific heat anomaly in 1933, leading to the development of the Gorter-Casimir theory of the thermodynamics of superconductors in 1934.[1] The Meissner-Ochsenfeld flux-exclusion effect,[2] discovered in 1933, led to the London theory of electromagnetic behavior in 1935.[3]

Theoretical developments in a third period of activity include the 1950 Ginzburg-Landau theory, which we discuss in Ch. 4, Pippard's nonlocal electrodynamics in 1953, the microscopic Bardeen, Cooper, Schrieffer (BCS) theory in 1957 and the prediction of the Josephson effect in 1962. Experimental developments during this period include the isotope effect, discovered in 1950 and the demonstration of an energy gap in the electronic

---

[1] C. J. Gorter and H. B. G. Casimir, *Phys. Z.* **35**, 963 (1934).

[2] W. Meissner and R. Ochsenfeld, *Naturwiss.* **21**, 787 (1933).

[3] F. and H. London, *Physica* **2**, 341 (1935); *Proc. Roy. Soc. (Lond.) A* **149**, 71 (1935).

excitation spectrum during the period 1953-60. Discussion of the 1961 discovery of flux quantization, of macroscopic quantum interference in 1964 and observation of the flux lattice in 1967 are postponed to Ch. 4.

We have mentioned very briefly in Ch. 1 the work of B. T. Matthias on intermetallic compounds and the development of oxide superconductors, which lead to the 1986 discovery by J. G. Bednorz and K. A. Müller of high-temperature superconductivity in copper oxide layer compounds.

2.2. The Two-Fluid Model

The independence of the superconductive state of previous history and the reversibility of the transition inferred from the Meissner-Ochsenfeld[2] effect led Gorter and Casimir[1] to develop a model of normal and superconducting electrons to account for the thermal properties of superconductors. Taking for the carrier concentration

$$n = n_n + n_s \tag{1}$$

with a concentration of normal carriers $n_n = fn$ and a concentration of superconducting carriers $n_s = (1 - f)n$, they were able to fit the heat capacity[4] with a Helmholz free energy

$$F = \sqrt{f}\, F_n(T) + (1 - f)\, F_s(T) \tag{2}$$

Note that for a mixed phase to be stable, Eq. (1) must be nonlinear in f.[5]

The thermodynamics of a free electron Fermi gas leads to an energy $U_n = {}^1\!/_2\, \gamma T^2$, a heat capacity $C_n = \gamma T$, an entropy $S_n = \int C\, dT/T = \gamma T$ and a consequent free energy

$$F_n = U_n - TS_n = -\frac{1}{2}\gamma T^2 \tag{3}$$

Assuming that the heat capacity of the superconducting fluid C_s is appreciable over only a limited range of temperature near T_c leads to the simplification $F_s \approx -\beta$. Minimizing Eq. (2) with respect to f for this choice of free energies yields for the fraction of normal carriers

$$f = (F_n/2F_s)^2 = t^4 \tag{4}$$

with $t = T/T_c$ and $T_c = 2(\beta/\gamma)^{1/2}$.

2.3. London Penetration

2.3.1. Theory

Following the discovery of the exclusion of magnetic flux by a superconductor–the Meissner-Ochsenfeld effect,[2] F. and H. London[3] together developed the appropriate electrodynamics from the expression for the rate of change of the lossless component of current[6]

---

[4]D. Shoenberg, *Superconductivity* (University Press, Cambridge, 1952) sec. 3.2.

[5]*ibid* sec. 6.3.

[6]R. Becker, G. Heller and F. Sauter, Z. *Phys.* **85**, 772 (1933).

$$\ell \frac{d}{dt}\mathbf{J} = \mathbf{E} \tag{5}$$

with the kinetic inductivity $\ell = m/n_s e^2$. With the definition of the vector potential

$$\mathbf{B} = \nabla \times \mathbf{A} \tag{6}$$

Faraday's law allows both electrostatic and magnetic contributions to the electric field

$$\mathbf{E} = -\nabla\phi\,\frac{\partial}{\partial t}\mathbf{A} \tag{7}$$

So long as there are no electrostatic fields, time-integration leads to the London equation

$$\ell\,\mathbf{J} = -\,\mathbf{A} \tag{8}$$

Differential equations equivalent to Eq. (8) are

$$\ell\,\nabla^2\mathbf{A} = -\,\mu\,\mathbf{A} \tag{9}$$

and similar equations with **A** replaced by **B** or **J** so long as these vectors are solenoidal (divergence-free as discussed in Ch. 3.)

At a plane surface, the field decays into the superconductor as

$$B(x) = \mu H_s \exp(-x/\lambda_L) \tag{10}$$

with the London penetration depth

$$\lambda_L = (\ell/\mu)^{1/2} \tag{11}$$

Current decays similarly into a superconductor with

$$\mu\,\lambda_L\,J(x) = -\,B(x) \tag{12}$$

The two-fluid model of Gorter and Casimir[1] suggests for the electrodynamics[7]

$$\mathbf{J} = \mathbf{J}_n + \mathbf{J}_s \tag{13}$$

With the relations

$$\rho\,\mathbf{J}_n = \mathbf{E} \qquad \ell\,d\mathbf{J}_s/dt = \mathbf{E} \qquad \mathbf{J} = \frac{1}{z}\mathbf{E} \tag{14}$$

where the resistivity is $\rho = m/fne^2\tau$, the kinetic inductivity is $\ell = m/(1-f)ne^2$ and the reciprocal of the complex specific impedance z at frequency ω is

[7] H. London, *Nature (London)* 133, 497 (1934).

$$\frac{1}{z}=\frac{1}{\rho}+\frac{1}{-i\omega\ell} \tag{15}$$

Substituting into the wave equation gives the dispersion relation in the two-fluid model

$$k^2 - \varepsilon\mu\omega^2 = i\omega\mu /z = i\omega\mu\sigma \ - \ \mu/\ell \tag{16}$$

or equivalently with the substitution of the London penetration depth

$$k^2 = - (1/\lambda_L)^2(1 - \omega^2\varepsilon\ell - i\omega\sigma\ell) \tag{17}$$

As discussed in Ch. 3, the surface impedance is

$$Z_s = R_s - i\omega L_s = E_s/H_s = (\mu/\varepsilon)^{1/2} = \omega\mu/k \tag{18}$$

At low frequency the surface inductance and resistance in the two-fluid model are then

$$L_s = \mu\lambda_L \tag{19}$$

$$R_s = - \tfrac{1}{2}\omega^2\mu\lambda_L\sigma\,\ell = -\tfrac{1}{2}\omega^2\mu^2\lambda_L{}^3\sigma \tag{20}$$

Calorimetric measurements of power absorption by H. London[8] demonstrated the presence of substantial loss just below T_c, indicating that σ is finite in a superconductor as expected from the two-fluid model with $\sigma \propto t^4$.

2.3.2. *Application*

The Londons[3] obtained for the average magnetization density of a sphere of radius a and penetration depth λ_L in a field H

$$M = - {}^3\!/_2\, H\ S(a/\lambda_L) \tag{21}$$

with

$$S(\zeta) = 1 - (3/\zeta) \coth \zeta + 3/\zeta^2 \tag{22}$$

The mean flux density within the sphere

$$B = H + {}^2\!/_3\, M = H[1 - S(\zeta)] \tag{23}$$

For $a >> \lambda_L$ we have $S(\zeta) \approx (1 - 1/\zeta)^3$. For $a << \lambda_L$ we have $S(\zeta) \approx \zeta^2/15$. For a magnetic field parallel to a flat plate of thickness t the magnetization density is

$$M = - H\ P(t/\lambda_L) \tag{24}$$

with

$$P(\zeta) = 1 - (2/\zeta) \tanh \zeta/2 \tag{25}$$

---

[8] H. London, *Proc. Roy. Soc. A* **176**, 522 (1940).

For a >> λ_L we have $P(\zeta) \approx 1 - 2/\zeta$ and for a << λ_L we have $P(\zeta) \approx \zeta^2/12$.

2.4. Pippard Electrodynamics

2.4.1. Anomalous skin effect

Normal metals

The measurements of H. London[3] of the surface resistance of Sn in the normal state at 3.8 K (just above the superconducting transition temperature) yielded a value three times that expected from the classical model of the rf surface impedance, which assumes a local relation between current and field of the form

$$\mathbf{J}(\mathbf{r}) = \sigma(\omega)\mathbf{E}(\mathbf{r}) \tag{26}$$

London attributed this anomalously large surface resistance to the fact that the electron mean free path l at low temperature becomes large compared with the classical penetration depth $\delta = \sqrt{2/\omega\sigma\mu}$. Under these conditions the current at position **r** arises from charge carriers that have passed through a wide range of electric fields between their last collision and their arrival at **r**. As a consequence, **J**(**r**) is not related locally to the electric field as expressed by Eq. (26), but depends of the electric fields over a region around **r** of extent l. This phenomenon leads to the anomalous skin effect in contrast to the classical skin effect, which is a consequence of Eq. (26).

From a series of experiments on pure metals at low temperature, which extended the work of H. London, Pippard[9] confirmed that the use of the local conductivity $\sigma(\omega)$ gives incorrect results and must be replaced by a nonlocal relation of which the simplest, neglecting retardation, is the Chambers formula[10,11]

$$\mathbf{J}(\mathbf{r}) = (3/4\pi l)\sigma(\omega)\int \boldsymbol{\rho}\, [\boldsymbol{\rho}\cdot\mathbf{E}(\mathbf{r} + \boldsymbol{\rho})]\, e^{-\rho/l}\, dV/\rho^4 \tag{27}$$

where $\boldsymbol{\rho}$ is the vector from **r** to the position at which current is induced. The induced current spreads and decays with the mean free path l. The theory with specular reflection of carriers from the surface was formalized by Reuter and Sondheimer.[12] The extension to diffuse surface scattering was performed by Dingle.[13]

Superconductors

In making careful measurements of the surface reactance of dilute alloys of Sn with varying amounts of In, Pippard[14,15] observed a strong dependence of the superconducting penetration depth λ on the carrier mean free path l even though the transition temperature

---

[9] A. B. Pippard, *Proc. Roy. Soc. A* **191**, 370, 385, 399 (1947).

[10] A. B. Pippard in *Advances in Electronics and Electron Physics*, ed. L. Marton (Academic Press, New York, 1954) Vol. 6, pp. 1-45.

[11] J. M. Ziman, *Principles of the Theory of Solids*, 2nd ed. (Cambridge Press, Cambridge, 1972) pp 283, 402.

[12] G. C. Reuter and E. H. Sondheimer, *Proc. R. Soc. London Ser. A* **195**, 336 (1948).

[13] R. B. Dingle, *Physica* **19**, 311 (1953).

[14] A. B. Pippard, *Proc. Roy. Soc. A* **203**, 98, 195 (1950).

[15] A. B. Pippard, *Proc. Roy. Soc. A* **216**, 547 (1953) .

and the critical field were little affected. This observation led Pippard to replace the mean free path in the exponent of Eq. (27) by a coherence length ξ to give for the current density

$$\mathbf{J}(\mathbf{r}) = -(3/4\pi\ell\xi_0)\int \boldsymbol{\rho}\,[\boldsymbol{\rho}\cdot\mathbf{A}(\mathbf{r}+\boldsymbol{\rho})]\,e^{-\rho/\xi}\,dV/\rho^4 \qquad (28)$$

where the coherence length ξ is given by $1/\xi = 1/\xi_0 + 1/l$ with ξ_0 an intrinsic coherence length. Note that $\mathbf{J}(\mathbf{r})$ is depressed by a factor ξ/ξ_0.

Although Eqs. (28) and (27) look rather similar, they are in fact quite different as Pippard[16] has pointed out. In the perfect conductivity limit of Eq. (27), l goes to infinity and flux gradually penetrates the medium. Eq. (28) with a finite coherence length maintains flux exclusion.

2.4.2. *Ineffectiveness*

Pippard suggested that in the extreme anomalous limit $kl >> 1$, only those carriers moving nearly parallel to the surface contribute effectively to $\mathbf{J}(\mathbf{r})$. This concept applied to normal conduction, leads to an effective conductivity

$$\sigma_{eff} = \sigma/|k\,l| \qquad (29)$$

which when substituted into Eq. (16) gives at low frequency

$$k^2 = i\omega\mu\sigma/|k\,l| \qquad \delta = (l/\omega\sigma\mu)^{1/3} \qquad (30)$$

Applied to a superconductor the concept suggests for the inductivity

$$\ell_{eff} = -\,i\,k\xi\,\ell \qquad (31)$$

which when substituted into Eq. (16) gives

$$k^2 = -\,\mu/k\xi\ell \qquad \lambda = (\xi\lambda_L^2)^{1/3} \qquad (32)$$

2.4.3. *Surface impedance*

Ginzburg[17] has discussed the superconducting surface impedance when the current is anomalous and concludes that it is always possible to define an *effective* dielectric function $\varepsilon(\omega, T)$ through the surface impedance as given by Eq. (18)

$$\varepsilon(\omega, T) = \mu/Z_s^2(\omega, T) \qquad (33)$$

For a normal conductor with

$$Z_s = \omega\mu/k = (-i\,\omega\mu/\sigma_{eff})^{1/2} \qquad (34)$$

we obtain from Eq. (30)

---

[16]A. B. Pippard, "The dynamics of conduction electrons," Sec. VI, in *Low-Temperature Physics*, eds. C. DeWitt, B. Dreyfus and P. G. DeGennes (Gordon and Breach, New York, 1962). Reprinted as *The Dynamics of Conduction Electrons* (Gordon and Breach, New York, 1965).

[17]V. L. Ginzburg, *Nuovo Cimento* **2**, 1234 (1953).

$$\sigma_{eff} = (\sigma^2 / \omega\mu\, l^2)^{1/3} \tag{35}$$

For a superconductor with

$$Z_s = \omega\mu/k = -\, i\omega(\mu\, \ell_{eff})^{1/2} \tag{36}$$

we obtain from Eq. (32)

$$\ell_{eff} = (\mu\xi^2\ell^2)^{1/3} \tag{37}$$

2.5. The BCS Microscopic Theory

2.5.1. Introduction

The theory of Bardeen, Cooper and Schrieffer (BCS)[18] has provided a model of the superconducting state in which paired charge-carriers interact over substantial distances to produce a coherent state.[19] This theory provides the quantum description essential for a microscopic understanding of the superconducting properties of metals, including the surface impedance.

L. N. Cooper[20,21] had discovered that two electrons with an attractive interaction always, in the presence of a filled Fermi sphere, form a bound pair, no matter how small their interaction, with a wave function that decays algebraically—as $1/r^2$ at large distances. This important step suggested that the normal state of a superconductor is unstable at sufficiently low temperature, leading to a stable phase in which pairs of electrons are bound.

The removal of a pair of carriers from the BCS state requires an energy 2Δ that depends exponentially on the parameter $N(\varepsilon_F)V$, where $N(\varepsilon_F)$ is the density of states at the Fermi surface and V is a typical matrix element of the interaction. Because the size of a Cooper pair must be of the order of Pippard's coherence length ξ (about 1 μm) and large compared with the average distance between electrons, the condensed state can not realistically be be described as a dilute gas of bound pairs.

The breakthrough came in 1957 when BCS succeeded in constructing a variational wave function for the superconducting ground state that takes many-body correlations into account and separated from the band of single-particle excitations by the energy gap 2Δ.[22] The BCS theory has been found to satisfactorily describe all superconductive phenomena in weakly coupled superconductors, which have $2\Delta \ll \varepsilon_F$.

In weak-coupling BCS theory the energy gap is related to the transition temperature T_c by

$$\Delta(0) = 1.764\, k_B T_c = 2\hbar\omega_D \exp[-1/N(\varepsilon_F)V] \tag{38}$$

where ω_D is the lattice Debye frequency. Near T_c the half-gap is expected to decrease as

$$\Delta(T) \approx 1.74\, \Delta(0)(1 - t)^{1/2} \tag{39}$$

---

[18] J. Bardeen, L. N. Cooper and J. R. Schrieffer, *Phys. Rev.* **106**, 162; **108**, 1175 (1957) .

[19] J. Bardeen, Nobel Lecture, *Science* **181**, 1209 (1973); *Physics Today* **26**, 41 (1973).

[20] L. N. Cooper, *Phys. Rev.* **104**, 1189 (1956).

[21] L. N. Cooper, Nobel Lecture, *Science* **181**, 908 (1973); *Physics Today* **26**, 31 (1973).

[22] J. R. Schrieffer, Nobel Lecture, *Science* **180**, 1243 (1973); *Physics Today* **26**, 28 (1973),

BCS have shown that the Meissner effect[2] is a consequence of a gap in the energy spectrum and obtain an intrinsic coherence length

$$\xi_0 = 2\hbar/\pi\delta p = \hbar v_F/\pi\Delta(0) \tag{40}$$

with just the electrodynamics anticipated by Pippard.[11,12]

A tabulation by Turneaure, Halbritter and Schwettman[23] of the London penetration depth $\lambda_L(0)$, the Pippard coherence length ξ_0 and other superconducting parameters is given in Table 1.

Table 1. Superconducting Parameters

Superconductor	$\lambda(0)$ (nm)	ξ_0 (nm)	$2\Delta(0)/k_BT$	T_c (K)
Al	16	1500	3.40	1.18
In	25	400	3.50	3.3
Sn	28	300	3.55	3.7
Pb	28	110	4.10	7.2
Nb	32	39	3.5 - 3.85	8.95 - 9.2
Ta	35	93	3.55	4.46
Nb_3Sn	50	6	4.4	18
NbN	50	6	4.3	≤ 17
$YBa_2Cu_3O_7$[a]	140	1.5	4.5	90

[a]These data were obtained with the use of a superconducting transmission line. See ref. [44] and the discussion of Sec. 2.6.3.

For typical metals, the coherence length ranges between 0.1 and 1 μm. The new copper oxide superconductors with large Δ and small v_F have unusually short coherence lengths– of the order of atomic distances.

2.5.2. *Electron-electron interaction*

In order to see how a charge moving through a lattice leaves a polarization wave behind it (with which a second charge can interact) we examine the following simple model. We consider the line of atoms shown in Fig. 1 separated by a distance a and harmonically bound with a resonance at the Debye frequency. We imagine a charge q moving along the line with velocity v. To each atom on the line the charge imparts an impulse $F\Delta t = Fa/v$.

The solution to this problem is a displacement wave that travels with the velocity of the charge. For $z < vt$ the transverse displacement is

$$x = (Fa/M\omega_D v) \sin \omega_D(z/v - t) \tag{41}$$

[23]J. P. Turneaure, J. Halbritter and H. A. Schwettman, *J. Superconduct.* 4, 341 (1991).

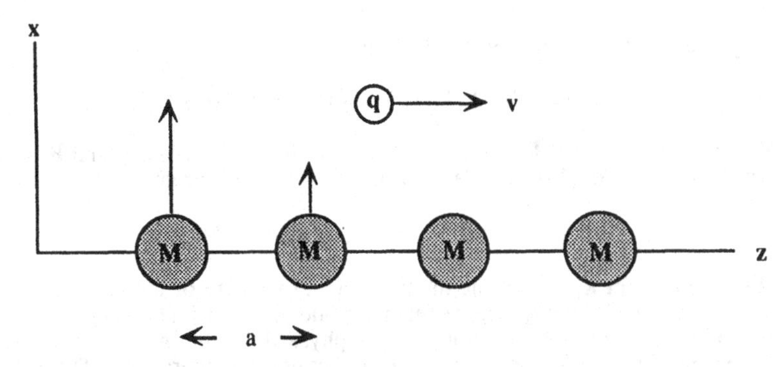

Figure 1. Line of atoms displaced by a moving charge q.

Ahead of the charge with $z > vt$ there is, of course, no displacement. A second charge couples to the displacement wave with energy

$$U = - Fx = - (F^2a/Mv\omega_D) \sin \omega_D(z/v - t) \tag{42}$$

Note that the maximum polarization lags behind the first electron by a distance

$$z = (\pi/2)(v/\omega_D) = \pi a\, v/v_s \tag{43}$$

where v_s is the velocity of sound in the medium. Thus, if electron velocities are of the order of the sound velocity we might expect an appreciable attractive interaction between electrons and the possibility of realizing this lowering of the energy without the electrons getting so close that Coulomb repulsion becomes important.

A more realistic treatment of the problem would be carried out in three (or possibly two) dimensions and would include interactions between atoms, leading to a Debye spectrum. The effect of these changes is to concentrate the polarization wave just behind the electron rather than to leave the infinite wake that develops in one dimension.

2.5.3. *Cooper pairs*

Although the process that we have just discussed leads to an attractive interaction between electrons, we do not expect that in normal metals it will produce much lowering of the energy. There are two reasons for this. First, electrons moving with the Fermi velocity produce very little lattice displacement. Second, a pair of electrons couple only if their velocities are quite close—essentially within the sound velocity v_s. This means that each electron couples only to a fraction $(v_s/v_F)^3 \approx 10^{-9}$ of the electrons. As we shall see, it is possible by producing correlated pairs to increase this fraction to $v_s/v_F \approx 10^{-3}$.

The following discussion of electron correlation draws on a paper by L. N. Cooper,[24] written to present the physical ideas underlying the BCS theory. We consider a pair of electrons in states represented by $k_1\uparrow$ and $k_2\downarrow$. The customary product wave function for uncorrelated electrons is written as

$$\Phi = k_1\uparrow(\mathbf{r}_1) \cdot k_2\downarrow(\mathbf{r}_2) \tag{44}$$

[24] L. N. Cooper, *Am. J. Phys.* **28**, 91 (1960).

A symmetrized wave function has the form

$$\Phi = k_1\uparrow(r_1) \cdot k_2\downarrow(r_2) + k_1\uparrow(r_2) \cdot k_2\downarrow(r_1) \tag{45}$$

Writing $\mathbf{r} = \mathbf{r}_1 - \mathbf{r}_2$ and $\mathbf{R} = \mathbf{r}_1 + \mathbf{r}_2$ together with $\mathbf{k} = \frac{1}{2}(\mathbf{k}_1 - \mathbf{k}_2)$ and $\mathbf{K} = \frac{1}{2}(\mathbf{k}_1 + \mathbf{k}_2)$, we obtain from Eq. (45) with plane wave states for the electrons

$$\Phi = 2\cos \mathbf{k}\cdot\mathbf{r}\, e^{2i\mathbf{K}\cdot\mathbf{R}} \tag{46}$$

We note that for $\mathbf{k}_1 \approx -\mathbf{k}_2$, the momentum and velocity of the correlated pair are very small. The kinetic energy is in the relative motion of the pair about their center of mass and not in the motion of the center of mass. A physical way of understanding this is given by the chemical notion of resonance between electronic configurations. The wave function of Eq. (46) is a linear combination of the two configurations shown in Figs. 2(a) and 2(b).

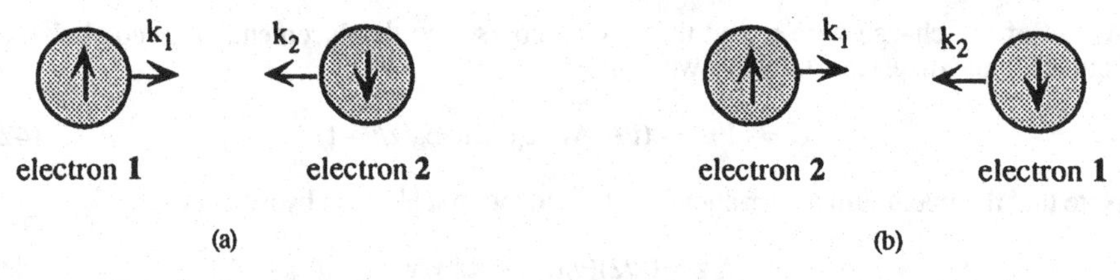

Figure 2. Cooper pairs in one configuration (a) and with the electrons exchanged (b).

If we look at either electron, we see that its motion resonates between $\mathbf{k}_1$ and $\mathbf{k}_2$. Although its mean-square momentum may be large, its average momentum is nearly zero.

The wave function of Eq. (46) is a spin singlet. Alternatively, we could have paired the electrons in a spin triplet. Then the space part of the wave function would have been antisymmetric on exchange of $\mathbf{r}_1$ and $\mathbf{r}_2$ and the wave function would vary as $\sin \mathbf{k}\cdot\mathbf{r}$ rather than as $\cos \mathbf{k}\cdot\mathbf{r}$. It is expected that the energy of the singlet pair will be lower than that of the triplet pair because of the lattice-induced attraction. The dominant part of the condensation energy arises from interaction *between* pairs, however, and this interaction depends little on whether we have singlet or triplet pairing.

We now see that pairs can take better advantage of the lattice interaction than can individual electrons. We may construct pairs with arbitrarily low velocity and without the restrictions of the exclusion principle on the pair momentum **K**. Thus we may achieve the full attractive energy limited only by the additional kinetic energy required to produce correlation between pairs.

2.5.4. Energy of the superconducting state

Before discussing pair energies, we introduce a modified notation. The probability that a one-electron state is occupied is given by $f(\varepsilon)$, the Fermi distribution function.[25] The zero of energy is taken at the Fermi surface as shown in Fig. 3. Then the energy required to promote an electron from a state with energy $-\varepsilon_2$ to a state with energy $+\varepsilon_1$ is simply given by

[25] Charles Kittel, *Introduction to Solid State Physics*, 6th ed. (Wiley, New York, 1986) ch.6.

$$\varepsilon = \varepsilon_1 + \varepsilon_2 \tag{47}$$

We may regard Eq. (47) as the energy required to create an electron above the Fermi surface and a hole below.

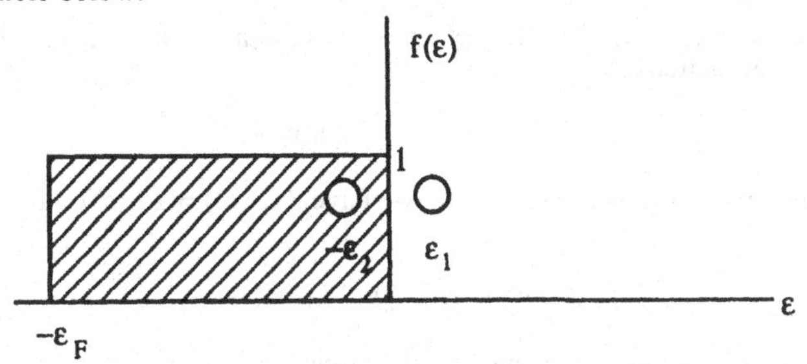

Figure 3. Fermi distribution function f(e).

Let us now consider correlated electron pairs instead of single electrons. The probability that a pair state of energy 2ε (energy ε per electron) is occupied is written as h(ε). A remarkable characteristic of this approach is that the pairs behave formally as bosons with regard to different states but as fermions with regard to the same states. This form of mixed statistics is an innovation of BCS theory. We show in Fig. 4 an arbitrary pair distribution h(e).

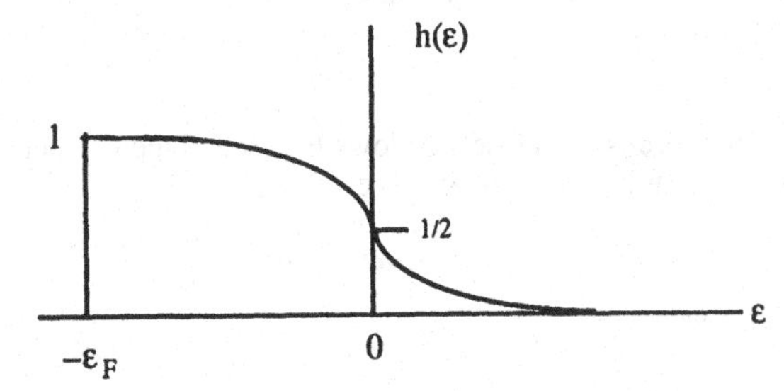

Figure 4. Arbitrary pair distribution function h(e).

Kinetic energy

The kinetic energy of correlated pairs is written in terms of the sum over filled states above the Fermi surface and empty correlated states below the Fermi surface:

$$W_{KE} = 2\,\Sigma_{\varepsilon>0}\ \varepsilon\, h(\varepsilon) + 2\,\Sigma_{\varepsilon<0}\ (-\varepsilon)\,[1 - h(\varepsilon)] \tag{48}$$

Interaction energy

The development in this section leads to an expression for the interaction energy. It is clear that if we are to achieve a lowering of the energy as a result of the interaction between pairs, the electron pairs must be correlated. That is, the states must both be empty or both filled in a *correlated* way. The correlation function is given by

$$\langle \Delta h(\varepsilon_1)\, \Delta h(\varepsilon_2)\rangle = \langle [h(\varepsilon_1) - \langle h(\varepsilon_1)\rangle]\ [h(\varepsilon_2) - \langle h(\varepsilon_2)\rangle]\rangle$$

$$= \langle h(\varepsilon_1)\ h(\varepsilon_2)\rangle - \langle h(\varepsilon_1)\rangle \langle h(\varepsilon_2)\rangle \quad (49)$$

If $h(\varepsilon)$ is a slowly varying function of energy, we may write the rms fluctuation in occupation (or correlation) as

$$\Delta h(\varepsilon) = [\langle h^2(\varepsilon)\rangle - \langle h(\varepsilon)\rangle^2]^{1/2} \quad (50)$$

Since $h(\varepsilon)$ has only the values zero or one, we have $\langle h^2(\varepsilon)\rangle = \langle h(\varepsilon)\rangle$ and Eq. (50) may be simplified to

$$\Delta h(\varepsilon) = \{h(\varepsilon)[1 - h(\varepsilon)]\}^{1/2} \quad (51)$$

where for simplicity of notation the average signs have been dropped. Assuming the maximum correlation between pairs allowed by the increased kinetic energy, we write the interaction energy as

$$W_I = -V\ \Sigma_{\varepsilon,\varepsilon'}\ \{h(\varepsilon)[1 - h(\varepsilon)]\ h(\varepsilon')[1 - h(\varepsilon')]\}^{1/2} \quad (52)$$

and the total electronic energy is the sum of Eqs. (48) and (52)

$$W = W_{KE} + W_I \quad (53)$$

Ground state

The remainder of the discussion closely follows BCS[18] and the Cooper article.[24] To obtain the function $h(\varepsilon)$ that minimizes the total energy we set

$$\partial W/\partial h(\varepsilon) = 0 \quad (54)$$

and obtain the expression

$$h(\varepsilon) = \frac{1}{2}(1 - \varepsilon/\sqrt{\varepsilon_0^2 + \varepsilon^2}\,) \quad (55)$$

sketched in Fig. 5(a) with

$$\varepsilon_0 = V\ \Sigma_{\varepsilon'}\ \{h(\varepsilon')[1 - h(\varepsilon')]\}^{1/2} \quad (56)$$

The rms fluctuation amplitude

$$\Delta h(\varepsilon) = \{h(\varepsilon)[1 - h(\varepsilon)]\}^{1/2} = \varepsilon_0/2\sqrt{\varepsilon_0^2 + \varepsilon^2} \quad (57)$$

is sketched in Fig. 5(b).

Substituting into Eq. (56) for $h(\varepsilon)$ from Eq. (55) gives

$$2/V = \Sigma_\varepsilon\ 1/\sqrt{\varepsilon_0^2 + \varepsilon^2} \quad (58)$$

which is the relationship between the interaction energy V and the energy parameter ε_0.

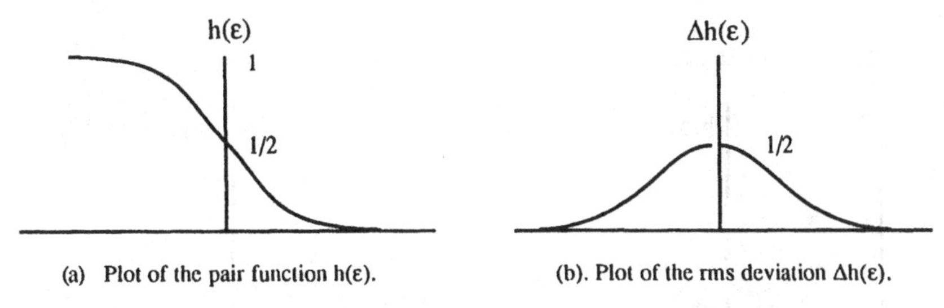

(a) Plot of the pair function h(ε). (b). Plot of the rms deviation Δh(ε).

Figure 5

Note that the sum given in Eq. (58) would diverge if taken over all states and must be limited to energies less than the Debye energy. This limitation gives a fraction of interacting pairs of order $\hbar\omega_D/\varepsilon_F = 2v_s k_D/v_F k_F \approx v_s/v_F \approx 10^{-3}$ as remarked earlier.

Excited states

In this section we calculate the energy required to break up a single correlated pair. We represent the excited state by a pair distribution function h(ε) (in which we are precluded from forming pair states at the single particle energies) and a single particle distribution function f(ε), which represents two uncorrelated electrons. These two functions are sketched in Figs. 6(a) and 6(b).

The increased kinetic energy is given by

$$\Delta W_{KE} = \varepsilon' + \varepsilon'' - 2\,[\,\varepsilon' h(\varepsilon') + \varepsilon''\, h(\varepsilon'')] \tag{59}$$

Substituting from Eq. (55), we obtain for the increased kinetic energy

$$\Delta W_{KE} = \varepsilon'^2/\sqrt{\varepsilon_0^2 + \varepsilon'^2} + \varepsilon''^2/\sqrt{\varepsilon_0^2 + \varepsilon''^2} \tag{60}$$

Note that if the number of electrons is to remain constant we must have on average

$$h(\varepsilon') + h(\varepsilon'') = 1 \tag{61}$$

The reduced interaction energy is given by

$$\Delta W_I = 2V\,\Sigma_\varepsilon\, \{h(\varepsilon)[1 - h(\varepsilon)]\}^{1/2}\, [\{h(\varepsilon')[1 - h(\varepsilon')]\}^{1/2} + \{\, h(\varepsilon'')[1 - h(\varepsilon'')]\}^{1/2}] \tag{62}$$

Again, substituting from Eq. (55) for h(ε) we obtain

$$\Delta W_I = \varepsilon_0^2/\sqrt{\varepsilon_0^2 + \varepsilon'^2} + \varepsilon_0^2/\sqrt{\varepsilon_0^2 + \varepsilon''^2} \tag{63}$$

Finally, adding Eqs. (60) and (63) we obtain for the increased energy

$$\Delta W = \sqrt{\varepsilon_0^2 + \varepsilon'^2} + \sqrt{\varepsilon_0^2 + \varepsilon''^2} \tag{64}$$

and we see that the minimum energy required to break up a correlated pair is given by

$$\Delta W = 2\,\varepsilon_0 \tag{65}$$

which is the energy gap 2Δ.

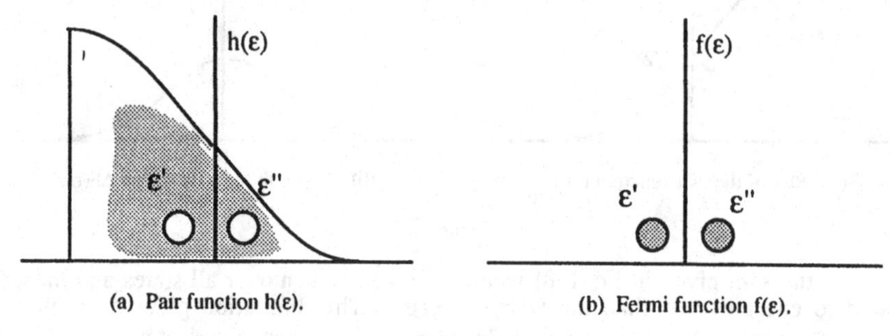

(a) Pair function $h(\varepsilon)$. (b) Fermi function $f(\varepsilon)$.

Figure 6

BCS write the free energy $F = W(T) - TS$ and obtain the thermal properties of the superconducting state. Their arguments follow directly from the expressions given above for the energy.

2.5.5. Coherence length

As a consequence of the presence of a gap in the energy spectrum, the BCS theory has a built-in coherence length

$$\xi_0 = \frac{\hbar v_F}{\pi\Delta(0)} \tag{66}$$

States within the gap $2\Delta(0)$ at the Fermi surface occupy an interval in wavenumber δk_F with

$$2\Delta(0) = \delta\varepsilon_F = v_F\,\hbar\,\delta k_F \tag{67}$$

Taking $\delta k_F = (2/\pi\xi_0)$ leads to Eq. (66).

2.6. Mattis-Bardeen Electrodynamics

2.6.1. Theory

Mattis and Bardeen[26] and independently Abrikosov, Gor'kov and Khalatnikov[27] have extended the BCS derivation of the Meissner effect to $\omega > 0$ through the development of a quantum mechanical form of the Chambers integral[10,11] with diffuse surface scattering. This work has recently been reviewed by Turneaure, Halbritter and Schwettman in one of a series of papers commemorating Bardeen's contributions to the understanding of superconducting and normal metals.[23] The following discussion draws substantially on this review.

---

[26] D. C. Mattis and J. Bardeen, *Phys. Rev.* **111**, 412 (1958) .

[27] A. A. Abrikosov, L. P. Gor'kov and I. M. Khalatnikov, *Sov. Phys.-JETP* **8**, 182 (1959).

Mattis and Bardeen begin by developing a relation between the current density and an ac applied field well within a conductor, where surface scattering may be neglected. They introduce the time-dependent Schrödinger equation

$$\mathcal{H}_0\Phi + \mathcal{H}_{ext}\Phi = i\hbar\frac{\partial\Phi}{\partial t} \tag{68}$$

Here $\mathcal{H}_0$ is the Hamiltonian is the absence of external fields but includes lattice scattering. $\mathcal{H}_{ext}$ is the Hamiltonian of the electromagnetic interaction, which is treated in first-order time-dependent perturtbation theory, yielding a current density for which only terms linear in the applied field are retained. Random scattering is included in the expression for the current density through the physically plausible factor exp $(-\rho/l)$. This approach leads to the following expression for the current density

$$J(r,t) = \frac{e^2N(0)v_F}{2\pi^2\hbar}\int\frac{\rho[\rho\cdot A_\omega(r')e^{-i\omega t}]I(\omega,\rho,T)e^{-\rho/l}}{\rho^4}dV' \tag{69}$$

which is similar to Pippard's expression as given by Eq. (28). The quantity N(0) is the one-spin electron density of states at the Fermi surface for the normal conducting metal. The kernel $I(\omega, \rho, T)$ in Eq. (69) is

$$\begin{aligned} I(\omega,\rho,T) = -\pi i \int_{\Delta-\hbar\omega}^{\infty}[1-2f(\varepsilon+\hbar\omega)][g(\varepsilon)\cos\alpha\varepsilon_2 - i\sin\alpha\varepsilon_2]e^{i\alpha\varepsilon_1}d\varepsilon \\ + \pi i\int_{\Delta}^{\infty}[1-2f(\varepsilon)][g(\varepsilon)\cos\alpha\varepsilon_1 + i\sin\alpha\varepsilon_1]e^{i\alpha\varepsilon_2}d\varepsilon \end{aligned} \tag{70}$$

with $\varepsilon_1 = (\varepsilon^2 - \Delta^2)^{1/2}$, $\varepsilon_2 = [(\varepsilon + \hbar\omega)^2 - \Delta^2]^{1/2}$, $g(\varepsilon) = (\varepsilon^2 + \Delta^2 + \hbar\omega\varepsilon)/\varepsilon_1\varepsilon_2$ and $\alpha = \rho/\hbar v_F$. This kernel is an extension of the BCS kernel[18] to nonzero frequencies. In the limit $\Delta \to 0$ Eqs. (69) and (70) reduce to Chamber's expression Eq. (27) for the current density of a normal conducting metal in agreement with Reuter and Sondheimer.[12]

Eqs. (69) and (70) for the current density are complex and involve integration over four variables. The current density is a function of the material parameters $\Delta(0)$, N(0), v_F and l as well as the applied field at angular frequency ω and the temperature T. The related parameters $\lambda_L(0)$ and ξ_0 may be taken in place of N(0) and v_F.

In developing their expression for the current density, Abrikosov *et al.*[27,28] start from the Green's function method, which allows a versatile and elegant treatment of perturbations such as scattering, characterized by an elastic mean free path, in contrast to a transport mean free path or a quasi-particle lifetime associated with the Boltzmann transport equation. This elastic mean free path l is related to the momentum uncertainty $\Delta p \approx \hbar/l$ and for large l yields nonlocal effects. Otherwise, the expression for the surface impedance given by Abrikosov *et al.* is equivalent to that given by Mattis and Bardeen.

Theories of the BCS surface impedance have been extended to include the effects of anisotropy and strong coupling. For conventional superconductors, inclusion of anisotropy usually leads to small corrections to the isotropic BCS surface impedance

---

[28] I. M. Khalatnikov and A. A. Abrikosov, *Adv. Phys.* 8, 45 (1958).

through averaged material parameters.[29] Strong-coupling corrections to the superconducting state are usually obtained with the Eliashberg[30] equations. Nam[31] has provided a general framework for calculating the surface impedance that includes the effects of strong coupling. For $\hbar\omega << 2\Delta$, Blaschke and Blocksdorf[32] have performed calculations that include strong-coupling effects for a number of superconductors including Sn, Pb and Nb. They compared their calculations with the weak-coupling BCS surface impedance calculated with the same material parameters. For Pb and Nb there are small but significant differences between the strong- and weak-coupling calculations but, as expected, none for Sn.

2.6.2. Numerical calculation

Miller[33] has performed numerical calculations of the Mattis-Bardeen surface impedance for finite ξ_0 in the Pippard limit $l \rightarrow \infty$ for diffuse scattering at the surface. As Turneaure *et al.* comment, these calculations show for Sn and Al that the inclusion of accepted values of material parameters for these superconductors gives surface resistances that are in good agreement with experiment and are significantly larger than those calculated from Eq. (28) in the Pippard limit. Corrections to the surface reactance are smaller.

Programs that evaluate the surface impedance numerically were developed by Turneaure[34,35] using the Mattis-Bardeen expression for the current density and independently by Halbritter[36,37] using the Green's function formalism of Abrikosov, Gor'kov and Dzaloshinskii.[38] Turneaure, following the approach of Miller, reduced the expression for the Mattis-Bardeen surface impedance with finite l to a form tractable for numerical integration with either diffuse or specular surface scattering.

2.6.3. Comparison with experiment

Samples with small values of residual resistance[34] allowed an early comparison of experiment with theory, demonstrating the characteristic exponential temperature dependence of surface resistance on the energy gap $R_s \propto \exp(-\Delta/k_BT)$. More recently, surface impedance measurements have been made and compared with the BCS surface impedance for a variety of superconductors including Pb, Nb, NbN, and Nb_3Sn and still more recently, the high-T_c superconductors. This experimental work provides surface impedance measurements over a wide range of material parameters and of temperature and frequency. Table 1, given earlier, lists a number of these superconductors and material parameters deduced from surface impedance and other measurements.

---

[29] R. Blaschke, J. Askenazi, O. Pictet, D. D. Koelling, A. T. von Kessel and F. M.Muller, *J. Phys. F* **14**, 175 (1984).

[30] G. M. Eliashberg, *Sov. Phys.-JETP* **11**, 696 (1960).

[31] S. B. Nam, *Phys. Rev.* **156**, 470 and 487 (1967).

[32] R. Blaschke and R. Blocksdorf, *Z. Phys. B* **49**, 99 (1982).

[33] P. B. Miller, *Phys. Rev.* **118**, 928 (1960).

[34] J. P. Turneaure, Ph.D. Dissertation, Stanford University, 1967 unpublished.

[35] J. P. Turneaure and I. Weissman, *J. Appl. Phys.* **39**, 4417 (1968).

[36] J. Halbritter, *Reports 3/69-2 and 3/70-6* (Kernforschungszentrum, Karlsruhe, 1969 and 1970).

[37] J. Halbritter, *Z. Phys.* **266**, 209 (1974).

[38] A. A. Abrikosov, L. P. Gor'kov and I. Yu. Dzaloshinskii, *Quantum Field Theoretical Methods in Statistical Physics* (Pergamon Press, New York, 1965).

Temperature dependence

Comparisons of the measured temperature dependence of the surface impedance with theory incorporate the material parameters in various ways. The quantities $\lambda_L(0)$ and v_F, which are averages over the Fermi surface, are obtained from other experiments or from theoretical analysis. The gap parameter $\Delta(0)$ is typically left as a free parameter in fitting the theory to experiment because the surface impedance is exponentially dependent on this parameter for $T/Tc < 0.5$. The temperature dependence of $\Delta(T)/\Delta(0)$ is obtaned from theory. The mean free path l may be left as a free parameter as well.

Frequency dependence

Measurements of the surface impedance with frequency have focused on the surface resistance because it is difficult to detect small differences in $\lambda(T, \omega) - \lambda(T, 0)$ expected for frequencies below $2\Delta/\hbar$. Such measurements of the surface resistance typically require several cavities excited in one or more modes to span a wide frequency range. Experimental data on high quality superconducting materials[39,40] are in good agreement with theory[41] when account is taken of energy-gap anisotropy.

Dependence on material parameters

To study the dependence of the surface impedance on the material parameters $\lambda_L(0)$, v_F, $\Delta(0)$ and l, measurements should be made over as wide a parameter range as possible. Generally, experimental values of $\lambda_L(0)$ and v_F can be deduced from other types of measurements such as the normal-state surface resistance in the vicinity of the anomalous limit, the electronic specific heat, and the magnetic properties of the superconductor. Because the surface impedance is so strongly dependent on $\Delta(0)$, this parameter is usually left free. The mean free path l may be deduced from measurements such as the low-temperature electrical conductivity or the normal-state surface resistance so long as the sample is not in the anomalous limit. These values are not always appropriate, however, because the mean free path is often reduced in the vicinity of the surface by additional scattering. For these reasons. the mean free path is often left as a free parameter as well.

In BCS theory, scattering reduces the momentum lifetime, leading to an increase in penetration depth

$$\lambda(T,l) = \sqrt{1+\pi\xi_0/2l}\;\lambda(T,\infty) \tag{72}$$

This dependence on mean free path has been verified through stripline measurements.[42]

The dependence of the surface resistance on mean free path is more complex. With decreasing l, the surface resistance at first decreases reaching a minimum for $l \approx \xi_0$ and then increases. Flécher[43] has examined this effect for Pb alloyed with varying amounts of Bi. Given the experimental uncertainties in l, the measurements appear to be in satisfactory agreement with theory.

---

[39] U. Klein, Ph.D. Dissertation, University of Wuppertal, 1981 unpublished.

[40] G. Müller, Ph.D. Dissertation, University of Wuppertal, 1983 unpublished.

[41] R. Blaschke, U. Klein and G. Müller, *Verhandl. DPG (VI)* **17**, 988 (1982).

[42] W. H. Henkels and C. J. Kircher, *IEEE Trans. Magn.* **13**, 63 (1977).

[43] P. Flécher, Ph.D. Dissertation, University of Karlsruhe, 1970 unpublished.

High-T_c superconductors

The surface impedance of ceramic samples and thick films of the high-T_c superconductors is dominated by grain boundaries. As discussed in Chs. 11 and 12, grain boundaries may exhibit Josephson phenomena, leading to field-dependent loss and penetration. Further, such grain boundaries lead to transient behavior in swept magnetic fields, as discussed in Ch. 13 and power-dependence as discussed in Ch. 14.

High quality thin films show little effect of the presence of grain boundaries. When such films are patterned into narrow filaments, however, strong evidence may appear of the presence of grain boundaries at filament edges. In very narrow filaments, grain boundaries may completely cross a filament. These boundaries then behave very much like grain boundaries in thick films or bulk ceramics, leading to an undesirable sensitivity to magnetic fields and to elevated power.

Anlage *et al.*[44] have determined the material parameters $\lambda_L(T)$, ξ_0, and $\Delta(T)$ from the temperature dependence of the frequency of a stripline resonator. Their value of 4.5 for $2\Delta(0)/k_BT_c$ is close to the BCS value 3.52. Other investigators, however, have found substantially larger gap values.[45]

Glass and Hall[46] have analyzed the microwave transmission data of Kobrin *et al.*[47] through thin epitaxial films of $YBa_2Cu_3O_7$. Values of σ_2 are obtained down to about 45 K below which there is insufficient transmission. Accurate values of σ_1 are obtained only down to 80 K because of the rapid drop in σ_1 with decreasing temperatures. These authors fit to the weak-coupling BCS expression for the complex conductivity in the local limit. A fit is obtained from 80 up to 88 K with the mean free path l and the carrier effective mass m* regarded as free parameters. Above 88 K the films become granular, requiring the application of effective medium theory. An interesting result of the analysis below 88 K is that the mean free path appears to increase rapidly with decreasing temperature.

---

[44] S. M. Anlage, B. W. Langley, G. Deutscher, J. Halbritter and M. R. Beasley, in press.

[45] Z. Schlesinger, R. T. Collins, F. Holtzberg, C. Field, G. Koren and A. Gupta, *Phys. Rev. B* **41**, 11 237 (1990).

[46] N. E. Glass and W. F. Hall, *Phys. Rev. B* **44**, 4495 (1991).

[47] P. H. Kobrin, J. T. Cheung, W. W. Ho, N. Glass, J. Lopez, I. S. Gergis, R. E. DeWames and W. F. Hall, *Physics C* **176**, 121 (1991).

3
ELECTRODYNAMICS

3.1. Microscopic Maxwell Equations

My purpose in this chapter is to show how electrodynamics may be used to analyze fields in magnetic and superconducting materials. Over fifteen years ago I wrote a textbook that tried to apply electrodynamics to solid state physics.[1] This chapter is largely a distillation of that text. I don't think I discovered anything that Maxwell didn't know and in some sense my approach to developing an electromagnetism course for solid state physicists was a return to Maxwell's ideas.

The issue that concerns the solid state physicist is the relation between an electrodynamic system and its environment. This problem doesn't arise in the physics of interacting particles since everything is described microscopically. The problem does arise, however, in a solid where there are mechanical, chemical and thermal processes that are *outside* electrodynamics. One needs a way to properly describe the way in which the electrodynamic system is coupled to the outside world. My thesis is that, as Maxwell understood very well, the constructed fields **D** and **H** do precisely this. The Maxwell equations in these fields provide a *macroscopic* formulation of electrodynamics.

The microscopic Maxwell equations are

$$\nabla \cdot \mathbf{E} = \rho/\varepsilon_0 \tag{1}$$

$$\nabla \times \mathbf{E} = -\partial \mathbf{B}/\partial t \tag{2}$$

[1] A. M. Portis, *Electromagnetic Fields, Sources and Media* (Wiley, New York, 1974).

$$\nabla \times \mathbf{B} = \mu_0(\mathbf{J} + \varepsilon_0 \partial \mathbf{E}/\partial t) \tag{3}$$

$$\nabla \cdot \mathbf{B} = 0 \tag{4}$$

$$\mathbf{F} = \rho \mathbf{E} + \mathbf{J} \times \mathbf{B} \tag{5}$$

where the Lorentz force per unit volume **F** defines **E** and **B**.

3.1.1. Current and charge conservation

Taking the divergence of Eq. (3) and using Eq. (1) we obtain

$$\nabla \cdot \mathbf{J} + \partial \rho/\partial t = 0 \tag{6}$$

This is a statement of the conservation of electric charge and is built into the Maxwell equations, which also lead to conservation equations of this form for potentials, for energy and for momentum.[2]

3.1.2. Vector and scalar potentials

Maxwell recognized that the equations of electrodynamics could be written in terms of potentials rather than fields. We digress for a moment to discuss sources of vector fields. Any vector **U** may be characterized by its sources of divergence and circulation

$$\nabla \cdot \mathbf{U} = \nabla^2 W \qquad \nabla \times \mathbf{U} = \nabla \times (\nabla \times \mathbf{V}) \tag{7}$$

from which we may write

$$\mathbf{U} = \nabla W + \nabla \times \mathbf{V} \tag{8}$$

with W and **V** the scalar and vector potentials of **U**. The vector ∇W is described as irrotational since its curl is zero. The vector $\nabla \times \mathbf{V}$ is described as solenoidal since its divergence is zero. We may read Eq. (4) as a statement that **B** is solenoidal with

$$\mathbf{B} = \nabla \times \mathbf{A} \tag{9}$$

where **A** is called the vector potential.[3] If we now substitute Eq. (9) into Eq. (2) we obtain the source of circulation of **E**

$$\nabla \times \mathbf{E} = -(\partial/\partial t)\, \nabla \times \mathbf{A} \tag{10}$$

We may integrate Eq. (10) to obtain

$$\mathbf{E} = -\nabla\phi - \partial \mathbf{A}/\partial t \tag{11}$$

---

[2] Such equations may all be written as the four-divergence of a four-vector, which gives them elegance but may obscure the physics.

[3] We note that Eq. (8) is a statement of the sources of circulation of **A**. Nothing has been said thus far about the sources of divergence of **A**.

Substituting Eqs. (9) and (11) into the Maxwell equations we have two equations in the potentials instead of four in the fields

$$\nabla^2\phi + (\partial/\partial t)\nabla\cdot\mathbf{A} = -\rho/\varepsilon_0 \quad (12)$$

$$\nabla(\nabla\cdot\mathbf{A}) - \nabla^2\mathbf{A} + \mu_0\varepsilon_0(\partial/\partial t)\,[\nabla\phi + (\partial/\partial t)\,\mathbf{A}] = \mu_0\,\mathbf{J} \quad (13)$$

3.1.3. Gauge and the divergence of A

Although we may have succeeded in reducing the four Maxwell equations to two, we have not simplified things very much. One problem is that we have an unspecified divergence of **A** floating around. We now examinine two choices that we might make for the divergence of **A**, which through Eq. (8) may be regarded as gauge transformations.

Coulomb gauge

The simplest choice is to take **A** solenoidal

$$\nabla\cdot\mathbf{A} = 0 \quad (14)$$

which gives for Eq. (12)

$$\nabla^2\phi = -\rho/\varepsilon_0 \quad (15)$$

This choice makes ϕ the *instantaneous* Coulomb potential and for that reason is called the Coulomb gauge.

Lorentz gauge

A very elegant solution to the gauge problem is to take

$$\nabla\cdot\mathbf{A} + \mu_0\varepsilon_0\,\partial\phi/\partial t = 0 \quad (16)$$

which looks a lot like Eq. (6), suggesting a kind of conservation of potential with **A** current-like and ϕ charge-like. Substituting into Eqs. (12) and (13) gives

$$\nabla^2\phi - \mu_0\varepsilon_0\,\partial^2\phi/\partial t^2 = -\rho \quad (17)$$

$$\nabla^2\mathbf{A} - \mu_0\varepsilon_0\,\partial^2\mathbf{A}/\partial t^2 = -\mu_0\,\mathbf{J} \quad (18)$$

The Maxwell equations in this form have a very beautiful symmetry with decoupled wave equations for ϕ and **A** whose sources are coupled by Eq. (6).[4]

London gauge

The London gauge, which is useful for superconductivity, takes $\nabla\phi = 0$ and also assumes **A** solenoidal as in Eq. (14). London's motivation was to scale solenoidal currents with the vector potential.

---

[4] By combining **A** and ϕ into a four-vector and the operator $\nabla^2 - \partial^2/\partial t^2$ into the four-Laplacian, we can reduce all of electrodynamics to a statement that the four-Laplacian of the four-potential is the four-current!

3.1.4. *Energy conservation and flow*

We now return to the Maxwell equations in their original form and note that if we take the divergence of a vector

$$\mathcal{S} = (1/\mu_0)\,\mathbf{E}\times\mathbf{B} \tag{19}$$

we obtain

$$\nabla\cdot\mathcal{S} + \partial u/\partial t = w \tag{20}$$

$$\partial u/\partial t = \varepsilon_0 E\,\partial E/\partial t + (1/\mu_0)B\,\partial B/\partial t \tag{21}$$

$$w = -\,\mathbf{E}\cdot\mathbf{J} \tag{22}$$

How do we interpret Eq. (20)? The clue is in the form of w. If $\rho\mathbf{E}$ is the force on the charge density ρ, then w must be the rate at which work is done on a unit volume of the electromagnetic field. If we then regard u as an energy density and **S** as energy flow, Eq. (20) becomes a statement of energy conservation.

3.1.5. *Momentum conservation and flow*

Let's now try to write an equation for momentum conservation. Taking our cue from the statement of energy conservation we might expect to write

$$\frac{\partial}{\partial t}\frac{\mathbf{dp}}{\mathrm{dV}} + \nabla\cdot\ddot{\mathcal{P}} = \mathbf{F}_{ext} \tag{23}$$

with **dp**/dV the momentum density, $\ddot{\mathcal{P}}$ the momentum flow tensor[5] and $\mathbf{F}_{ext}$ a (possibly fictitious) mechanical force that equilibrates the Lorentz force density and the electromagnetic pressure

$$\mathbf{F}_{ext} = -\,\rho\mathbf{E} - \mathbf{J}\times\mathbf{B} + \nabla u \tag{24}$$

with u given by Eq. (20). Working through the Maxwell equations confirms Eqs. (23) and (24) with $\ddot{\mathbf{I}}$ the unit tensor.

$$\mathbf{dp}/\mathrm{dV} = \varepsilon_0\,\mathbf{E}\times\mathbf{B} \tag{25}$$

$$\ddot{\mathcal{P}} = \left(\varepsilon_0 E^2 + \frac{1}{\mu_0}B^2\right)\ddot{\mathbf{I}} - \left(\varepsilon_0\mathbf{EE} + \frac{1}{\mu_0}\mathbf{BB}\right) \tag{26}$$

[5]Momentum flow must be represented by a tensor because there are two directions, the direction of the momentum and the direction of the flow.

3.2. Macroscopic Maxwell Equations

As elegant as is the microscopic formulation of the Maxwell equations, it does not tell us anything about the response of ρ and **J** to other than electrodynamic forces and, in fact, the theory only works when there are no other forces and the charges and currents are free. Maxwell's solution to this problem was to introduce special charges and currents subject to forces outside electrodynamics. For example, we may polarize a dielectric by an electric field produced by charges on capacitor plates with potentials established by an electrochemical cell. Or we may magnetize a material with a current flowing from a generator driven by a steam turbine. The behavior of the electrochemical cell and the turbine are outside usual electromagnetic theory and we do well to exclude their electrochemistry and thermodynamics.

3.2.1. External charge and current

We now distinguish between polarizing charges and *polarization* charge, on the one hand, and magnetizing currents and *magnetization* current on the other. We begin by separating the charge density ρ into two components

$$\rho = -\nabla \cdot \mathbf{P} + \rho_{ext} \tag{27}$$

where **P** is called the polarization density and describes the charge displacement resulting from electrodynamic forces. The second term represents charge that in some way is not subject to the usual forces of electrodynamics but is still conserved in the usual way

$$\nabla \cdot \mathbf{J}_{ext} + \partial\rho_{ext}/\partial t = 0 \tag{28}$$

This may all sound unphysical. It may be thought of as an artifice for asking something like the following question: if there were a charge density $\rho_{ext}(\mathbf{r})$ what would be the induced polarization density?

Substituting Eq. (27) into Eq. (6) using Eq. (28) gives

$$\nabla \cdot \mathbf{J} = \nabla \cdot \partial\mathbf{P}/\partial t + \nabla \cdot \mathbf{J}_{ext} \tag{29}$$

We call $\partial\mathbf{P}/\partial t$ the polarization current and characterize not only dielectric currents but also metallic currents by the rate of change of polarization.[6] Integrating Eq. (29), we introduce the magnetization density **M** whose curl gives the magnetization current.

$$\mathbf{J} = \partial\mathbf{P}/\partial t + \nabla \times \mathbf{M} + \mathbf{J}_{ext} \tag{30}$$

We now substitute Eqs. (28) and (30) into the Maxwell equations and obtain

$$\nabla \cdot (\varepsilon_0\mathbf{E} + \mathbf{P}) = \rho_{ext} \tag{31}$$

$$\nabla \times [(1/\mu_0)\mathbf{B} - \mathbf{M}] = \mathbf{J}_{ext} + (\partial/\partial t)(\varepsilon_0\mathbf{E} + \mathbf{P}) \tag{32}$$

Following Maxwell, we introduce two new fields, the displacement

$$\mathbf{D} = \varepsilon_0\mathbf{E} + \mathbf{P} \tag{33}$$

---

[6]This is the justification for the use of the dielectric function $\varepsilon(\omega, \mathbf{k})$ in describing metallic conductivity.

and the magnetizing field

$$\mathbf{H} = (1/\mu_0)\mathbf{B} - \mathbf{M} \tag{34}$$

The Maxwell equations may now be written with external sources as

$$\nabla \cdot \mathbf{D} = \rho_{ext} \tag{35}$$

$$\nabla \times \mathbf{E} + \partial\mathbf{B}/\partial t = 0 \tag{36}$$

$$\nabla \times \mathbf{H} - \partial\mathbf{D}/\partial t = \mathbf{J}_{ext} \tag{37}$$

$$\nabla \cdot \mathbf{B} = 0 \tag{38}$$

with the force

$$\mathbf{F} = \rho\mathbf{E} + \mathbf{J} \times \mathbf{B} + \mathbf{F}_{ext} \tag{39}$$

where $\mathbf{F}_{ext}$ is a force outside electrodynamics of electrochemical or thermoelectric origin, for example. The external force may be integrated around a closed contour to give

$$\oint \mathbf{F}_{ext} \cdot d\mathbf{r} = \rho\mathcal{E}_{ext} \tag{40}$$

where ρ is the charge density on which $\mathbf{F}_{ect}$ acts and $\mathcal{E}_{ext}$ is an external electromotive force. Maxwell called the quantity $\partial\mathbf{D}/\partial t$ the displacement current. Equations (9) and (11) may still be used to write **E** and **B** in terms of potentials.

3.2.2. Energy and work

What is meant by electromagnetic energy is now changed because of the presence of external charges, currents and forces that may do work on the system from outside. Taking the divergence of the Poynting vector

$$\mathcal{S} = \mathbf{E} \times \mathbf{H} \tag{41}$$

gives Eq. (20) with the rate of energy change and the rate of work redefined

$$\partial u/\partial t = \mathbf{E} \cdot \partial\mathbf{D}/\partial t + \mathbf{H} \cdot \partial\mathbf{B}/\partial t + \mathbf{F}_{ext} \cdot \mathbf{v} \tag{42}$$

$$w = -\mathbf{E} \cdot \mathbf{J}_{ext} + \mathbf{F}_{ext} \cdot \mathbf{v} \tag{43}$$

with **v** the velocity of the charges on which $\mathbf{F}_{ext}$ acts. Equation (42) is not in general integrable because of the presence of dissipation. We see from Eq. (43) that *external* currents and *external* forces are now the mechanisms for doing work on a system.

3.2.3. Momentum density and flow

The macroscopic Maxwell equations lead to macroscopic momentum conservation as given by Eq. (23) but with the momentum density and flow redefined as

$$dp/dV = \mathbf{D} \times \mathbf{B} \tag{44}$$

$$\tilde{\mathcal{P}} = (\mathbf{D} \cdot \mathbf{E} + \mathbf{B} \cdot \mathbf{H})\tilde{\mathbf{I}} - (\mathbf{DE} + \mathbf{BH}) \tag{45}$$

The force per unit volume is given by

$$\mathbf{F} = - \rho_{ext} \mathbf{E} - \mathbf{J}_{ext} \times \mathbf{B} + \nabla u + \mathbf{F}_{ext} \tag{46}$$

with

$$\nabla u = \tfrac{1}{2}[(\nabla \mathbf{D}) \cdot \mathbf{E} + (\nabla \mathbf{B}) \cdot \mathbf{H} + (\nabla \mathbf{E}) \cdot \mathbf{D} + (\nabla \mathbf{H}) \cdot \mathbf{B}] \tag{47}$$

Field momentum

The sum of the kinetic and field momenta of a charge at **r** is

$$\mathbf{p} = m\mathbf{v} + \int dV\, \mathbf{D} \times \mathbf{B} = m\mathbf{v} + \int dV\, \mathbf{D} \times (\nabla \times \mathbf{A}) \tag{48}$$

where we have used Eq. (45) for the field momentum. In order to determine the field momentum we substitute into Eq. (48) the identity

$$\mathbf{D} \times (\nabla \times \mathbf{A}) = \nabla(\mathbf{A} \cdot \mathbf{D}) - \nabla \cdot (\mathbf{AD} + \mathbf{DA}) + \mathbf{A}\nabla \cdot \mathbf{D} + \mathbf{D}\nabla \cdot \mathbf{A} - \mathbf{A} \times (\nabla \times \mathbf{D}) \tag{49}$$

with **A** solenoidal and **D** irrotational so that the last two terms are identically zero. Integrating over all space, the first two terms on the right of Eq. (49) integrate to zero and the momentum of the pair becomes

$$\mathbf{p}(\mathbf{r}) = m\, \mathbf{v}(\mathbf{r}) + q\, \mathbf{A}(\mathbf{r}) \tag{50}$$

3.3. Wave Properties

3.3.1. Dielectric function

Because magnetic fields are screened in a distance that is not always small compared with sample size, it is safest to treat magnetic screening currents explicitly. We use the two-fluid model with some carriers conducting as normal electrons

$$\mathbf{J}_n = \sigma \mathbf{E} \tag{51}$$

and other carriers providing a supercurrent that screens the interior

$$\ell\, \mathbf{J}_s = - \mathbf{A} \tag{52}$$

In addition we have to allow for the possibility of polarization current associated with core electrons

$$\mathbf{J}_{core} = (\partial/\partial t)\, \mathbf{P}_{core} \tag{53}$$

For a transverse electric field

$$\mathbf{E}(\mathbf{r}, t) = \mathbf{E}\, e^{i(\mathbf{k} \cdot \mathbf{r} - \omega t)} \tag{54}$$

E and **k** are orthogonal and the divergence of the electric field

$$\nabla \cdot \mathbf{E}(\mathbf{r}, t) = i\mathbf{k} \cdot \mathbf{E}(\mathbf{r}, t) \tag{55}$$

is zero. From Eq. (1) there can be no charge density and we may take $\phi = 0$. From Eq. (15) with $\phi = 0$ Eq. (11) becomes

$$\mathbf{E} = i\omega\, \mathbf{A} \tag{56}$$

Adding Eqs. (51) and (52), the total polarization current is

$$\partial \mathbf{P}/\partial t = - i\omega \mathbf{P} = -i\, \omega \mathbf{P}_{core} + \sigma \mathbf{E} - (1/\ell)\, \mathbf{A} \tag{57}$$

or using Eq. (55)

$$\mathbf{P} = \mathbf{P}_{core} + (i\sigma/\omega - 1/\omega^2 \ell)\mathbf{E} \tag{58}$$

The dielectric function $\varepsilon(\omega)$ is defined by

$$\mathbf{D} = \varepsilon_0\, \mathbf{E} + \mathbf{P} = \varepsilon(\omega)\, \mathbf{E} \tag{59}$$

Substituting Eq. (58) into (59) gives

$$\varepsilon(\omega) = \varepsilon_{core} - 1/\omega^2 \ell + i\sigma/\omega \tag{60}$$

and the dispersion relation

$$\omega = k/\sqrt{\mu_{core}\, \varepsilon(\omega)} \tag{61}$$

or what is equivalent

$$\omega^2 + i\omega\sigma/\varepsilon_{core} = 1/\ell\varepsilon_{core} + k^2/\mu_{core}\, \varepsilon_{core} \tag{62}$$

3.3.2. Penetration depth

Substituting the total current from Eq. (30) with the polarization from Eq. (58) into Eq. (3) gives in the absence of external current and charge

$$(1/\mu_0)\, \nabla \times \mathbf{B} = (\sigma - i\omega\varepsilon_{core})\, \mathbf{E} - (1/\ell)\, \mathbf{A} + \nabla \times \mathbf{M} \tag{63}$$

With the understanding that only core currents are considered to contribute to the magnetization, we write

$$\mathbf{B} = \mu_0(\mathbf{H} + \mathbf{M}) = \mu_{core}\, \mathbf{H} \tag{64}$$

Eq. (63) simplifies to

$$\nabla \times \mathbf{B} = \mu_{core}\,[(\sigma - i\omega\varepsilon_{core})\,\mathbf{E} - (1/\ell)\,\mathbf{A}] \tag{65}$$

Taking the curl of Eq. (64) and using Eq. (4) gives

$$\nabla^2\mathbf{B} - \mu_{core}\,[(\sigma - i\omega\varepsilon_{core})\,\partial\mathbf{B}/\partial t - (1/\ell)\,\mathbf{B}] = 0 \tag{66}$$

For propagation of a transverse plane wave into the superconductor

$$\mathbf{B}(\mathbf{r}, t) = \mathbf{B}\, e^{i(\mathbf{k}\cdot\mathbf{r} - \omega t)} \tag{67}$$

we obtain for the complex propagation vector

$$k^2 = \omega^2\,\mu_{core}\,\varepsilon(\omega) \tag{68}$$

which is equivalent to Eq. (61). This relation is plotted in Fig. 1.

3.3.3. London plasma frequency

A longitudinal electrostatic wave propagates through a medium at frequencies above the plasma frequency. Because the divergence of the displacement vector **D** is zero, the requirement for propagation of a plasma wave is $\varepsilon(\omega) = 0$. Setting Eq. (60) to zero and neglecting conduction gives for the London plasma frequency

$$\omega_L = 1/\sqrt{\varepsilon_{core}\ell} \tag{69}$$

At frequencies below ω_L, the permittivity $\varepsilon(\omega)$ is negative and $k = i\kappa$ is imaginary as shown in Fig. 1.

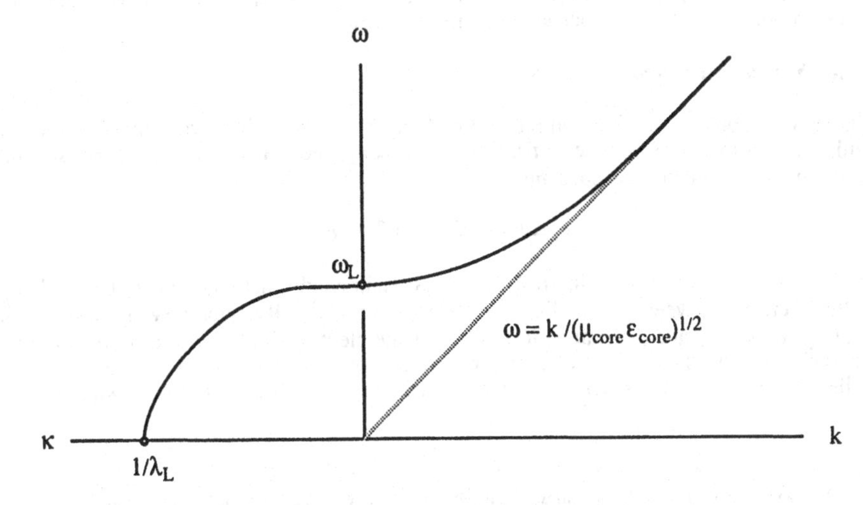

Figure 1. The dispersion relation of a lossless London plasma. At frequencies above the London plasma frequency, k is real and a transverse wave propagates through the superconductor. At frequencies below ω_L, the wave amplitude decays exponentially with attenuation constant κ.

3.3.4. Characteristic impedance

The characteristic impedance Z_c of a homogeneous medium may be defined in terms of the load presented at z by the medium beyond z

$$Z_c = E(z)/\int_z J\,dz = E(z)/H(z) = \omega\mu/k = [\mu_{core}/\varepsilon(\omega)]^{1/2} \tag{70}$$

where the final equality follows from Eq. (68).

3.3.5. Surface impedance

The surface impedance is the ratio of the tangential electric field at the surface E_s to the total current—polarization and magnetization currents—integrated *from the surface* into the medium

$$\int_s \mathbf{J}\,dz = \int (\nabla \times \mathbf{H})\,dz = \mathbf{n} \times \mathbf{H}_s \tag{71}$$

which gives

$$Z_s = R_s - iX_s = E_s/H_s \tag{72}$$

where we have used Eq. (2). The surface reactance $X_s = \omega L_s$, where L_s is called the surface inductance, is positive and arises from the inertia of the supercarriers. The surface resistance R_s is positive or zero and is non-vanishing only if there are normal or quasi-normal carriers that generate microwave loss. We see later that microwave loss may arise from the motion of viscously damped magnetic flux, driven by superconducting currents.

The surface impedance Z_s of a homogeneous medium is normally identical to the characteristic impedance Z_c. In Section 3.6 we consider an exception to this rule that arises when volume currents are associated with magnetization.

3.3.6. Surface electromagnetic waves

Under conditions of total internal reflection of an electromagnetic wave from the interface with a medium of lower index of refraction $n_1 < n_2$, the wave decays into the second medium with an attenuation constant[7]

$$\kappa_1 = (n_2^2 \sin^2\theta_2 - n_1^2)^{1/2}\,\omega/c \tag{73}$$

for angles of incidence $\theta_2 > \sin^{-1} n_1/n_2$. Under certain conditions a wave may be localized at the interface between two media, attenuating into both.[8,9] By symmetry, the wave must propagate with either the magnetic field **H** or the electric field **E** in the wavefront and parallel to the interface. The first case is called transverse magnetic (TM) and the second is called transverse electric (TE). The interface and coordinate directions are shown in Fig. 2.

---

[7] M. Born and E. Wolf, *Principles of Optics: Electromagnetic Theory of Propagation, Interference and Diffacction of Light*, 6th ed. (Pergamon, Oxford, 1980) sec. 1.5.4.

[8] R. F. Wallis in *Electromagnetic Surface Excitations*, eds. R. F. Wallis and G. I. Stegeman (Springer, Berlin, 1986).

[9] H. Raether, *Surface Plasmons on Smooth and Rough Surfaces and on Gratings* (Springer, Berlin, 1988).

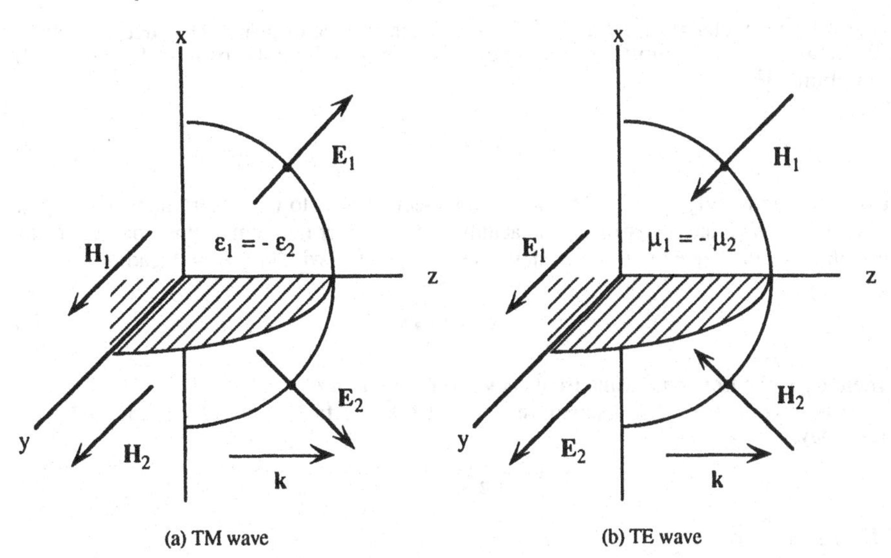

Figure 2. Electric and magnetic fields of a surface wave. The interface is in the yz-plane. The wavevector **k** is taken along the z-direction. a) A TM surface wave propagates with the magnetic field along the y-direction and the electric fields in the xz-plane. b) A TE surface wave propagates with the electric field along the y-direction and the magnetic fields in the xz-plane.

TM surface wave

The magnetic field **H** is parallel to both the interface, which is in the yz-plane and to the wavefront, which is in the xy-plane. In the upper medium the magnetic field is

$$\mathbf{H}_1 = \hat{\mathbf{y}} H \exp(ikz - \kappa_1 x) \tag{74}$$

and in the lower medium the magnetic field is

$$\mathbf{H}_2 = \hat{\mathbf{y}} H \exp(ikz + \kappa_2 x) \tag{75}$$

For the wave to attenuate in both media, both κ_1 and κ_2 must be positive. The Ampère-Maxwell law gives for the electric fields in the two media

$$i\omega\varepsilon_1\mathbf{E}_1 = (ik\hat{\mathbf{x}} + \kappa_1\hat{\mathbf{z}})H \exp(ikz - \kappa_1 x) \tag{76}$$

$$i\omega\varepsilon_2\mathbf{E}_2 = (ik\hat{\mathbf{x}} - \kappa_2\hat{\mathbf{z}})H \exp(ikz - \kappa_2 x) \tag{77}$$

For E_z continuous across the interface we must have $\varepsilon_1/\varepsilon_2 = -\kappa_1/\kappa_2$ and the dielectric permittivities are thus required to be opposite in sign in the two media.

In the short wavelength limit a plasma wave at the surface of a superconductor is nearly electrostatic with a limiting frequency given approximately by the Stern-Ferrell condition[10,11,12]

$$\omega^2 = \frac{1}{2}\omega_p^2 \tag{78}$$

This result is readily obtained by taking the electric field to be the gradient of a scalar potential.[13] With the upper medium vacuum and the lower medium a superconductor, the condition for the wave to be electrostatic is $k^2 = \kappa_1^2 = \kappa_2^2$ with $\varepsilon_1 + \varepsilon_2 = 0$ leading to

$$\omega^2 = \frac{1}{2}\omega_L^2 \tag{79}$$

where ω_L is the London plasma frequency given by Eq. (69).

It is also possible to propagate a wave at the interface between conductors[16] at a frequency

$$\omega^2 = \frac{1}{2}(\omega_{p1}^2 + \omega_{p2}^2) \tag{80}$$

TE surface wave

The electric field **E** is parallel to both the interface, which is in the xz-plane and to the wavefront, which is in the xy-plane. In the upper medium the electric field is

$$\mathbf{E}_1 = \hat{\mathbf{y}} E \exp(ikz - \kappa_1 x) \tag{81}$$

and in the lower medium the electric field is

$$\mathbf{E}_2 = \hat{\mathbf{y}} E \exp(ikz + \kappa_2 x) \tag{82}$$

For the wave to attenuate in both media, both κ_1 and κ_2 must be positive. The Faraday law gives for the magnetic fields in the two media

$$i\omega\mu_1\mathbf{H}_1 = -(ik\hat{\mathbf{x}} + \kappa_1\hat{\mathbf{z}})E \exp(ikz - \kappa_1 x) \tag{83}$$

$$i\omega\mu_2\mathbf{H}_2 = -(ik\hat{\mathbf{x}} - \kappa_2\hat{\mathbf{z}})E \exp(ikz - \kappa_2 x) \tag{84}$$

For H_z continuous across the interface we must have $\mu_1/\mu_2 = -\kappa_1/\kappa_2$ and the magnetic permeabilities are thus required to be opposite in sign in the two media. This condition may be achieved in resonant magnetic materials.[14,15]

---

[10] R. A. Ferrell, *Phys. Rev.* **111**, 1214 (1958).

[11] E. A. Stern, *Phys. Rev. Lett*, **8**, 7 (1961).

[12] R. A. Ferrell and E. A. Stern, *Am. J. Phys.* **30**, 810 (1962).

[13] C. Kittel, *Introduction to Solid State Physics*, 6th ed. (Wiley, New York, 1986) ch. 10, probs. 1 and 2.

[14] E. F. Sarmento and D. R. Tilley, in *Electromagnetic Surface Modes*, ed. A. D. Boardman (Wiley, Chichester, 1982).

[15] D. R. Tilley in *Electromagnetic Surface Excitations*, eds. R. F. Wallis and G. I. Stegeman (Springer, Berlin, 1986).

Attenuated total reflection

Surface electromagnetic waves may be observed through their coupling to bulk electromagnetic waves using the technique of attenuated total reflection.[16,17] Resonant absorption is observed when the dispersion curves of the two waves intersect.

3.4. Superconducting Currents

3.4.1. Polarization and magnetization current

Josephson[18] has discussed the relationship between magnetization current $\nabla \times \mathbf{M}$ and polarization current $\partial \mathbf{P}/\partial t$ in metals. Irrotational current must arise from changing polarization and can not be associated with magnetization because of the vector identity

$$\nabla \cdot (\nabla \times \mathbf{M}) = 0 \tag{85}$$

Solenoidal current, on the other hand, may be associated either with magnetization or with polarization. By making use of physical arguments that relate the current to the magnetic field **B**, Josephson finds for normal metal in equilibrium that the current is to be associated with magnetization and not with polarization. Josephson has also analysed the penetration of flux quanta into a superconductor. We take up this part of his argument in Ch.4.

In superconductors, where the current is given by the London Equation

$$\ell \mathbf{J} + \mathbf{A} = 0 \tag{86}$$

the current may be associated either with the time-derivative of a polarization density **P** or with the curl of a magnetization density **M**. The two formulations are found to give different expressions for the permittivity ε and the permeability μ. They must, however, give the same expressions for the wavenumber $k(\omega)$ and for the surface impedance $Z_s(\omega)$, since these are both measurables. To examine this requirement, we work through the two formulations.

Polarization current

The electric field $\mathbf{E} = i\omega\mathbf{A}$ with $\mathbf{J} = -i\omega\mathbf{P}$ give for the permittivity

$$\varepsilon(\omega, k) = \varepsilon_{core} + P/E = \varepsilon_{core} - 1/\omega^2\ell \tag{87}$$

The permeability arises only from localized atomic currents

$$\mu = \mu_{core} \tag{88}$$

where the subscript *core* stands for all contributions except from the supercarriers. The wavevector is

---

[16] G. Kovacs in *Electromagnetic Surface Modes*, ed. A. D. Boardman (Wiley, Chichester, 1982).

[17] F. Abeles in *Electromagnetic Surface Excitations*, eds. R. F. Wallis and G. I. Stegeman (Springer, Berlin, 1986).

[18] B. D. Josephson, *Phys. Rev.* **152**, 211 (1966).

$$k(\omega) = \omega\sqrt{\varepsilon\mu} = \sqrt{\mu_{core}}\,(\omega^2\varepsilon_{core} - 1/\ell)^{1/2} \tag{89}$$

The wavevector is a measurable quantity since the phase of a classical field may be determined. The surface impedance

$$Z_s(\omega) = E_s/H_s = \sqrt{\mu/\varepsilon} = \omega\sqrt{\mu_{core}}\,(\omega^2\varepsilon_{core} - 1/\ell)^{-1/2} = \omega\mu_{core}/k \tag{90}$$

may be determined either from a determination of the electric and magnetic fields at the interface with vacuum or from the surface reflection coefficient. Note that Z_s is *increased* by the supercurrents.

Magnetization current

Alternatively, we take the superconducting current to be associated with magnetization

$$\mathbf{J} = \nabla \times \mathbf{M} \tag{91}$$

which leads in the medium to $\mathbf{J} = i\mathbf{k} \times \mathbf{M}$ and with $\mathbf{B} = i\mathbf{k} \times \mathbf{A}$ gives for the permeability of the medium

$$\mu = B/H = B/(B/\mu_0 - M_{core} - M) = \mu_{core}/\,(1 + \mu_{core}/k^2\ell\,) \tag{92}$$

and for the permittivity

$$\varepsilon = \varepsilon_{core} \tag{93}$$

The wavevector is

$$k(\omega) = \omega\sqrt{\varepsilon\mu} = \omega\sqrt{\varepsilon_{core}\mu_{core}}\,\,(1 + \mu_{core}/k^2\ell\,)^{-1/2} \tag{94}$$

Although Eqs. (89) and (94) look different, they are in fact the same as may be verified by solving Eq. (94) explicitly for k.

We may expect for the surface impedance

$$Z_c(\omega) = E/H = \sqrt{\mu/\varepsilon} = \sqrt{\mu_{core}/\varepsilon_{core}}\,\,(1 + \mu_0/k^2\ell\,)^{-1/2} = k(\omega)/\omega\varepsilon_{core} \tag{95}$$

which is seen to be *reduced* by the supercurrents. Since H differs in the two formulations, the characteristic impedances need not be the same. The surface impedance, which determines the reflection and transmission coefficients at the interface with vacuum, must be the same, however. We are forced to conclude that Eqs. (90) and (95) for Z_s are incompatible and that we must have made a mistake.

The error that we have made is to apply Eq. (86) not only in the medium but at the interface as well. The problem with associating the supercurrent with magnetization is that the magnetization drops discontinuously to zero at the interface with a nonphysical surface current from Eq. (91)

$$\mathbf{I}_s = \mathbf{n} \times \mathbf{M}_s \tag{96}$$

3.4.2. *Kinetic inductivity*

The impedance of an inductor wound from a superconductor has three components

$$Z = -i\omega L + R - i\omega L_k \tag{97}$$

as shown in Fig. 3. The first term is the usual inductance $L = N\Phi/I$ where N is the number of turns carrying a current I and enclosing the flux Φ. R is the resistance of the windings and L_k is a new term called the *kinetic* inductance of the superconducting carriers.

The work done to establish a current I is the stored energy

$$U = {}^1\!/_2 L\, I^2 + {}^1\!/_2 L_k I^2 \tag{98}$$

The first term as usual represents energy stored in the magnetic field

$$U_{mag} = \int dV \int \mathbf{H} \cdot d\mathbf{B} \tag{99}$$

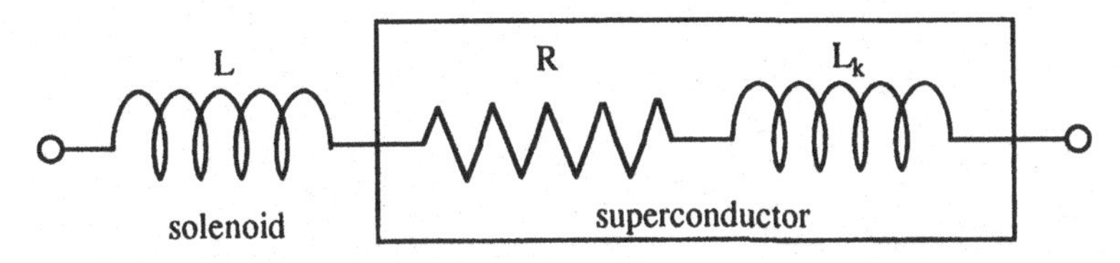

Figure 3. Three lumped-element components of a solenoid wound from a superconductor. The inductance $L = N\Phi/I$ is the usual magnetic inductance of the windings. The resistance R and the kinetic inductivity L_k arise from the carriers in the superconductor.

What about the second term? From the London point of view, the superconducting carriers are accelerated by the electric field and thus this term must be considered to be electric energy

$$U_{elec} = \int dV \int \mathbf{E} \cdot d\mathbf{D} = \int dV \int \mathbf{E} \cdot \mathbf{J} dt = -\int dV \int \mathbf{J} \cdot d\mathbf{A} \tag{100}$$

With the London equation Eq. (86) we obtain the simple result for the electric energy

$$U_{elec} = \frac{1}{2\ell} \int J^2 dV \tag{101}$$

Using $\ell = m/nq^2$ one may verify that this is just the expression for the kinetic energy of the carriers. For the London case of static magnetic fields screened in the London penetration depth $\lambda_L = \sqrt{\ell/\mu}$ the electric and magnetic contributions to the energy are just equal and each is one-half the total.

The total energy may be written in a particularly simple form by transforming Eq. (99) through use of the identity

$$\nabla \cdot (\mathbf{H} \times d\mathbf{A}) = d\mathbf{A} \cdot \mathbf{J} - \mathbf{H} \cdot d\mathbf{B} \tag{102}$$

to obtain

$$U_{mag} = \int dV \int \mathbf{J} \cdot d\mathbf{A} - \oint d\mathbf{S} \cdot \int \mathbf{H} \times d\mathbf{A} \tag{103}$$

Adding Eqs. (100) and (103) gives the total energy in terms of the boundary contribution

$$U_{tot} = -\oint d\mathbf{S} \cdot \int \mathbf{H} \times d\mathbf{A} \tag{104}$$

4
SUPERCONDUCTING PHASE AND FLUX QUANTA

4.1. Superconducting Phase

We begin this chapter with a discussion of the application of Schrödinger theory to the quantum mechanics of superconductivity.[1] Electromagnetic theory, which we discussed in Ch. 3, and Schrödinger theory are about all that we will need for a discussion of the Josephson effect and flux quantization as well as a quantum theory of the Meissner effect. The nicest treatment that I have seen of the application of Schrödinger theory to these problems is in Vol. III of *The Feynman Lectures on Physics*.[2] This was Feynman's final lecture in Cal Tech's two-year introductory physics course and he described it as "only for entertainment." We shall miss Richard Feynman.

In Schrödinger theory the state of a superconductor is represented by a wave function $\Psi(\mathbf{r}, t)$, which is a solution of the time-dependent Schrödinger equation

$$\mathcal{H}\,\Psi(\mathbf{r}, t) = i\hbar\,(\partial/\partial t)\,\Psi(\mathbf{r}, t) \tag{1}$$

where $|\Psi(\mathbf{r}, t)|^2$ is an observable that has been subject to varying interpretations.

4.1.1. Charge and current density

[1] F. London, *Phys. Rev.* 74, 562 (1948); *Superfluids*, Vol. 1 (Wiley, New York, 1950) part E.

[2] Richard P. Feynman, Robert B. Leighton and Matthew Sands, *The Feynman Lectures on Physics, Quantum Mechanics*, Vol. III, (Addison-Wesley, Reading, MA, 1965.) secs. 21-1 through 21-9.

The local charge density in Schrödinger theory is

$$\rho(\mathbf{r}, t) = q\,\Psi^*(\mathbf{r}, t)\,\Psi(\mathbf{r}, t) \tag{2}$$

From the BCS theory of superconductivity, the Schrödinger "elementary particle" must be pair of electrons with charge $q = -2e$ and mass $m = 2m_e$ in a spin singlet state with nearly paired momenta. For the high-temperature superconductors, this particle may in some sense be a boson with the superconducting state a Bose condensation of electron pairs. The bosons are acted on by electric and magnetic fields and by the chemical potential of the host. For a discussion of Bose condensation and chemical potential that is oriented toward solids, I suggest Kittel and Kroemer's *Thermal Physics*.[3]

The quantum current is most directly obtained from charge conservation

$$\nabla \cdot \mathbf{J} = -\partial\rho/\partial t = -q\,(\Psi^*\,\partial\Psi/\partial t + \Psi\,\partial\Psi^*/\partial t) \tag{3}$$

Substituting from Eq. (1) we have

$$\nabla \cdot \mathbf{J} = (q/i\hbar)\,[\Psi^*\,(\mathcal{H}\,\Psi) - \Psi\,(\mathcal{H}\,\Psi)^*] \tag{4}$$

If there is no magnetic field **B**, only the momentum operator $\mathbf{p} = -i\hbar\,\nabla$ contributes to Eq. (4) and we have

$$\nabla \cdot \mathbf{J} = -(q/2i\hbar m)\,[\Psi^*\,(p^2\,\Psi) - \Psi\,(p^2\,\Psi)^*] = (q/2m)\,\nabla \cdot [\Psi^*\,(\mathbf{p}\Psi) + \Psi\,(\mathbf{p}\,\Psi)^*] \tag{5}$$

which gives for the current density

$$\mathbf{J} = (q/2m)\,[\Psi^*\,(\mathbf{p}\,\Psi) + \Psi\,(\mathbf{p}\,\Psi)^*] \tag{6}$$

In a magnetic field **B**, the current is more complicated. This is because the momentum has two components, the particle momentum $m\mathbf{v}(\mathbf{r})$ and the field momentum $q\mathbf{A}(\mathbf{r})$ as discussed in Ch. 3.

$$\mathbf{p}(\mathbf{r}) = m\,\mathbf{v}(\mathbf{r}) + q\,\mathbf{A}(\mathbf{r}) \tag{7}$$

The Hamiltonian in a magnetic field is obtained by replacing $\mathbf{p}(\mathbf{r})$ by $\mathbf{p}(\mathbf{r}) - q\,\mathbf{A}(\mathbf{r})$. In place of Eq. (5) we have (skipping a lot of algebra)

$$\nabla \cdot \mathbf{J} = -(q/2ihm)\,\{\Psi^*\,[|\mathbf{p} - q\mathbf{A}|^2\,\Psi] - \Psi\,[|\mathbf{p} - q\mathbf{A}|^2\,\Psi]^*\}$$

$$= (q/2m)\,\nabla \cdot \{\Psi^*\,[(\mathbf{p} - q\mathbf{A})\,\Psi] + \Psi[(\mathbf{p} - q\mathbf{A})\,\Psi]^*\} \tag{8}$$

which gives for the current

$$\mathbf{J} = (q/2m)\,\{\Psi^*\,[(\mathbf{p} - q\mathbf{A})\Psi] + \Psi\,[(\mathbf{p} - q\mathbf{A})\,\Psi]^*\} \tag{9}$$

4.1.2. Schrödinger wave function

We write the wave function as

[3] C. Kittel and H. Kroemer, *Thermal Physics*, 2nd ed. (Freeman, New York, 1980). See also C. Kittel, *Introduction to Solid State Physics*, 6th ed. (Wiley, New York, 1986).

$$\Psi(\mathbf{r}, t) = \psi(\mathbf{r}, t)\, e^{i\phi(\mathbf{r}, t)} \tag{10}$$

where $\phi(\mathbf{r}, t)$ is the phase of the wavefunction and the amplitude $\psi(\mathbf{r}, t)$ is real. Substituting into Eq. (9), the current density is

$$\mathbf{J} = (1/m)\rho(\mathbf{r}, t)\, (\hbar\, \nabla\phi - q\, \mathbf{A}) \tag{11}$$

It is usual to write $\nabla\phi(\mathbf{r}, t)$ as the wavevector $\mathbf{k}(\mathbf{r}, t)$ and the phase is then

$$\phi(\mathbf{r}, t) = \int d\mathbf{r} \cdot \mathbf{k}(\mathbf{r}, t) \tag{12}$$

The amplitude of the wavefunction is the solution of a Schrödinger equation without the magnetic field

$$-(\hbar^2/2m)\, \nabla^2\psi + q\, V(\mathbf{r})\, \psi = [\mu(\mathbf{r}) + q\, V(\mathbf{r})]\psi \tag{13}$$

where $V(\mathbf{r})$ is the electrostatic potential and $\mu(\mathbf{r})$ is the chemical potential.[3] The sum $\mu(\mathbf{r}) + q\, V(\mathbf{r})$ is called the electrochemical potential.

4.1.3. Quantum fluid dynamics

Feynman[2] discusses charge flow in terms of the equations of motion of a fluid and obtains three equations

$$m\mathbf{v} = \hbar\nabla\phi - q\, \mathbf{A} \tag{14}$$

$$\nabla \cdot (\rho\mathbf{v}) + \partial\rho/\partial t = 0 \tag{15}$$

$$\hbar\, \partial\phi/\partial t = {}^1\!/_2\, mv^2 + qV(\mathbf{r}) + \mu(\mathbf{r}) \tag{16}$$

The first equation is equivalent to Eq. (7). Taking its curl we obtain

$$m\, \nabla \times \mathbf{v} = q\, \mathbf{B} \tag{17}$$

The second equation is the statement of charge conservation as given by Eq. (3). The third equation gives the hydrodynamics. By taking the gradient of Eq. (16) and introducing the convective derivative

$$d\mathbf{v}/dt = \partial\mathbf{v}/\partial t + (\mathbf{v} \cdot \nabla)\, \mathbf{v} \tag{18}$$

we obtain the more familiar form with the Lorentz force augmented by the gradient of the chemical potential

$$m\, d\mathbf{v}/dt = q\, (\mathbf{E} + \mathbf{v} \times \mathbf{B}) - \nabla\, \mu(\mathbf{r}) \tag{19}$$

4.1.4. Quantization of magnetic flux

The quantization of flux in the hollow core of a superconducting cylinder is a consequence of phase coherence within the superconductor. Flux quantization was predicted by H.

London and F. London[4] and first observed experimentally by Little and Parks.[5] For the wavefunction to be single-valued we require for any closed contour

$$\oint d\mathbf{r} \cdot \nabla\phi = 2\pi n \quad (20)$$

Substituting Eq. (14) into Eq. (20) with $\mathbf{J} = \rho\mathbf{v}$ gives

$$\hbar \oint d\mathbf{r} \cdot \nabla\phi = \oint d\mathbf{r} \cdot [m\mathbf{v} + q\,\mathbf{A}] = q\oint d\mathbf{r} \cdot [\ell\,\mathbf{J}_s + \mathbf{A})] = nh \quad (21)$$

with the kinetic inductivity $\ell = m/\rho q$. Using $\mathbf{B} = \nabla \times \mathbf{A}$ Eq. (21) leads to

$$\oint d\mathbf{r} \cdot \mathbf{A} = \Phi = nh/q - \ell \oint d\mathbf{r} \cdot \mathbf{J}_s \quad (22)$$

with $q = -2e$. Within a contour over which $\mathbf{J}$ vanishes, the enclosed flux must be a multiple of the flux quantum $\Phi_0 = h/q$. This result applies not only to the core of a superconducting cylinder as in the Little-Parks experiment[5] but also to normal filaments within otherwise homogeneous superconductors and to the interstices of granular superconductors, which are discussed in Ch. 11.

4.1.5. London penetration depth

Isotropic screening

The screening of magnetic fields by superconductors arises from the constancy of the phase of the superconducting state. This requires

$$\hbar\,\mathbf{k} = q\,(\ell\,\mathbf{J} + \mathbf{A}) = 0 \quad (23)$$

which leads to the London equation

$$\ell\,\mathbf{J} + \mathbf{A} = 0 \quad (24)$$

As we saw in Ch. 2, this equation gives the decay of the vector potential into a superconductor

$$A(x) = A(0)\exp(-x/\lambda_L) \quad (25)$$

where $\lambda_L = \sqrt{\ell/\mu}$ is the London penetration depth.

Anisotropic screening

Clem[6] has discussed the systematics of magnetic screening by the high-temperature superconductors, which are highly anisotropic. For a uniaxial superconductor we introduce band masses m_{xy} and m_z and the band energy is

---

[4] H. London and F. London, *Physica* **2**, 341 (1935); *Proc. Roy. Soc. (Lond.)* **A149**, 71 (1935).
[5] W. A. Little and R. D. Parks, *Phys. Rev. Lett.* **9**, 9 (1962); *Phys. Rev.* **133**, A97 (1964).
[6] J. R. Clem, *Physica C* **162-164**, 1137 (1989).

$$\varepsilon = \hbar^2(k_x^2 + k_y^2)/2m_{xy} + \hbar^2 k_z^2/2m_z \tag{26}$$

At low carrier density the wavevectors at the Fermi surface are

$$\hbar\, k_{F_{xy}} = (2m_{xy}\,\varepsilon_F)^{1/2} \qquad \hbar\, k_{Fz} = (2m_z\varepsilon_F)^{1/2} \tag{27}$$

The volume of the Fermi ellipsoid is

$$V_F = (4\pi/3)\, k_{xy}^2\, k_z = (4\pi/3\hbar^3)\, (2\varepsilon_F)^{3/2}\, (m_{xy}^2 m_z)^{1/2} \tag{28}$$

and the area of the Fermi surface is

$$S_F = (4\pi/\hbar^2)\; 2\varepsilon_F\, (m_{xy}^2 m_z)^{1/3} \tag{29}$$

With n carriers per unit volume the Fermi volume is $V_F = 4\pi^3 n$. From Eq. (28) the Fermi energy is

$$\varepsilon_F = (3\pi^2 n/2)^{2/3}\, \hbar^2\, (m_{xy}^2 m_z)^{-1/3} \tag{30}$$

The Fermi momenta are then

$$k_{Fz} = (2\pi^2 n/2)^{1/3}\, (8m_z/m_{xy})^{1/6} \tag{31}$$

$$k_{Fxy} = (2\pi^2 n/2)^{1/3}\, (8m_{xy}/m_z)^{1/6} \tag{32}$$

The high-temperature superconductors are planar with high conductivity in the plane and little conduction between planes. The band energy changes very little with k_z, which is represented by a large band mass m_z. For the band mass m_z sufficiently large the Fermi surface may contact the zone boundary. We then have $k_{Fz} \approx \pi/c$, the Fermi volume is $V_F \approx (2\pi/c)\pi k_{xy}^2$ and the free Fermi surface area is $S_F \approx 2\pi k_{xy}^2$.

For an anisotropic superconductor the scalar mass m may be replaced by the tensor $\ddot{\mathbf{m}}$ and Eq. (23) becomes

$$\hbar\, \mathbf{k} = (1/\rho)\, \ddot{\mathbf{m}} \cdot \mathbf{J} + q\, \mathbf{A} \tag{33}$$

In place of Eq. (24) we write

$$\ddot{\mathbf{m}} \cdot \mathbf{J} + \rho q\, \mathbf{A} = 0 \tag{34}$$

Using $\mu\mathbf{J} = \nabla \times \mathbf{B}$ and $\mathbf{B} = \nabla \times \mathbf{A}$, Eq. (34) becomes

$$\nabla \times (\ddot{\mathbf{m}} \cdot \nabla \times \mathbf{B}) + \rho q \mu\, \mathbf{B} = 0 \tag{35}$$

Expanding the triple vector product with $\nabla \cdot \mathbf{B} = 0$ we obtain the tensor London equation

$$(\nabla \cdot \ddot{\mathbf{m}} \cdot \nabla)\, \mathbf{B} = \rho q \mu\, \mathbf{B} \tag{36}$$

A magnetic field parallel to the *c*-axis is screened in the plane with

$$\lambda_{Lxy} = (m_{xy}/\rho q \mu)^{1/2} \tag{37}$$

A magnetic field perpendicular to the c-axis is screened in the ab-plane with

$$\lambda_{Lxy} = (m_{xy}/\rho q\mu)^{1/2} \quad (38)$$

and along the c-axis with

$$\lambda_{Lz} = (m_z/\rho q\mu)^{1/2} \quad (39)$$

4.2. Ginzburg-Landau Theory

4.2.1. Introduction

Ginzburg and Landau[7,8,9] have developed a phenomenological theory of superconductivity through the application of the Landau theory of phase transitions.[10,11] Where for ferroelectrics, the polarization density **P(r)** is the order parameter and for ferromagnets it is the magnetization density **M(r)**, for superconductors it has been shown to be the complex gap function $\Delta(\mathbf{r})$,[12] which in the Ginzburg-Landau theory is represented by the mean value of $\mathcal{V}\Psi_\downarrow^*(\mathbf{r})\Psi_\uparrow(\mathbf{r})$ V where $\mathcal{V}$ is the coupling energy and V is the sample volume.

The free energy density functional in a magnetic field takes the form

$$\mathcal{F}_{sB} = \mathcal{F}_{s0} - \int \mathbf{H} \cdot d\mathbf{B} + \frac{1}{2m} |(-i\hbar\nabla - q\,\mathbf{A})\Psi|^2 \quad (40)$$

The zero field functional $\mathcal{F}_{s0}$ at temperatures not too far below T_c is taken as

$$\mathcal{F}_{s0} = \mathcal{F}_{n0} - \alpha|\Psi|^2 + \frac{1}{2}\beta|\Psi|^4 \quad (41)$$

The Ginzburg-Landau theory contains the two free parameters, α and β together with the pair-charge $q = -2e$ and the pair- mass $m = 2m_e$. The coefficient of $|\Psi|^4$ turns out to be

$$\beta = \mu H_c^2(T)\, n^2(T) \quad (42)$$

where n(T) is the superconducting pair concentration and H_c is the thermodynamic critical field. The coefficient of $|\Psi|^2$ is

$$\alpha = (1 - t)\, n(0)\beta \quad (43)$$

with $t = T/T_c \leq 1$.

Minimizing $\mathcal{F}_{sB}$ with respect to the conjugate pseudo-wavefunction Ψ^* as well as with

---

[7] V. L. Ginzburg and L. D. Landau, *Zh. Eksp. Teor. Fiz.* **20**, 1064 (1950). An English translation is contained in *Men of Physics, Vol. 1: L. D. Landau*, ed. D. ter Haar (Pergamon, New York, 1965), pp. 138-167.

[8] V. L. Ginzburg and L. D. Landau, *Soviet Phys. JETP* **5**, 1442 (1957)

[9] V. L. Ginzburg, *Soviet Phys. JETP* **3**, 621 (1956); **7**, 78 (1958).

[10] L. D. Landau and I. M. Lifshitz, *Course of Theoretical Physics, Vol. 5: Statistical Physics*, (Pergamon, London, 1958) ch. 4.

[11] C. Kittel and H. Kroemer, *Thermal Physics*, 2nd ed. (Freeman, New York, 1980), pp. 298-304.

[12] L. P. Gor'kov, *Soviet Phys. JETP* **9**, 1364 (1959).

respect to the vector potential **A** leads to the two Ginzburg-Landau differential equations

$$\frac{1}{2m}(-i\hbar\nabla - q\,\mathbf{A})^2\,\Psi - \alpha\Psi + \beta|\Psi|^2\Psi = 0 \tag{44}$$

$$\nabla^2\mathbf{A} = (\,i\hbar\mu_0 q/2m)(\Psi^*\nabla\Psi - \Psi\nabla\Psi^*) + (\mu_0 q^2/m)|\Psi|^2\,\mathbf{A} \tag{45}$$

In obtaining Eq. (45) we have used the identity

$$\int dV\,\mathbf{H}\cdot d\mathbf{B} = \int dV\ \mathbf{J}\cdot d\mathbf{A} \tag{46}$$

with the integral over the volume of the sample. The expression given in Eq. (46) for the magnetic energy is simpler than that originally chosen by Ginzburg and Landau and leads as well to Eq. (45). The advantage of this form is that it can be applied to other magnetic problems.

With $\phi(\mathbf{r})$ the phase of the wavefunction Ψ, Eq. (45) assumes the expected form

$$\mathbf{J} = (q/m)|\Psi|^2\,(\hbar\nabla\,\phi - q\,\mathbf{A}) \tag{47}$$

The equation

$$\hbar\,\nabla\,\phi = m\,\mathbf{v} + q\,\mathbf{A} = q\,(\ell\,\mathbf{J} + \mathbf{A}) \tag{48}$$

with $\phi(\mathbf{r})$ constant, is the London equation, which applies only in weak fields. Ginzburg-Landau theory leads to a breakdown of the London equation in strong fields with a spatial variation in in both ψ and ϕ.

In place of α and β the parameters of the theory may be taken as two of the following:

(i) the thermodynamic critical field

$$H_c = \alpha/\sqrt{\beta\mu} \tag{49}$$

(ii) the London penetration depth

$$\lambda_L = \sqrt{m\beta/q^2\alpha\mu} \tag{50}$$

(iii) and the coherence length

$$\xi = \hbar/\sqrt{2\alpha m} \tag{51}$$

The Ginzburg-Landau parameter is

$$\kappa = \lambda_L/\xi = (m/hq)\sqrt{2\beta/\mu} = 2\pi\sqrt{2}\,\lambda_L^2\mu H_c/\Phi_0 \tag{52}$$

with $\Phi_0 = h/q$. Agreement with the observed properties of metals like Pb and Sn is obtained for $\kappa \approx 0.1$. Ginzburg and Landau noticed early that for $\kappa > 1/\sqrt{2}$ their differential equations developed anomalous behavior which was further investigated by Abrikosov.

4.2.2. Applications

Critical field of a thin film

For a film of thickness d comparable to or thinner than the penetration depth λ_L, the applied field uniformly penetrates the film. For $\kappa < 1$, the coherence length will exceed the film thickness with the gap function $\Delta(r)$ uniform through the film. Solution of the Ginzburg-Landau equations lead to a transition field

$$H_t = 2\sqrt{6}\,(\lambda_L/d)H_c \tag{53}$$

This value is $\sqrt{2}$ times that calculated from the diamagnetic increase in energy and indicates that the penetration depth in the Ginzburg-Landau domain has grown to

$$\lambda_{GL}(H_t) = \sqrt{2}\,\lambda_L \tag{54}$$

Critical current of a thin film

For a sufficiently thin film, magnetic field effects may be ignored and we may concentrate on the increase in kinetic energy. The transition current density above which superconductivity is destroyed is

$$J_c = H_c(T)/3\sqrt{6}\,\pi\lambda(T) \tag{55}$$

Bardeen has compared the predictions of Ginzburg-Landau theory with the results of the microscopic theory.[13]

Surface energy

It is usual to write the surface energy of a superconductor as

$$\varepsilon_s = \tfrac{1}{2}\mu H_c^2\delta \tag{56}$$

For $\kappa \ll 1$ the parameter δ is[14,15]

$$\delta = \frac{4\sqrt{2}}{3}\xi = 1.89\xi \tag{57}$$

and for $\kappa \gg 1$ the parameter δ is

$$\delta = -\tfrac{8}{3}\left(\sqrt{2}-1\right)\lambda_{GL} = -1.104\lambda_{GL} \tag{58}$$

The crossover from positive to negative surface energy is at $\kappa = 1/\sqrt{2}$ and is responsible for the distinct properties of what are called type II superconductors.

---

[13] J. Bardeen, *Rev. Mod. Phys.* **34**, 667 (1962).

[14] G. Rickayzen, *Theory of Superconductivity* (Interscience, New York, 1965) sec. 9.6.

[15] M. Tinkham, *Introduction to Superconductivity* (McGraw-Hill, New York, 1975); reprinted (Krieger, Malabar, FL, 1980) sec. 4-3.

Deutscher and Müller[16] have analyzed the reduction in gap function at interior surfaces associated with large-angle grain boundaries in the high-T_c superconductors. Because of the very short coherence length in the high-T_c superconductors, such interior boundaries may substantially degrade the properties of these superconductors.

High surface current densities

Lam *et al.*[17] model nonlinear current distributions at surfaces using the time-independent Ginzburg-Landau equations. Ginzburg-Landau theory was chosen because it is a well known and widely accepted formalism that deals with the nonlinear response to fields and currents strong enough to change the superconducting carrier density. The equations to be solved are

$$\nabla_s^2 A_z = u^2 A_z \tag{59}$$

$$\frac{1}{\kappa^2}\nabla_s^2 u = u\left(u^2 - 1 + \tfrac{1}{2}A_z^2\right) \tag{60}$$

where A_z is the vector potential with $u = \lambda_L/\lambda_{GL}$ and λ_{GL} the Ginzburg-Landau penetration depth.

4.3. Flux Quanta

Abrikosov[18,19] extended the Ginzburg-Landau theory to variable κ and concluded that there exist two distinct classes of superconductors: type I superconductors with $\kappa < 1/\sqrt{2}$ and type II superconductors with $\kappa > 1/\sqrt{2}$. Ginzburg and Landau had already found that the surface energy of a type II superconductor is negative, allowing for the stabilization of superconductivity to an upper critical field

$$H_{c2} = \sqrt{2}\,\kappa\, H_c \tag{61}$$

Abrikosov investigated the properties of type II superconductors at fields close to H_{c2} and found a regular lattice of flux lines to be stable with $|\Psi|^2 = 0$ at their cores.

Abrikosov showed that type II superconductors allow the nucleation of flux quanta

$$\Phi_0 = h/2e = 2.06783 \times 10^{-15}\ \text{Wb} \tag{62}$$

assuming $q = 2e$ with a threshold field

$$H_{c1} = (\Phi_0/4\pi\mu\lambda_L^2)\ln\kappa \tag{63}$$

At low fields, the flux lines are distinct and outside the core region satisfy the equation

$$\nabla^2 A - (1/\lambda_L)^2 A = (\Phi_0/2\pi\lambda_L^2)\,\nabla\phi \tag{64}$$

---

[16]G. Deutscher and K. A. Müller, *Phys. Rev. Lett.* **59**, 1745 (1987).

[17]C. W. Lam, D. M. Sheen, S. M. Ali and D. E. Oates, *IEEE Trans. Appl. Superconduc* **2**, 58 (1992).

[18]A. A. Abrikosov, *Dokl. Akad. Nauk., SSSR* **86**, 489 (1952).

[19]A. A. Abrikosov, *Soviet Physics JETP* **5** , 1174 (1957).

Well outside the core of a flux line, λ_L is constant as assumed in the London theory. The phase increases by 2π around the core with

$$\nabla \phi = (1 / r)\, \hat{\theta} \tag{65}$$

Taking the curl of Eq. (64) yields

$$\nabla^2 \mathbf{B} - (1/\lambda_L)^2 \mathbf{B} = - (\Phi_0/\lambda_L{}^2)\, \hat{\mathbf{z}}\, \delta(\mathbf{r}) \tag{66}$$

The solution of Eq. (66) is

$$\mathbf{B}(\mathbf{r}) = (\Phi_0/2\pi\lambda_L{}^2)\, K_0(r/\lambda_L)\, \hat{\mathbf{z}} \tag{67}$$

where $K_0(r/\lambda_L)$ is the modified Bessel function. For $r >> \lambda_L$ the function falls off exponentially as

$$K_0(r / \lambda_L) \approx (\pi\, \lambda_L/2\, r)^{1/2} \exp(- r / \lambda_L) \tag{68}$$

For $r << \lambda_L$ the function has a logarithmic singularity, which may be taken down to a core radius $\lambda_L/\kappa = \xi$. The flux enclosed within a cylinder of radius r is

$$\int d\mathbf{S} \cdot \mathbf{B} = \Phi_0 - \mu\lambda_L{}^2 \oint d\mathbf{r} \cdot \mathbf{J} \tag{69}$$

which approaches the quantum of flux Φ_0 for $r \to \infty$.

4.3.1. Characteristic lengths and fields

The Ginzburg-Landau theory assumes near T_c that α and as a result $|\Psi|^2$ go to zero as $(1 - t)$ with β constant where $t = T/T_c$ is the reduced temperature. With coefficients obtained from the BCS theory, the Ginzburg-Landau macroscopic theory leads near T_c to expressions for the coherence length and penetration depth[20]

$$\xi(t) = 0.74\, \xi_0(1 - t)^{-1/2} \tag{70}$$

$$\lambda_L(t) = 0.71\, \lambda_L(0)(1 - t)^{-1/2} \tag{71}$$

where ξ_0 is the Pippard coherence length. The critical fields[20] are

$$H_{c1}(T) \approx H_c{}^2(T)/H_{c2}(T) = 1.64\, (2\pi\xi_0{}^2/\Phi_0)\, H_c{}^2(0) \tag{72}$$

$$H_{c2}(T) = [\phi_0/2\pi\xi^2(T)] = 1.83\, (\Phi_0/2\pi\xi_0{}^2)(1 - t) \tag{73}$$

The theory also leads to a thermodynamic critical current (or depairing current)[20]

$$J_c = (1/3\pi\sqrt{6})\, H_c(T)/\lambda_L(T) \propto (1 - t)^{3/2} \tag{74}$$

In Ch. 10 are discussed the related critical currents in Josephson junctions. Studies of the

[20] M. Tinkham, *Introduction to Superconductivity*, (McGraw-Hill, New York, 1975; Krieger, Malabar, 1980).

temperature dependence of microwave absorption near T_c may be useful in identifying loss mechanisms since these mechanisms depend variously on characteristic lengths, fields and critical currents.

4.3.2. Observation of flux quanta

Essmann and Träuble[21,22] observed fluxon arrays emerging from Pb alloyed with a small amount of In. They employed a decoration technique in which very small ferromagnetic particles, which marked the emergence of flux, were observed with an electron microscope. A very similar technique has recently been employed to observe the emergence of quantized flux from single crystal $YBa_2Cu_3O_7$.[23,24] Fluxons have also been observed emerging from ceramic $YBa_2Cu_3O_{7-\delta}$ by means of similar decoration technique.[25,26]

Neutron studies

Neutrons are sensitive to magnetic fields because of their magnetic moment. In a spatially varying magnetic field, each neutron has a Zeeman energy that is a function of position, giving rise to diffraction of the neutron beam. The diffraction angle θ is of the order of d/λ where $d = \sqrt{\Phi_0/B}$ is the distance between vortices and $\lambda \approx 10$ Å is the wavelength of cold neutrons, leading to $\theta \approx 1°$. Neutron diffraction studies of classical superconductors have been reviewed by Brandt.[27] The technique has only recently been applied to the high-temperature superconductors.[28,29,30]

Tunneling studies

Scanning tunneling microscopy (STM) on a surface of superconducting 2H-$NbSe_2$ reveals features of the electronic structure of the cores of flux lines.[31,32]

4.3.3. London equation with sources

In this section we augment the London equation with external current sources and write

21 U. Essmann and H. Träuble, *Phys. Lett. A* **24**, 526 (1967).

22 H. Traüble and U. Essman, *J. Appl. Phys.* **39**, 4052 (1968). For a recent review, see E. H. Brandt and U. Essmann, *Phys. Stat. Sol. B* **144**, 13 (1987).

23 P. L. Gammel, D. J. Bishop, G. J. Dolan, J. R. Kwo, C. A. Murray, L. F. Schneemeyer and J. V. Waszczak, *Phys. Rev. Lett.* **59**, 2592 (1987).

24 Z. H. Mai and X. Chu, *Solid State Commun.* **65**, 877 (1988) have used the Bitter decoration method with finely powdered $YBa_2Cu_3O_{7-\delta}$ in place of magnetic powder.

25 C. J. Jou, E. R. Weber, J. Washburn and W. A. Soffa, Appl. Phys. Lett. **52**, 326 (1988).

26 A. Ourmazd, J. A. Rentschler, W. J. Skocpol and D. W. Johnson, Jr., *Phys. Rev. B* **36**, 8914 (1987).

27 E. H. Brandt, *Phys. Rev. B* **169**, 107 (1991).

28 E. M. Forgan, D. McK. Paul, H. A. Mook, P. A. Timmins, H. Keller, S. Sutton and J. S. Abell, *Nature* **343**, 735 (1990).

29 E. M. Forgan, *Physica B* **169**, 107, (1991).

30 E. M. Forgan, D. McK. Paul, H. A. Mook, S. L. Lee, R. Cubitt, J. S. Abell, F. Gencer and P. Timmins, *Physica C* **185-189**, 247 (1991).

31 H. F. Hess, R. B. Robinson, R. C. Dynes, J. M. Valles Jr.and J. V. Waszczak, *Phys. Rev. Lett.* **62**, 214 (1989).

32 See the review of A. Khurana in "Search & Discovery," *Physics Today* **43** (6), 17 (1990).

$$\ell\,\mathbf{J} + \mathbf{A} = (\hbar/q)\,\nabla\,\phi = \ell\,\mathbf{J}_{ext} \tag{75}$$

Taking the curl of Eq. (75) we obtain

$$\nabla \times (\nabla \times \mathbf{B}) + (\mu/\ell)\,\mathbf{B} = \mu\,\nabla \times \mathbf{J}_{ext} \tag{76}$$

We regard $\mathbf{J}_{ext}$ as the quantum currents that establish the fluxon $\Phi_0 = h/q$. These currents are distributed around the fluxon core , which we idealize by shrinking the core area to zero at **R** with

$$\ell\,\nabla \times \mathbf{J}_{ext} = \Phi_0\,\delta(\mathbf{r} - \mathbf{R})\,\hat{\mathbf{z}} \tag{77}$$

where $\delta(\mathbf{r} - \mathbf{R})$ is a two-dimensional delta function. Integrating Eq. (77) over a circle of radius r in a plane normal to the unit vector $\hat{\mathbf{z}}$ and centered at **R**, we obtain with Stokes' theorem

$$\ell \oint d\mathbf{r} \cdot \mathbf{J}_{ext} = 2\pi r\,\ell\,J_{ext} = \Phi_0 \tag{78}$$

or equivalently with $\ell = \mu\,\lambda_L^2$

$$J_{ext} = \Phi_0/(2\pi\mu\,r\,\lambda_L^2) \tag{79}$$

Flow that falls off as 1/r has angular momentum density independent of radius that is characteristic of a vortex, a name commonly used for the fluxon. We have for Eq. (76)

$$\lambda_L^2\,\nabla \times (\nabla \times \mathbf{B}) + \mathbf{B} = \Phi_0\delta(\mathbf{r} - \mathbf{R})\,\mathbf{z} \tag{80}$$

with $\lambda_L = (\ell/\mu)^{1/2}$. The solution to Eq. (80) is the magnetic field of a charge-vortex, screened by diamagnetic currents

$$\mathbf{B} = (1/2\pi\lambda_L^2)\,K_0(r/\lambda_L)\,\Phi_0\,\mathbf{z} \tag{81}$$

with $K_0(\rho)$ the *modified Bessel function* . For ρ large we have

$$K_0(\rho) \approx (\pi/2\rho)^{1/2}\,(1 - 1/8\rho + \cdots)\exp(-\rho) \tag{82}$$

For ρ small the function has a logarithmic singularity

$$K_0(\rho) \approx \ln(2/\rho) \tag{83}$$

which is to be taken down to the core radius $\rho \approx \kappa = \xi/\lambda_L$. For $r < \xi$ the core is in the normal state with an internal field of the order of 2 H_{c1}. B as a function of distance from the core is shown in Fig. 1.

We may obtain the flux enclosed within a circle of radius r by integrating Eq. (80)

$$\ell \int d\mathbf{S} \cdot \nabla \times \mathbf{J} + \int d\mathbf{S} \cdot \mathbf{B} = \Phi_0 \tag{84}$$

where **J** is the superconducting current density. The second integral is the enclosed flux. Converting the first integral gives for the enclosed flux

$$\Phi = \Phi_0 - \ell \oint d\mathbf{r} \cdot \mathbf{J} \tag{85}$$

At densities less than $1/\xi^2$, vortices may be treated independently and the London equation for **B** may be written with multiple sources as

$$\lambda_L^2\, \nabla \times (\nabla \times \mathbf{B}) + \mathbf{B} = \Phi_0 \Sigma_{\mathbf{R}}\, \delta(\mathbf{r} - \mathbf{R})\, \hat{\mathbf{z}} \tag{86}$$

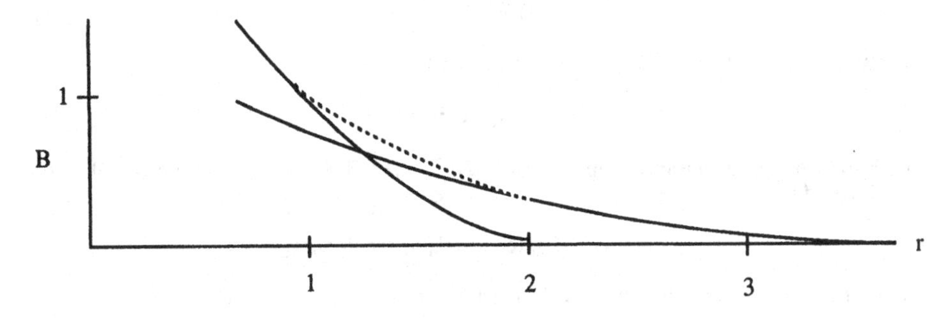

Figure 1. For $r < \xi$ the core is in the normal state with an internal field of the order of 2 H_{c1}. B as a function of distance from the core is shown.

4.3.4. Energy of a vortex line

The vortex line energy has two contributions, the magnetic field energy and the kinetic energy of the carriers. In the limit $\kappa >> 1$ the computed energy per unit length is[33]

$$\varepsilon_l = \frac{1}{2}\mu H_c^2\,(4\pi\xi^2)\ln\kappa \tag{87}$$

Lower critical field

The lower critical field is obtained as

$$H_{cl} = \frac{\varepsilon_l}{\Phi_0} = \frac{H_c}{\sqrt{2}\kappa}\ln\kappa \tag{88}$$

4.4. Types of Superconductivity

We begin this section by contrasting the magnetic properties of type I and type II superconductors. These materials are sometimes designated as soft and hard, respectively. The names make the analogy with soft and hard magnetic materials. Soft magnetic materials have high permeability with low coercive fields and low remanent magnetization and thus little hysteresis or energy loss. Hard magnetic materials have properties that are the opposite of soft materials with high remanence and coercivity and large hysteretic loss.

---

[33] M. Tinkham, *Introduction to Superconductivity* (McGraw-Hill, New York, 1975; Krieger, Malabar, 1980) sec 5-1.2.

4.4.1. *Type I superconductors*

Flux penetration

The penetration of magnetic flux into a type I superconductor is described by the London equation, giving

$$(\ell/\mu)\,\nabla^2 \mathbf{B} - \mathbf{B} = 0 \tag{89}$$

For an external magnetic field parallel to a plane surface we obtain

$$\mathbf{B} = \mu_0\,\mathbf{H}_{ext}\,\exp[-x/\lambda_L] \tag{90}$$

with the London penetration depth $\lambda_L = (\ell/\mu)^{1/2}$. The flux density within a flat plate of thickness d is

$$B(x) = \mu_0\,H_{ext}\,(\cosh x/\lambda_L)\,/\,(\cosh d/2\lambda_L) \tag{91}$$

and for a cylinder of radius a with the field parallel to the axis the flux density is

$$B = \mu_0\,H_{ext}\,I_0(r/\lambda_L)/I_0(a/\lambda_L) \tag{92}$$

where $I_0(r/\lambda_L)$ is the *associated Bessel function.*

For a sphere, the London result for the magnetic moment μ is[34,35]

$$\mu = -\frac{1}{2}a^3\,H_{ext}\,P(a/\lambda_L) \tag{93}$$

with $P(x) = 1 - (3/x)(\coth x) + 3/x^2$. For $x << 1$ we obtain $P \approx x^2/15$ and $x >> 1$ gives $P \approx (1 - 1/x)^3$.

4.4.2. *Type II superconductors*

Penetration of flux

For internal fields less than the lower critical field H_{c1}, penetration of flux into a type II superconductor is reversible as described for a type I superconductor and there is little ac loss well below T_c. For $H_{int} > H_{c1}$ vortices penetrate the sample. The mean concentration of vortices increases until at the upper critical field H_{c2}, the mean flux density in the sample is $\langle B\rangle = H_{ext}$. At fields higher than H_{c2} the material is normal.

For an ideal type II superconductor without pinning, the penetrating flux is a function H_{int} and is uniform. We have London screening of the residual flux in place of Eq. (89)

$$(\ell/\mu)\,\nabla^2(B - n\Phi_0) - (B - n\Phi_0) = 0 \tag{94}$$

Hysteresis

[34] F. London, *Superfluids Volume I , Macroscopic Theory of Superconductivity* (Wiley, New York, 1960).

[35] D. Shoenberg, *Proc. Roy. Soc. (London)* A **175**, 49 (1940); *Superconductivity*, (Cambridge Press, London, 1960) ch. 5 and app.

To exhibit hysteresis, a sample is first cooled below T_c in zero field. The field H_{int} is raised from 0 to $H_{c1} < H_0 < H_{c2}$, reduced to $-H_0$ and then raised again to H_0. In Fig. 2 we show a plot of the magnetization density[36] around such a cycle[37] for an epitaxial film of $YBa_2Cu_3O_{7-\delta}$. We first observe that the penetration of flux is not reversible, showing hysteresis. What in a ferromagnet would be called remanent induction is here called trapped flux. The second observation is that energy must be dissipated around such a cycle. The work done per unit volume around a cycle is

$$w = \oint H_{ext}\, dM \tag{95}$$

The picture that we develop is that vortices become pinned by defects and magnetic work must be done to depin them. Such processes exist at the surface and inhibit the nucleation of vortices. Similar processes take place within the superconducting volume and inhibit the motion of flux into and out of the interior. When we discuss granular superconductors in Ch. 11 we will encounter additional dissipation processes arising from flux-slip in clusters that contain Josephson junctions.

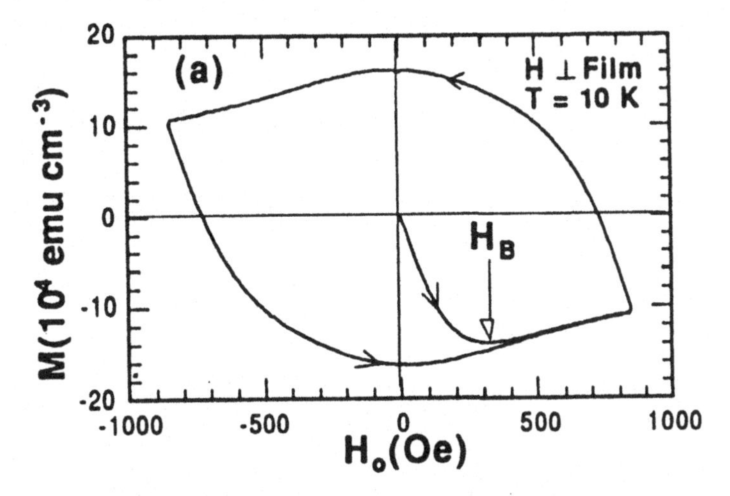

Figure 2. Plot of the magnetization density M around a cycle for an epitaxial film of $YBa_2Cu_3O_{7-\delta}$ deposited on (100) $SrTiO_3$. Flux reaches the center of the film in a perpendicular field of about 300 Oe [36].

It is usual to write M for increasing (↑) and decreasing (↓) fields as the sum of reversible and irreversible parts

$$M^{\uparrow} = M_{rev} + M_{irrev} \qquad M^{\downarrow} = M_{rev} - M_{irrev} \tag{96}$$

[36] C. Schlenker, J. Dumas, C. J. Liu and S. Revenaz in *Chemistry of High Temperature Superconductors*, ed. C. N. R. Rao (World Scientific, Singapore, 1991).

[37] It is customary to treat flux penetration in terms of an effective sample magnetization. As we have commented in Ch. 3, one must be careful since the magnetization arises from macroscopic polarization currents and not from microscopic magnetization currents.

which gives

$$M_{rev} = \frac{1}{2}(M\uparrow + M\downarrow) \qquad M_{irrev} = \frac{1}{2}(M\uparrow - M\downarrow) \tag{97}$$

Equations (96) and (97) define M_{rev} and M_{irrev}. The reversible magnetization need not be identical with the equilibrium magnetization M_{eq} obtained by cooling the superconductor in the field.

ac loss

Magnetic fields are associated with the flow of current in a superconducting wire or film. If the current is alternating, flux flows into and out of the superconductor. If the current is sufficiently small that H at the surface is always less than H_{c1}, the flux density is reversible. At higher current densities vortices are nucleated and penetrate the sample. Hysteretic losses associated with the entry and exit of vortices are fed by the current source with the mechanism for energy transfer the electric field generated by vortex motion.

5
MAGNETIC RESONANCE AND RELAXATION

5.1. Nuclear Magnetic Resonance

5.1.1. Local magnetic fields

Local fields in magnetic materials

Local magnetic fields at spins have been measured in magnetic materials by nuclear magnetic resonance, Mössbauer emission, perturbed angular correlation, muon free precession and nuclear ordering.[1] The field may typically be written as the sum of five terms[2]

$$\mathbf{B}_{loc} = \mu_0(\mathbf{H}_{ext} - \mathbf{H}_{demag} + \mathbf{H}_{Lorentz}) + \mathbf{B}_{dipolar} + \mathbf{B}_{core} \tag{1}$$

A detailed discussion of local fields may be found in solid state physics texts. We have already discussed the source of the external and demagnetizing fields. The Lorentz field arises from the field $\mathbf{B}_{dipolar}$ from a spherical region surrounding the spin site. The Lorentz field is the negative of the demagnetizing field from the surface of the spherical region. The core field may usually be divided into three or more components

$$\mathbf{B}_{core} = \mathbf{B}_{contact} + \mathbf{B}_{dipolar} + \mathbf{B}_{orbital} + \cdots \tag{2}$$

The field $\mathbf{B}_{orbital}$ arises from currents near the central site. The field $\mathbf{B}_{contact}$ arises from electron spin density at the central site and $\mathbf{B}_{dipolar}$ from spins near the central site. Writing the central magnetization density

---

[1] A. J. Freeman and R. E. Watson, "Hyperfine interactions in magnetic materials," in *Magnetism*, eds. G. T. Rado and H. Suhl (Academic Press, New York, 1965) vol. IIA, ch. 4.

[2] W. Marshall, *Phys. Rev.* **110**, 1280 (1958).

$$M = (g\mu_B S)\, \Psi^*(0)\, \Psi(0) \tag{3}$$

the contact and dipolar fields for a spherical spin distribution are

$$\mathbf{B}_{contact} + \mathbf{B}_{dipolar} = \mu_0(\mathbf{H} + 4\pi\, \mathbf{M}) = \frac{8\pi}{3}\, \mu_0 \mathbf{M} = \frac{8\pi}{3}\, \mu_0 g \mu_B S\, \Psi^*(0)\, \Psi(0) \tag{4}$$

Local fields in superconductors

For an ideal type I superconductor the magnetic flux density **B** is zero and we expect no local field. Let us imagine that the Meissner effect is incomplete because of normal regions that allow flux to penetrate. The mean permeability is usually written as $\mu = f\mu_0$ or equivalently $\chi = -(1 - f)$ where f is the fraction of normal material. We may ask for the mean magnetic field $\langle \mathbf{B} \rangle$ in the material as well as the fields $\mathbf{B}_s$ and $\mathbf{B}_n$ in the superconducting and normal regions.[3] We write

$$\langle \mathbf{B} \rangle = \mu \mathbf{H} = (1 - f)\, \mathbf{B}_s + f\, \mathbf{B}_n \tag{5}$$

We write for the field **H**

$$\mathbf{H} = \mathbf{H}_{ext} - N\, \mathbf{M} = \mathbf{H}_{ext} + (1 - f) N\, \mathbf{H} \tag{6}$$

which gives

$$\mathbf{H} = \mathbf{H}_{ext}/[1 - (1 - f)N] \tag{7}$$

Taking $\mathbf{B}_s = 0$ in Eq. (5) and using Eq. (7) we obtain for the field in the normal phase

$$\mathbf{B}_n = \mu \mathbf{H}_{ext}/[1 - (1 - f)N] \tag{8}$$

At f = 0 with no normal material, $\mathbf{B}_n$ is the field $\mu\mathbf{H}$ that we have already obtained for a superconductor. For f = 1 when all the material is in the normal phase, we obtain $\mathbf{H}_{ext}$ as expected.

5.1.2. Experimental survey

La_2CuO_4

Lütgemeier and Pieper[4] have observed the NMR spectrum of ^{139}La in pure and doped La_2CuO_4 at 4.2 K at a fixed frequency of 27.1 MHz with conclusions very similar to those of Furó and Jánossy.[5] Analysis of their spectrum leads to a quadrupole splitting ν_Q = 6.388 MHz and a hyperfine field H = 0.921 T with Cu moments along the direction of the Cu-O-Cu chains in the ab-planes that connect the CuO_6 octahedra. Pulsed Cu NMR has

---

[3] M. Mehring, in *Earlier and Recent Aspects of Superconductivity,*, eds. J. G. Bednorz and K. M. Müller (Springer, Berlin, 1990).

[4] H. Lütgemeier and M. W. Pieper, *Solid State Commun.* **64**, 267 (1987) .

[5] I. Furò and A. Jànossy, *Jpn. J. Appl. Phys.* **26**, L1307 (1987).

been observed by M. Lee *et al.*[6] in $La_{2-x}Sr_xCuO_4$ with x = 0.17.

$YBa_2Cu_3O_7$

Lütgemeier and Pieper[4] observed both Cu NMR and NQR in $YBa_2Cu_3O_{7-\delta}$ and concluded that there were two active Cu sites. They were unable to detect magnetic ordering in their sample, which was insufficiently depleted in oxygen for the establishment of antiferromagnetic order. They examined the ^{63}Cu and ^{65}Cu NQR spectrum as a function of frequency in zero magnetic field and the NMR spectrum as a function of field at 63.5 MHz. The NQR frequencies at site 1 are 20.4 and 22.0 MHz respectively for the two isotopes. The NQR frequencies at site 2 are 29.1 and 31.5 MHz. Kitaoka *et al.*[7] have obtained similar results.

Spin-lattice relaxation studies of Cu nuclei in rare-earth substituted $YBa_2Cu_3O_{7-\delta}$ make it is possible to identify site 1 with nuclei in CuO chains Cu(1) and site 2 with nuclei in the Cu_2O planes Cu(2). Pennington *et al.*[8] have observed Cu NMR in a sample of thirty oriented small single crystals and draw the same conclusion. Other studies support this assignment.[9]

5.2. Nuclear Relaxation

5.2.1. *Studies of the vortex lattice*

Early studies of the magnetic field of a vortex lattice were undertaken by Rossier and MacLaughlin[10] and by Pincus *et al.*[11] who observed the broadening of the nuclear magnetic resonance (NMR) in superconducting vanadium. Later NMR studies[12,13,] have provided additional details. Flux flow was shown by Delrieu[14] to narrow the resonance.

Hentsch[15] and Mehring *et al.*[3,16,17] have reported the magnetic field distribution in superconducting $Tl_2Ba_2CuO_6$. The characteristic lineshape expected for a triangular vortex lattice is observed and leads to a low-temperature penetration depth $\lambda = 180$ nm.

Extensive muon studies of vortex fields in classical type II superconductors and more recently in the high-T_c superconductors are described in Sec. 5.2.

---

[6]M. Lee, M. Yudkowsky, W. P. Halperin, J. Thiel, S.-J. Hwu and K. R. Poeppelmeier, *Phys. Rev. B* **36**, 2378 (1987).

[7]Y. Kitaoka, S. Hiramatsu, K. Ishida, T. Kohara, Y. Oda, K. Amaya and K. Asayama, *Physica* **148B**, 298 (1987).

[8]C. H. Pennington, D. J. Durand, D. B. Zax, C. P. Slichter, J. P. Rice and D. M. Ginsberg, *Phys. Rev. B* **37**, 7944 (1988).

[9]Y. Kitaoka, S. Hiramatsu, T. Kohara, K. Asayama, K. Oh-Ishi, M. Kikuchi and N. Kobayashi, *Jpn. J. Appl. Phys.* **26**, L397 (1987) .

[10]D. Rossier and D. E. MacLaughlin, *Phys. Kondens. Materie* **11**, 66 (1960).

[11]P. Pincus, A. C. Gossard, V. Jaccarino and J. H. Wernick, *Phys. Lett.* **13**, 21 (1964).

[12]A. G. Redfield, *Phys. Rev. Lett.* **17**, 381 (1966); *Phys. Rev.* **162**, 367 (1967).

[13]J. M. Delrieu and J. M. Winter, *Solid State Commun.* **4**, 545 (1966).

[14]J. M. Delrieu, *J. Low Temp. Phys.* **6**, 197 (1972).

[15]F. Hentsch, N. Winzek, M. Mehring, Hj. Mattausch and A. Simon, *Physica C* **158**, 137 (1989).

[16]M. Mehring, F. Hentsch, Hj. Mattausch and A. Simon, *Z. Phys. B.* **77**, 355 (1989).

[17]M. Mehring, F. Hentsch, Hj. Mattausch and A. Simon, "Magnetic field distribution of the vortex lattice in the superconductor $Tl_2Ba_2CuO_6$ (T_c = 85 K): ^{205}Tl NMR investigation," *Solid State Commun.* **75**, 753 (1990).

5.2.2. Longitudinal relaxation

Significant experimental support for the BCS theory had come from the experimental studies of Hebel and Slichter of longitudinal nuclear relaxation in superconducting aluminum.[18] It had been expected that the relaxation rate would decrease smoothly as the sample went superconducting and the normal fraction of electrons decreased. What was observed instead was a peak in the relaxation rate just below T_c followed by the expected decrease. Stimulated by this result, BCS calculated the longitudinal relaxation and found that their theory led to a "coherence peak" around T_c of the observed form. Subsequently, studies of the acoustic and electromagnetic properties around T_c also demonstrated evidence of coherence.[19]

Measurements of longitudinal relaxation in $YBa_2Cu_3O_7$ have given no evidence of a coherence peak below T_c.[20,21] The absence of such a peak has been attributed to strong pair-breaking around T_c, smearing out the BCS density of states.[22,23]

5.3. Muon Spin Rotation and Relaxation

Time-differential positive muon-spin rotation (μSR)[24] allows the determination of the spectrum of frequencies of a precessing muon spin from the time that the muon is detected entering the sample with spin antiparallel to its initial momentum until muon decay is detected by the emission of a positron, predominantly along the positive direction of the muon spin. The direction of the detected positron and the time interval are recorded for each of millions of muons, from which the muon frequency spectrum is computed. From the magneto-mechanical ratio of the muon $\gamma = \omega/B = 2\pi \times 135.5$ MHz/T internal magnetic fields are readily computed. In addition to internal fields, sensed by the muon either in motion or at rest in the sample, external fields along or transverse to the direction of the incoming beam (and thus along or transverse to the initial direction of the muon spin) may be applied. Such fields modify the frequency spectrum of the precessing muon and allow for the detection of local magnetization with the possibility of magnetic order.

By comparing the observed static internal field with the calculated dipolar field in planar antiferromagnet $Sr_2CuO_2Cl_2$, Le *et al.*[25] have concluded that muons are located in this material near the center of the CuO_2 planes. Comparison with muon studies of La_2CuO_4 and Nd_2CuO_4 suggests that that muon sites in the oxide superconductors are closely associated with oxygen atoms.[26] Boekema *et al.*[27] came to a similar conclusion in their muon study of α-Al_2O_3.

5.3.1. Experimental survey

---

[18] L. C. Hebel and C. P. Slichter, *Phys. Rev.* **113**, 1504 (1959).

[19] D. M. Ginsberg and L. C. Hebel in *Superconductivity*, ed. R. D. Parks (Dekker, New York, 1969).

[20] R. E. Walstedt, W. W. Warren, R. F. Bell, G. F. Brennert, G. P. Espinosa, R. J. Cava, L. F. Schneemeyer and J. V. Waszczak, *Phys. Rev. B* **38**, 9299 (1988).

[21] P. C. Hammel, M. Takigawa, R. H. Heffner, Z. Fisk and K. C. Ott, *Phys. Rev. Lett.* **63**, 1992 (1989).

[22] Y. Kuroda and C. M. Varma, *Phys. Rev.* **42**, 8619 (1990).

[23] L. Coffey, *Phys. Rev.. Lett.* **64**, 1071 (1990).

[24] A. Schenck, *Muon spin rotation spectroscopy: principles and applications in solid state physics* (Adam Hilger, Bristol, 1985).

[25] L. P. Le, G. M. Luke, B. J. Sternlieb, Y. J. Uemura, J. H. Brewer, T. M. Riseman, D. C. Johnston and L. L. Miller, *Phys. Rev. B* **42**, 2182 (1990).

[26] G. M. Luke *et al.*, *Physica C* **162-164**, 825 (1989).

[27] C. Boekema, K.C. Chan, R. L. Lichti, A. B. Denison, D. W. Cooke, R. H. Heffner, R. L. Hutson and M. E. Schillaci, *Hyp. Int.* **32**, 667 (1986).

We give primary attention in this section to muon studies of the magnetic ordering of La_2CuO_4 and $YBa_2Cu_3O_7$ and related compounds obtained either by depletion of oxygen or by partial replacement of trivalent La or Y by a divalent ion. We also review studies of $YBa_2Cu_3O_{7-\delta}$ in which Y has been replaced by a magnetic rare earth element (*RE*).

La_2CuO_4

Uemura, Kossler, *et al.*[28] performed muon spin relaxation (μSR) experiments in three polycrystalline $La_2CuO_{4-\delta}$ samples that as a result of different heat treatment had differing degrees of oxygen depletion. Sample 1 was heated to 950 °C and cooled in air and should have had the largest oxygen depletion. Sample 2 was heated in oxygen at atmospheric pressure and should have been intermediate. Sample 3 was heated in oxygen under pressure and should have had the least depletion.

All three samples showed evidence of freezing of magnetic moments with abrupt internal field onsets close in temperature to the measured maxima in the magnetic susceptibility. For samples 1 and 2 the transition was at 250 K and for sample 3 between 200 and 225 K. Spin-freezing was determined in zero magnetic field from time-histograms of μSR counting rates F(t) and B(t) from a set of forward counters and backward counters

$$F(t) = \exp[-t/\tau_\mu]\,[1 + A(t)] \tag{9}$$

$$B(t) = \exp[-t/\tau_\mu]\,[1 - A(t)] \tag{10}$$

where $\tau_\mu = 2.2$ μsec is the muon lifetime and A(t) is the muon-decay asymmetry

$$A(t) = [F(t) - B(t)]/[F(t) + B(t)] \tag{11}$$

The onset of spin-freezing was established from the amplitude of the asymmetry, which was fitted by a single frequency with a relaxation rate that increased from 0.5 μsec^{-1} at low temperatures to 2 μsec^{-1} at the transition with no indication of a divergence. This relaxation rate corresponds at most to a 5% line-broadening. The asymmetry represents about half the muons stopped in the sample with the other half presumably stopped at sites where the precession frequency is much lower. The temperature dependence of the frequency must be proportional to the magnetic field H_{loc} at the muon site $\omega = \gamma\, B_{loc}$ with $\gamma/2\pi = 135.5$ MHz/T. The mean local field $B_{loc} \approx 42$ mT is in turn proportional to the temperature-dependent magnetization M_S. The magnetization is found to rise abruptly at the transition to a value about half its ultimate low-temperature value. If the local field is taken to be the dipolar field at a site in the cuprate planes, a magnitude of 0.15 μ_B is estimated for the copper moments. Although the transition appeared to be first-order, this was not firmly established.

Budnick *et al.*[29] have also studied the magnetic freezing of La_2CuO_4 with very similar results. In addition, the doped compound $La_{2-x}Sr_xCuO_4$ was studied[30] from x = 0 to 0.05. At x = 0 a magnetic transition was observed at 250 K and dropped to 6 K at x = 0.05. Above x = 0.05 the sample became superconducting with no indication of spin-

---

[28] Y. J. Uemura, W. J. Kossler, X. H. Yu, J. R. Kempton, H. E. Schone, D. Opie, C. E. Stronach, D. C. Johnston, M. S. Alvarez and D. P. Goshorn, *Phys. Rev. Lett.* **59** 1045 (1987) .

[29] J. I. Budnick, A. Golnik, Ch. Niedermayer, E. Recknagel, M. Rossmanith, A. Weidinger, B. Chamberland, M. Filipkowski and D. P. Yang, *Phys. Lett. A* **124** 103 (1987).

[30] J. I. Budnick, B. Chamberland, D. P. Yang, Ch. Niedermayer, A. Golnik, E. Recknagel, M. Rossmanith and A. Weidinger, *Europhys. Lett.* **5** , 651 (1988).

freezing down to 6 K. Given the large reduction in the spin-freezing temperature, it is surprising that the precession frequency in the most highly doped sample was almost three-quarters of that in the undoped sample. The reduction in spin-freezing temperature must therefore be caused by a reduction in coupling between moments and not by a reduction in the moments. The depolarization rate was observed to increase with x and was attributed to inhomogeneity in the internal magnetic field at the muon site.

Harshman *et al.*[31] have performed zero-field (ZF) and longitudinal-field (LF) μSR measurements on both a flux-grown single crystal and sintered powders of $La_{2-x}Sr_xCuO_4$ with x = 0.00, 0.01, 0.02 and 0.05 with critical-slowing-down reported near the spin-freezing temperature. A summary of their data for sintered powders above and below the spin-freezing temperature is shown in Fig. 1.

Measurements in an external field parallel to the initial muon spin direction has led to a two-component model of muon-spin depolarization in which muons are bound to two types of negatively charged oxygen ion sites. One muon site is taken to be near those oxygen ions that provide superexchange paths within the cuprate planes. A second site is taken to be near oxygen sites out of the cuprate planes. For the moments of the copper ions ordered antiferromagnetically, muons near planar sites should experience nearly zero local field because of cancellation of contributions from the two sublattices. Disorder in which the spins of adjacent copper ions are parallel should lead to a large local field. Muons near oxygen ions out of the plane couple primarily to the moment of a single Cu ion and are relatively insensitive to disorder. This model fits the x-dependence of the muon depolarization quite well. An inhomogeneity in B_{loc} is observed with increasing x and attributed to ferromagnetically paired moments of Cu^{2+} ions.

Sternlieb *et al.*[32] have compared muon spin relaxation and neutron scattering in a large single crystal[33] of $La_{1-x}Sr_xCuO_4$ with x = .06. Neither superconductivity nor antiferromagnetism develops in this concentration range . Muon studies indicate that the copper electronic moments are frozen below 3.9 K while neutron scattering indicates freezing below 20 K. This difference arises because muons are sensitive to much lower fluctuation frequencies.

$YBa_2Cu_3O_7$

Studies by Nishida *et al.*[34,35] of $YBa_2Cu_3O_{7-\delta}$ with $\delta > 0.6$ have confirmed a transition to a magnetically frozen phase at a transition temperature that depends on oxygen depletion. Their most depleted sample showed muon precession at $\nu \approx 4$ MHz and a local magnetic field B ≈ 300 G up to 250 K. A less depleted sample showed precession at about the same frequency but only up to 10K. As with La_2CuO_4, even though the transition temperature is sensitive to oxygen depletion, the frozen magnetization is relatively insensitive.

Brewer *et al.*[36] have carefully measured the dependence of the magnetic transition

---

[31] D. R. Harshman, G. Aeppli, G. P. Espinosa, A. S. Cooper, J. P. Remeika, E. J. Ansaldo, T. M. Riseman, D. Ll. Williams, D. R. Noakes, B. Ellman and T. F. Rosenbaum, *Phys. Rev. B* **38** (3) (1988).

[32] B. J. Sternlieb, G. M. Luke, Y. J. Uemura, T. M. Riseman, J. H. Brewer, P. M. Gehring, K. Yamada, Y. Hidaka, T. Murakami, T. R. Thurston and B. J. Birgeneau, *Phys. Rev. B* **41**, 8866 (1990).

[33] Y. Hidaka, Y. Enomoto, M. Suzuki, M. Oda and T. Murakami, *J. Cryst. Growth* **85**, 581 (1987).

[34] N. Nishida, H. Miyatake, D. Shimada, S. Okuma, M. Ishikawa, T. Takabatake, Y. Nakazawa, Y. Kuno, R. Keitel, J. H. Brewer, T. M. Riseman, D. Ll. Williams, Y. Watanabe, T. Yamazaki, K. Nishiyama, K. Nagamine, E. J. Ansaldo and E. Torikai, *Jpn. J. Appl. Phys.* **26**, L1856 (1987).

[35] N. Nishida, H. Miyatake, D. Shimada, S. Okuma, M. Ishikawa, T. Takabatake, Y. Nakazawa, Y. Kuno, R. Keitel, J. H. Brewer, T. M. Riseman, D. Ll. Williams, Y. Watanabe, T. Yamazaki, K. Nishiyama, K. Nagamine, E. J. Ansaldo and E. Torikai, *J. Phys. Soc. Jpn.* **57**, 597 (1988) .

[36] J. H. Brewer, E. J. Ansaldo, J. F. Carolan, A. C. D. Chaklader, W. N. Hardy, D. R. Harshman, M. E. Hayden, M. Ishikawa, N. Kaplan, R. Keitel, J. Kempton, R. F. Kiefl, W. J. Kossler, S. R. Kreitzman, A. Kulpa, Y. Kuno, G.

temperature on oxygen depletion. Their starting material was fully oxygenated with an uncertainty in δ of 0.05. After the removal of amounts of oxygen controlled to 0.01 the

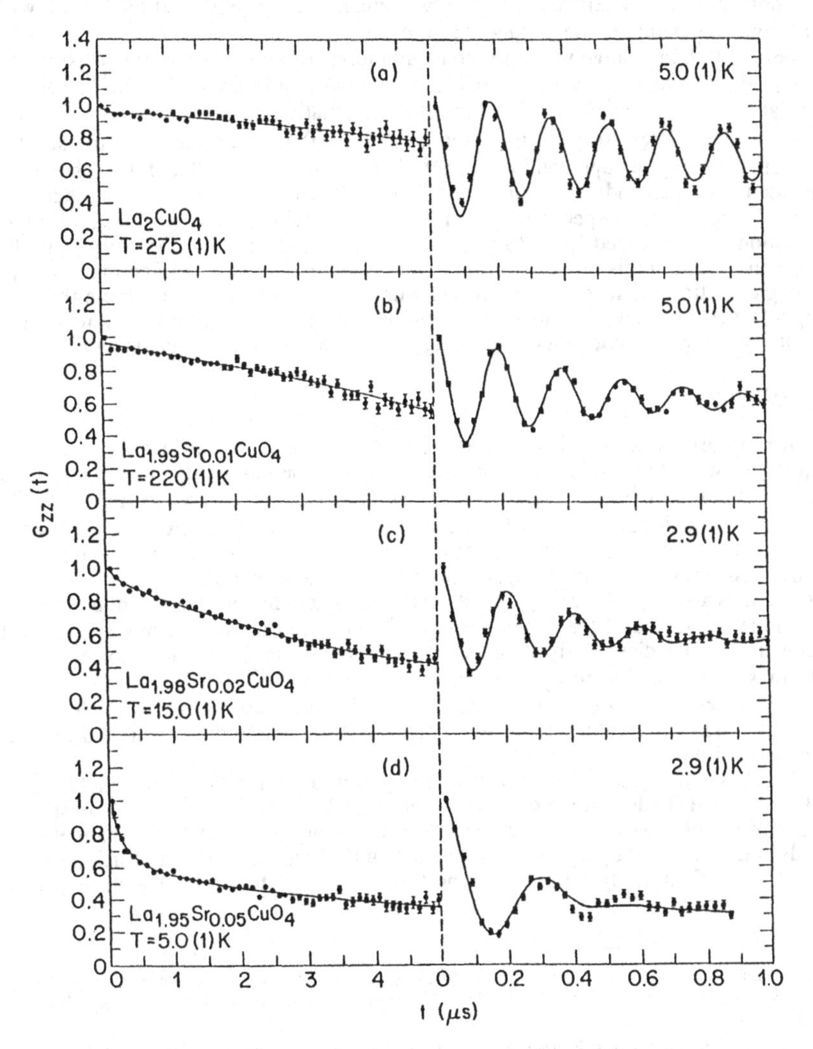

Figure 1. Summary of zero-field data of Harshman *et al.* [39] above and below the freezing-temperature for (a) La_2CuO_4 at 275 K and 5.0 K, (b) $La_{2-x}Sr_xCuO_4$ with x = 0.01 at 220 K and 5.0 K, (c) $La_{2-x}Sr_xCuO_4$ with x = 0.02 at 15.0 K and 2.9 K and (d) $La_{2-x}Sr_xCuO_4$ with x = 0.05 at 5.0 K and 2.9 K. The lines through the data are fits based on a function developed by the authors.

M. Luke, M. Miyatake, K. Nagamine, Y. Nakazawa, N. Nishida, K. Nishiyama, S. Ohkuma, T. M. Riseman, G. Roehmer, P. Schleger, D. Shimada, C. E. Stronach, T. Takabatake, Y. J. Uemura, Y. Watanabe, D. Ll. Williams, T. Yamazaki and B. Yang, *Phys. Rev.* **60**, 1073 (1988) .

samples were slowly annealed to establish a uniform oxygen concentration. The development of a magnetic moment at low temperature was observed for $\delta > 0.6$ with the transition temperature climbing to 450 K at $\delta = 1$.

For $\delta \approx 0.9$, in a range where the transition temperature is relatively insensitive to δ, a peak in the relaxation rate $1/T_1 = 0.4\ \mu s^{-1}$ was observed around 341 K,[37] leading to a correlation time $\tau_c = \omega^2/T_1 \approx 0.5$ ns for critical fluctuations.

Particularly interesting is the region around $\delta = 0.6$, where there is a transition from magnetic ordering to superconductivity.[34] Samples with $\delta = 0.60$ and 0.65 showed sharp transitions to superconductivity at 33 and 25 K and then the development of magnetization below 5 and 10 K, respectively. For the sample with $\delta = 0.60$, single-component relaxation was observed from 33 K down to 5 K, indicating that the entire sample was superconducting in this range. A reduction in asymmetry at 5 K marked the development of magnetization. It was not established whether the portion of the sample that became magnetically ordered remained superconducting or became normal. An increase in frequency at the ordering temperature suggests a reduction in the superconducting fraction.

$LnBa_2Cu_3O_7$

As already discussed, the detection of a precessing muon signal in zero magnetic field requires slow spin relaxation if not spin-freezing with the possibility of magnetic ordering. Such a signal has been observed in $GdBa_2Cu_3O_{7-\delta}$ by several groups.[38,39,40] The Konstanz group[35] resolved two frequencies below 2.3 K. The lower frequency $\nu = 4.7$ MHz (corresponding to a local field H = 33 mT) was dominant while the intensity of the higher frequency was sample-dependent and was associated with a defect muon stopping site, such as an oxygen vacancy, which could be present in variable amounts from sample to sample. The data of the Tokyo group[36] is shown in Fig. 4. The observation of these precession frequencies clearly indicates spin-freezing below 2.3 K and is consistent with an AF transition as indicated by magnetic susceptibility.[41,42,43] The Los Alamos group[37] performed zero-field (ZF) and transverse-field (TF) μSR studies between 5.8 and 300 K. They obtained temperature-dependent depolarization, which they attributed to Korringa relaxation of the Gd ion.

Various symmetric sites in the cuprate as well as in the Gd planes yield calculated values of local fields that are consistent with the observed frequencies. To establish the origin of the observed muon precession it is necessary to calculate the magnetic dipolar fields at the muon stopping sites, which unfortunately are not known. Such a calculation might be performed by minimizing the total energy of the system with respect to the

---

[37] G. M. Luke, R. F. Kiefl, J. H. Brewer, T. M. Riseman, D. Ll. Williams, J. R. Kempton, S. R. Kreitzman, E. J. Ansaldo, N. Kaplan, Y. J. Uemura, W. N. Hardy, J. F. Carolan, M. E. Hayden and B. X. Yang, *Physica* **C153-155**, 759(1988).

[38] A. Golnik, Ch. Niedermayer, E. Recknagel, M. Rossmanith, A. Weidinger, J. I. Budnick, B. Chamberland, M. Filipkowsky, Y. Zhang, D. P. Yang, L. L. Lynds, F. A. Otter and C. Baines, *Phys. Lett. A* **125**, 71 (1987).

[39] N. Nishida, H. Miyatake, S. Okuma, Y. Kuno, Y. Watanabe, T. Yamazaki, S. Hikami, M. Ishikawa, T. Takabatake, Y. Nakazawa, S. R. Kreitzman, J. H. Brewer and C-Y. Huang, *Jpn. J. Appl. Phys.* **27**, L94 (1988).

[40] D. W. Cooke, R. L. Hutson, R. S. Kwok, M. Maez, H. Rempp, M. E. Schillaci, J. L. Smith, J. O. Willis, R. L. Lichti, K-C. B. Chan, C. Boekema, S. P. Weathersby, J. A. Flint and J. Oostens, *Phys. Rev. B* **37**, 9401 (1988).

[41] C. Jiang, Y. Mei, S. M. Green, H. L. Luo and C. Politis, *Z. Phys. B* **68**, 15 (1987).

[42] T. Kaneko, H. Toyoda, H. Fujita, Y. Oda, T. Kohara, K. Ueda, Y. Yamada, I. Nakada and K. Asayama, *Jpn. J. Appl. Phys.* **26**, L1956 (1987).

[43] I. Oguro, T. Tamegai and Y. Iye, *Physica* **148B**, 456 (1987).

magnetic configuration, as shown by Sen,[44] or simply by assuming a magnetic configuration. Using the extended Luttinger-Tisza method, Sen has shown that an antiferromagnetic state is always the configuration of lowest energy. He has found that the magnetic fields at the Cu sites are smaller than the measured upper critical fields,[45] which might account for the observed coexistence of superconductivity and magnetic ordering.

The situation appears to be rather different in $HoBa_2Cu_3O_{7-\delta}$. The observations of Nishida *et al.*[36] indicate spin-freezing of the Ho moments below 3 K but no explicit local field for muon precession. The slow decay of muon asymmetry above 3 K is attributed to the fluctuating local fields of paramagnetic Ho moments.

Nishida *et al.*[36] have reported that the relaxation rate increased monotonically with decreasing temperature down to 7 K, where the relaxation function changed, indicating a transition from paramagnetism to a frozen phase. These results have been confirmed by Graboy *et al.*[46] This change was characterized both by an increase in the muon relaxation rate and by a change in the form of the relaxation function itself--from root exponential to Gaussian. At still lower temperatures the μSR signal could be described by Kubo-Toyabe (KT) depolarization in a Gaussian distribution of local fields, indicating a nearly static field distribution with a width of about 13.5 mT at 2.35 K. This width is large compared with the field at the muon site in $GdBa_2Cu_3O_{7-\delta}$, which orders antiferromagnetically near 2 K.

A subsequent experiment performed by Barth *et al.*[47] at temperatures down to 39 mK have been analyzed in terms of a rapidly damped oscillatory component arising from a broadened distribution of static fields. The most probable field value approached zero as the temperature was increased to 2.5 K.

Measurement of the Van Vleck second moment of the muon depolarization can be used in an attempt to establish the muon site. This moment was determined from transverse field (TF) μSR in nonmagnetic $YBa_2Cu_3O_{7-\delta}$, where the depolarization is caused entirely by the Cu nuclear moments.[44] The TF second-moment computed at a site half-way between two copper ions in the cuprate planes was found to be in agreement with the observed TF second-moment in $YBa_2Cu_3O_{7-\delta}$. The TF second-moment computed at this site from the rare earth moments in $GdBa_2Cu_3O_{7-\delta}$ and $HoBa_2Cu_3O_{7-\delta}$ was also found to be in agreement with the muon data.

5.3.2. *Fourier expansion of vortex fields*

The London penetration depth has been computed in the oxide superconductors from the

---

[44] H. Sen, *Phys. Lett. A* **129** (1988) 131.

[45] H. C. Hamaker *et al. Phys. Lett.* **A81** (1981) 91.

[46] I. E. Graboy, V. G. Grebinnik, I. I. Gurevich, V. N. Duginov, V. A. Zhukov, A. R. Kaul, B. F. Kirillov, E. P. Krasnoperov, A. B. Lazarev, B. A. Nikolsky, V. G. Olshevsky, A. V. Pirogov, V. Yu. Pomjakushin, A. N. Ponomarev, V. A. Suetin and S. N. Shilov, to be published.

[47] S. Barth, P. Birrer, F. N. Gygax, B. Hitti, J. Hulliger, E. Lippelt, H. R. Ott and A. Schenck, *Physica* **C153-155**, 767 (1988).

rate of muon spin relaxation resulting from a magnetic field distribution.[48,49,50,51,52,53,54] The analysis requires the expansion of **B**(**r**) in a Fourier series and the computation of the second moment of the field distribution.

Direct lattice

A two-dimensional periodic lattice (more properly a net) is characterized by primitive translation vectors **a** and **b** such that a lattice vector

$$\mathbf{R} = u\,\mathbf{a} + v\,\mathbf{b} \tag{12}$$

with u and v integers joins all equivalent points.

Reciprocal lattice

A two-dimensional reciprocal lattice is written as

$$\mathbf{G} = h\,\mathbf{A} + k\,\mathbf{B} \tag{13}$$

For the lattice of points **G** to be reciprocal to the lattice of points **R** we require

$$\mathbf{a}\cdot\mathbf{A} = 2\pi \qquad \mathbf{a}\cdot\mathbf{B} = 0 \qquad \mathbf{b}\cdot\mathbf{A} = 0 \qquad \mathbf{b}\cdot\mathbf{B} = 2\pi \tag{14}$$

which is satisfied by

$$\mathbf{A} = 2\pi\,(\mathbf{b}\times\mathbf{z})/(\hat{\mathbf{z}}\cdot\mathbf{a}\times\mathbf{b}) \qquad \mathbf{B} = 2\pi\,(\hat{\mathbf{z}}\times\mathbf{a})/(\hat{\mathbf{z}}\cdot\mathbf{a}\times\mathbf{b}) \tag{15}$$

where $\hat{\mathbf{z}}$ is a unit vector normal to the ab-plane. Using Eq. (14) we obtain from Eqs. (12) and (13)

$$\mathbf{G}\cdot\mathbf{R} = 2\pi\,(uh + vk) = 2\pi N \tag{16}$$

where N is an integer. The unit cell in the direct lattice is the region about a lattice point bounded by lines that are the perpendicular bisectors of all lattice vectors **R**. The area of the unit cell is

---

[48] G. Aeppli, R. J. Cava, E. J. Ansaldo, J. H. Brewer, S. R. Kreitzman, G. M. Luke, D. R. Noakes and R. F. Kiefl, *Phys. Rev. B* **35**, 7129 (1987).

[49] W. J. Kossler, R. J. Kempton, X. H. Yu, H. E. Schone, Y. J. Uemura, A. R. Moodenbaugh, M. Suenaga and C. E. Stronach, *Phys. Rev. B* **35**, 7133 (1987).

[50] F. N. Gygax, M. Hitti, E. Lippelt, A. Schenck, D. Cattani, J. Cors, M. Decroux, Ø. Fischer and S. Barth, *Europhys. Lett.* **4**, 473 (1987).

[51] D. R. Harshman, G. Aeppli, E. J. Ansaldo, B. Batlogg, J. H. Brewer, J. F. Carolan, R. J. Cava, M. Celio, A. C. D. Chaklader, W. N. Hardy, S. R. Kreitzman, G. M. Luke, D. R. Noakes and M. Senba, *Phys. Rev. B* **36**, 2386 (1987).

[52] B. Pümpin, H. Keller, W. Kündig, W. Odermatt, I. M. Slavic, J./ W. Schneider, H. Simmler, P. Zimmerman, J. G. Bednorz, Y. Maeno, K. A. Müller, C. Rossel, E. Kaldis, S. Rusiecki, W. Assmus and J. Kowalewski, *Physica C* **162-164**, 151 (1989).

[53] D. R. Harshman, L. F. Schneemeyer, J. V. Wasczak, G. Aeppli, R. J. Cava, B. Batlogg, L. W. Rupp, E. J. Ansaldo and D. Ll. Williams, *Phys. Rev. B* **39**, 851 (1989).

[54] E. J. Ansaldo, B. Batlogg, R. J. Cava, D. R. Harshman, L. W. Rupp, T. M. Riseman and D. Ll. Williams, *Physica C* **162-164**, 259 (1989).

$$A_c = \hat{\mathbf{z}} \cdot \mathbf{a} \times \mathbf{b} \tag{17}$$

Similarly the unit cell in the reciprocal lattice is the region about a reciprocal lattice point bounded by perpendicular bisectors of all reciprocal lattice vectors **G**. The area of the unit cell in reciprocal space is

$$\Omega_c = \hat{\mathbf{z}} \cdot \mathbf{A} \times \mathbf{B} \tag{18}$$

Using Eq. (15) we obtain for the relation between Eqs. (17) and (18)

$$\Omega_c = (2\pi)^2/A_c = 4\pi^2 n \tag{19}$$

where n is the number of sites in the direct lattice.

Field expansion and second moment

Any function, and in particular a magnetic field, periodic in the direct lattice may be expanded as a Fourier series in reciprocal lattice vectors

$$B(\mathbf{r}) = \Sigma_G B_G e^{i\mathbf{G} \cdot \mathbf{r}} \tag{20}$$

Replacing **r** in Eq. (20) by **r** + **R** and using Eq. (16) we establish the periodicity of B(**r**). The mean magnetic field over the unit cell (or any sufficiently large region of space) is simply given by the **G** = 0 Fourier component of Eq. (20)

$$\langle B(\mathbf{r})\rangle = B_0 \tag{21}$$

Taking Eq. (20) for **B**(**r**), the London equation leads to

$$B_G = n\Phi_0/(1 + \lambda_L^2 G^2) \tag{22}$$

The second moment is defined as

$$(\Delta B)^2 = \langle(B - \langle B\rangle)^2\rangle = \langle B^2\rangle - \langle B\rangle^2 = (n\Phi_0)^2 \,\Sigma' (1 + \lambda_L^2 G^2)^{-2} \tag{23}$$

where $n = B/\Phi_0 = 1/A_c$ is the vortex density and the **G** = 0 term is excluded from the summation. With the definitions

$$x = 4\pi n\, \lambda_L^2 \qquad\qquad \Delta x = 4\pi \Delta n\, \lambda_L^2 \tag{24}$$

the second moment may be written in dimensionless form as

$$(\Delta x)^2 = x^2 \,\Sigma' (1 + \pi\, x\, G^2/\Omega_c)^{-2} \tag{25}$$

We next find the second moments of the square and triangular lattices.

The square lattice.

The primitive vectors of the square lattice are

$$\mathbf{a} = d\,\hat{\mathbf{x}} \qquad \mathbf{b} = d\,\hat{\mathbf{y}} \tag{26}$$

with $n=1/d^2$ and $\hat{\mathbf{x}}$ and $\hat{\mathbf{y}}$ orthogonal unit vectors. From Eq. (15) the reciprocal lattice vectors are

$$\mathbf{A} = (2\pi/d)\,\hat{\mathbf{x}} \qquad \mathbf{B} = (2\pi/d)\,\hat{\mathbf{y}} \tag{27}$$

Summing over the first twenty-four reciprocal lattice vectors we obtain for the second moment using Eq. (25)

$$(\Delta x)^2 = 4x^2[(1+\pi x)^{-2} + (1+2\pi x)^{-2} + (1+4\pi x)^{-2} + 2(1+5\pi x)^{-2} + (1+8\pi x)^{-2}] \tag{28}$$

At high density $x >> 1$ we obtain from Eq. (28)

$$(\Delta x)^2 = (4/\pi^2)\,(1 + {}^1/_4 + {}^1/_{16} + {}^2/_{25} + {}^1/_{64}) \tag{29}$$

which leads to

$$\Delta B \approx 1.1866\,(2/\pi)(\Phi_0/4\pi\lambda_L^2) = 0.7554\,\Phi_0/4\pi\lambda_L^2 \tag{30}$$

To obtain the low density limit of Eq. (25) we have to include terms at least out to $G \approx 1/\lambda_L$. This is best done by summing the first N terms explicitly and integrating over the remainder

$$(\Delta B)^2 \approx (n\Phi_0)^2 \sum (1+\lambda_L^2G^2)^{-2} + \int 2\pi G dG/\Omega_c(1+\lambda_L^2G^2) \tag{31}$$

with $\pi G_N^2 = (N+1)\Omega_c$. Performing the integration and using Eq. (25) the second moment is

$$(\Delta x)^2 = x^2 \sum{}' (1+\pi x G^2/\Omega_c)^{-2} + x/[1+(N+1)x] \tag{32}$$

which gives for $x = 4\pi n\lambda_L^2 << 1$

$$\Delta x \approx \sqrt{x} \tag{33}$$

The triangular lattice

The primitive translation vectors of the triangular lattice are

$$\mathbf{a} = d\,\hat{\mathbf{x}} \qquad \mathbf{b} = \tfrac{1}{2} d\,(\hat{\mathbf{x}} + \sqrt{3}\,\hat{\mathbf{y}}) \tag{34}$$

with $n = 2/\sqrt{3}\,d^2$. The reciprocal lattice vectors are

$$\mathbf{A} = (2\pi/d)\,[\,\hat{\mathbf{x}} - (1/\sqrt{3})\,\hat{\mathbf{y}}\,] \qquad \mathbf{B} = (2\pi/d)\,(2/\sqrt{3})\,\hat{\mathbf{y}} \tag{35}$$

We sum over the lowest six Fourier components of the triangular lattice and integrate over the remainder to obtain

$$(\Delta x)^2 \approx 6x^2/(1+2\pi x/\sqrt{3})^2 + x/(1+7x) \tag{36}$$

The high density limit of Eq. (36) is

$$\Delta B \approx (9/2\pi^2 + 1/7)^{1/2}\, \Phi_0/4\pi\lambda_L^2 = 0.7738\, \Phi_0/4\pi\lambda_L^2 \qquad (37)$$

In Fig. 2 we plot Eq. (36).

The second moment is reduced slightly by including additional terms as we have done for the square lattice. Summing over the lowest thirty-six Fourier components we obtain in place of Eq. (37)

$$(\Delta x)^2 = 6x^2[(1+2\pi x/\sqrt{3})^{-2} + (1+6\pi x/\sqrt{3})^{-2} + (1+8\pi x/\sqrt{3})^{-2} + 2(1+14\pi x/\sqrt{3})^{-2} + (1+54\pi x/\sqrt{3})^{-2}] \qquad (38)$$

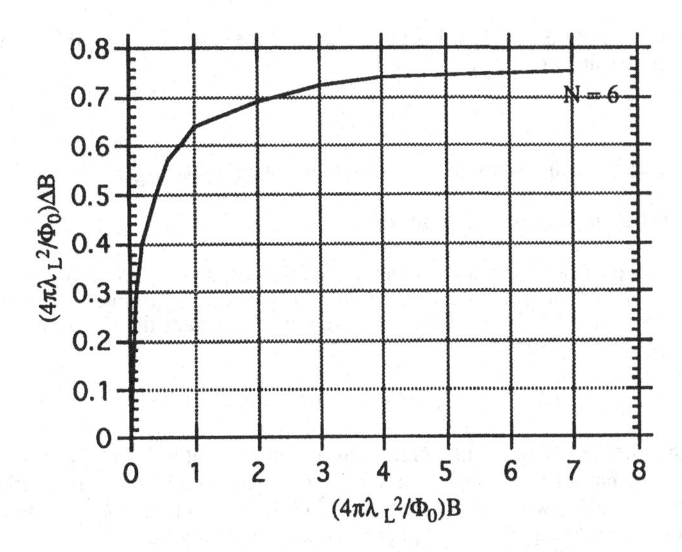

Figure 2. Second moment of a triangular lattice obtained by summing over the six lowest Fourier components and integrating over the balance.

The high density limit of Eq. (38) is

$$\Delta B = 0.7657\, \Phi_0/4\pi\lambda_L^2 \qquad (39)$$

Laminar flux distribution

A rectangular $a \times b$ lattice in which one distance is held fixed makes a transition as a function of field from a vortex lattice at low fields to a laminar flux distribution at high fields. We take the distance between laminar planes to be a and the distance between vortices in a plane to be b. The vortex density is then $n = 1/ab$. As the density increases, the vortex lattice begins as rectangles elongated parallel to the planes, passes through a square lattice and and finally forms rectangles elongated normal to the laminar planes. For the distance a between planes not much larger than λ_L, the second moment goes through a pronounced maximum. The second moment does not saturate at higher fields, but continues to grow, the strong indicator of a laminar flux distribution.

Amorphous flux distribution at high density

We now use the Fourier expansion method to obtain the second moment of an amorphous distribution of vortices. We allow all Fourier components of fluxon position except that we require fixed density over areas larger than the unit cell A_c. We accomplish this by excluding Fourier components within the first Brillouin zone of area $\Omega_c = (2\pi)^2/A_c$. We treat G as a continuous quantity through the integral portion of Eq. (31)

$$(\Delta B)^2 = (n\Phi_0)^2 \int_{G_{min}} (2\pi G\, dG)/\Omega_c(1 + \lambda_L{}^2G^2)^2$$

$$= (n\Phi_0{}^2/4\pi\lambda_L{}^2)/(1 + \lambda_L{}^2G_{min}{}^2) = (n\Phi_0{}^2/4\pi\lambda_L{}^2)/(1 + 4\pi n\lambda_L{}^2) \quad (40)$$

where we have used $\pi G_{min}{}^2 = \Omega_c = 4\pi^2 n$ as the area of the central cell. Taking Eq. (40) to the limit of high density we obtain

$$\Delta B \approx \Phi_0/4\pi\lambda_L{}^2 \quad (41)$$

independent of vortex concentration as for lattices but about 50% larger.

Amorphous flux distribution at low density

Finally, we apply the Fourier expansion method to an amorphous distribution of vortices at low density, where we can regard the distribution of vortices as completely random. We obtain the second moment by lifting the restriction of constant density beyond A_c and integrate Eq. (40) to $G = 0$ obtaining

$$\Delta B = (n/4\pi\lambda_L{}^2)^{1/2}\, \Phi_0 = (\Phi_0 B/4\pi\lambda_L{}^2)^{1/2} \quad (42)$$

We see that the second moment increases as the square root of B at all fields. We may understand this behavior as arising from density fluctuations within an area $4\pi\lambda_L{}^2$. The mean number of vortices within this area is $4\pi\lambda_L{}^2 n$ and the variance is $4\pi\lambda_L{}^2 n$ leading to a variation in density $(n/4\pi\lambda_L{}^2)^{1/2}$ and the spread in field given by Eq. (42).

5.3.3. *Diffusive narrowing*

We now treat narrowing of the second moment as the result of muon diffusion, which has been reported in Nb.[21,22] In our treatment of diffusive narrowing we utilize the diffusion equation, which represents the diffusing muons by a density $\rho(\mathbf{r}, t)$ and finds the phase of precessing muons $\phi(\mathbf{r}, t)$ for such a distribution. This procedure is appropriate if a muon at $(\mathbf{r}, t)$ has sampled the entire space weighted by $\rho(\mathbf{r}, t)$ for previous times. This condition will not be satisfied as the muon begins to diffuse. After the muon has diffused for some time and has taken a large number of steps of length λ the approximation should be appropriate. It does require that the jump time τ be much shorter than the muon life time τ_μ and even then misses early depolarization at times $t < \tau$.

Muon diffusion equation

The diffusing muon current is related to the gradient of its density by Fick's law

$$\mathbf{J} = -D\,\nabla\rho \tag{43}$$

where $D \approx \lambda^2/3\tau$ is called the diffusion constant. Coupling Fick's law with the condition for particle conservation

$$\nabla \cdot \mathbf{J} = -\partial\rho/\partial t \tag{44}$$

we obtain the diffusion equation

$$D\,\nabla^2\rho(\mathbf{r}, t) = \partial\rho(\mathbf{r}, t)/\partial t \tag{45}$$

For a particle initially at $\mathbf{r} = 0$ the solution to the diffusion equation is

$$\rho(r, t) = (4\pi Dt)^{-3/2}\exp(-r^2/4Dt) \tag{46}$$

Field narrowing

We now consider a periodic field

$$B(\mathbf{r}) = B_G\,e^{iGx} \tag{47}$$

For a muon initially at $\mathbf{r} = 0$ the mean field at time t is

$$\langle B(\mathbf{r}, t)\rangle = (4\pi Dt)^{-3/2}\int B_G\exp(-|\mathbf{r}'-\mathbf{r}|^2/4Dt)\,e^{iGx'}dx'dy'dz' \tag{48}$$

Integrating over y' and z' we obtain

$$\langle B(\mathbf{r}, t)\rangle = (4\pi Dt)^{-1/2}\int H_G\exp[-(x'-x)^2/4Dt]\,e^{iGx'}dx' \tag{49}$$

Integrating Eq. (49) by completing the square in the exponent we obtain

$$\langle B(\mathbf{r}, t)\rangle = B_G\,e^{iGx}\exp(-DG^2t) \tag{50}$$

and we see that the apparent periodic field decays exponentially at a characteristic rate

$$1/\tau_G = DG^2 \approx \lambda^2G^2/3\tau \tag{51}$$

To obtain the phase of a precessing muon we integrate Eq. (50) over all previous times

$$\phi(\mathbf{R}, t) = \int \gamma H_G e^{iGx}e^{-t/\tau_G}dt = \gamma H_G\tau_G e^{iGx}\left(1 - e^{-t/\tau_G}\right) \tag{52}$$

5.3.4. Disordered flux lattices

In this section we obtain the second moment for small and random vortex displacements with the field macroscopically uniform. To obtain an expression for the second moment we must first develop some of the mathematics for treating random variables. The magnetic field at **r** from a vortex lattice is

$$\mathbf{B}(\mathbf{r}) = \Sigma_{\mathbf{R}}\, \mathbf{B}(\mathbf{r} - \mathbf{R}) \tag{53}$$

where $\mathbf{B}(\mathbf{r} - \mathbf{R})$ is the field at **r** from a vortex at **R**. Expressing $\mathbf{B}(\mathbf{r} - \mathbf{R})$ as a Fourier integral

$$\mathbf{B}(\mathbf{r} - \mathbf{R}) = \int dV_k\, \mathbf{B}_k \exp\{i\mathbf{k}\cdot(\mathbf{r} - \mathbf{R})\} \tag{54}$$

and substituting into (37) we obtain the Fourier components

$$\mathbf{B}_k = n\Phi_0/(1 + \lambda_L^2\, k^2) \tag{55}$$

Substituting (55) into (54), the field is

$$\mathbf{B}(\mathbf{r}) = \Sigma_{\mathbf{R}} \int dV_k\, \mathbf{B}_k \exp\{i\mathbf{k}\cdot(\mathbf{r} - \mathbf{R})\} \tag{56}$$

For **R** periodic the sum over **R** gives the Dirac delta function $\delta(\mathbf{k} - \mathbf{G})$. For the disordered case we replace **R** by $\mathbf{R} + \delta\mathbf{R}$. Assuming $\mathbf{k}\cdot\delta\mathbf{R}$ small we expand the exponential to obtain

$$\mathbf{B}(\mathbf{r}) = \Sigma_{\mathbf{R}} \int dV_k\, \mathbf{B}_k\, [1 - i\mathbf{k}\cdot\delta\mathbf{R} - \tfrac{1}{2}(\mathbf{k}\cdot\delta\mathbf{R})^2] \exp\{-i\mathbf{k}\cdot(\mathbf{r} - \mathbf{R})\} \tag{57}$$

The mean square field is

$$\langle \mathbf{B}(\mathbf{r})^2 \rangle = (1/V) \int dV\, \Sigma_{\mathbf{R}\mathbf{R}'} \int dV_{k'}\, dV_k\, \mathbf{B}_{k'} \mathbf{B}_k\, [1 - i\mathbf{k}\cdot\delta\mathbf{R} - \tfrac{1}{2}(\mathbf{k}\cdot\delta\mathbf{R})^2]$$

$$\times [1 + i\mathbf{k}'\cdot\delta\mathbf{R}' - \tfrac{1}{2}(\mathbf{k}'\cdot\delta\mathbf{R}')^2] \exp\{i\mathbf{k}\cdot(\mathbf{r} - \mathbf{R}) - i\mathbf{k}'\cdot(\mathbf{r} - \mathbf{R}')\} \tag{58}$$

Integrating over V and summing over **R** and **R'** leads to

$$\langle \mathbf{B}(\mathbf{r})^2 \rangle = \Sigma_G\; \mathbf{B}_G^2\, [1 - (\mathbf{G}\cdot\delta\mathbf{R})^2] \tag{59}$$

which gives a *reduction* in the second moment and is the usual motional narrowing result.

6
FLUX PINNING, CREEP AND FLOW

6.1. Flux Pinning

6.1.1. Forces on vortices

We identify three forces that act on vortices. The first is that exerted by the flow of external current. The second force on vortices is from the repulsion of other vortices, which may be regarded as a magnetic pressure. The third force is from structural defects that, if large enough, will pin vortices. These defects may be localized, such as small inclusions of normal material. There are also extended defects, such as the filamentary structure obtained when a superconductor is forced into vicor glass as well as granular structure, which is discussed in Ch. 11.

Lorentz force

External currents exert a force on vortices that is the equilibrant of the Lorentz force

$$\mathbf{F} = \Phi_0\, \hat{\mathbf{z}} \times \mathbf{J}_{ext} \tag{1}$$

Vortex motion in turn produces an electric field

$$\mathbf{E} = -\,\mathbf{v} \times \mathbf{B} \tag{2}$$

against which the current must be driven, requiring electrical work. If there is an irreversible component of vortex motion over a cycle, net electrical work will be done.

Magnetic pressure

The force per unit length on a vortex from other vortices is

$$\mathbf{F} = \int dS\, \mathbf{J}_{ext} \times \mathbf{B} \tag{3}$$

where $\mathbf{J}_{ext}$ is the vortex current density and $\mathbf{B}$ is the magnetic field of the other vortices. We use the vector potential to transform Eq. (3) to

$$\mathbf{F} = \int dS\, \mathbf{J}_{ext} \times (\nabla \times \mathbf{A}) = -\int dS\, (\nabla \times \mathbf{J}_{ext}) \times \mathbf{A} \tag{4}$$

plus terms that can be dropped. Using the London equation

$$\ell\, \mathbf{J} + \mathbf{A} = 0 \tag{5}$$

and the integral of the curl of $\mathbf{J}_{ext}$ from Eq. (5-31)

$$\int dS\, (\nabla \times \mathbf{J}_{ext}) = (1/\ell)\int dS\, \Phi_0\, \hat{\mathbf{z}}\, \delta(\mathbf{r}) = (1/\ell)\Phi_0\, \hat{\mathbf{z}} \tag{6}$$

we obtain for the magnetic force

$$\mathbf{F} = \Phi_0\, \hat{\mathbf{z}} \times \mathbf{J} \tag{7}$$

where $\mathbf{J}$ is the induced current.

Pinning forces

We finally consider the force that defects exert on vortices. As we have seen, forces on vortices arise from currents. We choose to regard pinning currents as external currents and write them as $\nabla \times \mathbf{H}$. The force on a vortex may then be written as

$$\mathbf{F} = \Phi_0\, \hat{\mathbf{z}} \times \mathbf{J}_{ext} = \Phi_0\, \hat{\mathbf{z}} \times (\nabla \times \mathbf{H}) \tag{8}$$

where $\mathbf{H}$ is a microscopic field.

Equilibration of current

Bean's critical state model[6] assumes that for vortices to enter or leave a medium, a critical force per unit volume

$$\mathbf{F}_c = \mathbf{J}_c \times \mathbf{B} \tag{9}$$

must be equilibrated. The critical force arises from inhomogeneities in the medium and is expressed in terms of a critical current density $\mathbf{J}_c$, which may be a function of the flux density $\mathbf{B}$. The force generated by inhomogeneities is equilibrated by a magnetic pressure gradient arising from the variation in flux density

$$\mathbf{F} = -\nabla\, \mathcal{U} = -(\nabla\, \mathbf{B}) \cdot \mathbf{H} = (\nabla \times \mathbf{B}) \times \mathbf{H} - (\mathbf{H} \cdot \nabla)\, \mathbf{B} \tag{10}$$

which must be associated with a volume current density

$$\nabla \times \mathbf{B} = \mu\, \mathbf{J} \tag{11}$$

From Eqs. (9), (10) and (11) we obtain $\mathbf{J} = -\mathbf{J}_c$ and the magnetic current just compensates the current arising from inhomogeneities.

6.1.2. Critical phenomena

A number of investigators have noticed that the strength of vortex pinning may be characterized by a macroscopic critical current density J_c. To move flux, a critical pressure-gradient BJ_c is required. Associated with this pressure-gradient is a magnetic field-gradient $dB/dz = \mu J_c$. The magnetic field-gradient may be detected by bulk magnetization measurements or by direct measurement of critical shielding.

Bean critical state

Bean[1] computed the mean magnetization of a superconducting cylinder on the assumption of a constant critical current J_c. His results are in good agreement with measurements of the mean magnetization of Nb_3Sn cylinders of radii a = 0.12 and 0.24 cm in fields up to 7 kOe. The critical current density J_c was found to be 1.2×10^5 A/cm^2.

We introduce in addition to the lower critical field H_{c1} an effective field $H^* = aJ_c$. For external fields $H < H_{c1} = 1.6$ kOe, flux is largely excluded so long as the diameter of the cylinder is large compared with λ_L. The mean magnetization density of the cylinder for $H < H_{c1}$ is $-M \approx H$ and is shown in Fig. 1 a).

For fields $H_{c1} < H < (H_{c1} + H^*)$, the induction B drops linearly to H_{c1} at a radius $r_c = a[1 - (H - H_{c1})/H^*]$. For radii $r < r_c$ flux is completely excluded. The mean magnetization of the cylinder is shown in Fig. 1 b)

$$-M = \frac{1}{3}H^*\left[1+2\left(\frac{r_c}{a}\right)^3\right]+(H-H^*)\left(\frac{r_c}{a}\right)^2 \tag{12}$$

Finally for fields $H > (H_{c1} + H^*)$, B drops linearly to $(H - H^*) > H_{c1}$ on the axis, vortices penetrate the entire cylinder, and the mean magnetization density of the cylinder as shown in Fig. 1 c) is

$$-M = \frac{a}{3}J_c \tag{13}$$

Two other examples are a flat plate and a sphere. For the external field H in the plane of a flat plate of thickness w, the magnetization density at high fields is $-M = (w/4)J_c$. For a sphere of radius a, the magnetization density at high fields is $-M = (3\pi a/32)J_c$.

The high-temperature superconductors are axial and conduct largely in planes. A random distribution of anisotropic spherical grains exhibits a mean magnetization density

$$-M = \frac{1}{3}\times\frac{3\pi a}{32}J_c = \frac{\pi a}{32}J_c \tag{14}$$

Bean[2] has applied an alternating magnetic field to a synthetic type II superconductor of lead forced into a porous Vycor glass cylinder and has measured the harmonic content of the flux that penetrates the cylinder. As expected, only odd harmonics are observed. Good agreement is obtained with the critical field model and $J_c = 2.1 \times 10^5$ A/cm^2.

---

[1] C. P. Bean, *Phys. Rev. Lett.* **8**, 250 (1962).

[2] C. P. Bean, *Rev. Mod. Phys.* **36**, 31 (1964).

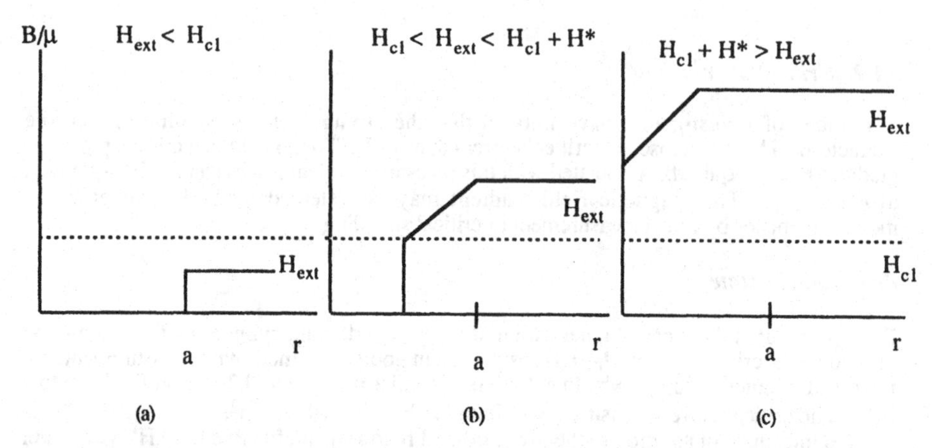

Figure 1. Mean magnetization of a superconducting cylinder in the critical state. a) For fields at the surface less than H_{c1}, flux is completely excluded. b) For fields at the surface greater than H_{c1} but less than $H_{c1} + H^*$ (defined in the text) the flux density drops linearly into the cylinder. At a critical radius r_c at which $H_{int} = H_{c1}$ the flux density drops almost discontinuously to zero. c) For fields at the surface greater than $H_{c1} + H^*$ the flux density drops linearly within the cylinder.

Bean assumed the following simple model:

(i) For $H < H_{c1}$ the magnetic field is screened within a very short distance of the surface and we may take $B = 0$ within the superconductor.

(ii) For $H > H_{c1}$ and J_c constant, flux penetrates the superconductor with a density that decreases linearly with distance down to H_{c1} at which point B drops to zero.

The surface flux density

$$\Phi_s = \int B\, dx \tag{15}$$

for an initially increasing field $H > H_{c1}$ has the form

$$\Phi_s = (1/J_c)\,(H^2 - H_{c1}^2) \tag{16}$$

If the field is increased to the value $+ H_s$ and then decreased toward $- H_s$, the surface flux density has the form

$$\Phi_s = (1/2J_c)\mu[H_s^2 - \tfrac{1}{2}(H_s - H)^2] \tag{17}$$

where for simplicity we have set $H_{c1} = 0$. For fields increasing from $-H_s$ toward H_s the surface flux density is

$$\Phi_s = -\,(1/2J_c)\mu[H_s^2 - \tfrac{1}{2}(H_s + H)^2] \tag{18}$$

For a material with hysteresis, the work performed per unit volume when the sample is taken around a *closed cycle* of applied field H is

$$W_v = \oint H\, dB \tag{19}$$

In terms of the surface flux density Eq. (15), the work per unit area performed around a cycle is

$$W_s = \oint H\, d\Phi_s \tag{20}$$

The work W_s performed around a cycle is one-third the area of the rectangle bounded by $H = \pm H_s$ and $\Phi_{max} = \pm \mu H_s^2/2J_c$

$$W_s = \frac{2}{3}\mu H_s^3/J_c \tag{21}$$

The power absorbed per unit area of surface is

$$P_s = W_s/T = \frac{1}{2\pi}\omega W_s \tag{22}$$

where T is the period and $\omega = 2\pi/T$ is the frequency. Substituting into Eq. (22) and equating to the definition of surface resistance

$$P_s = \frac{1}{2}R_s H_s^2 \tag{23}$$

the surface resistance is

$$R_S = \frac{2}{3\pi}\omega\mu(H_s/J_c) \tag{24}$$

For a plate of thickness d that the field penetrates from one side, the surface resistance passes through a maximum for $H_s > J_c d$ and drops off as $1/H_s$

$$R_s = \frac{2}{\pi}\omega\mu d\frac{J_c d}{H_s}\left(1 - \frac{2}{3}\frac{J_c d}{H_s}\right) \tag{25}$$

The surface reactance at low rf fields is

$$X_s = \frac{1}{2}\omega\mu d\frac{H_s}{J_c d} \tag{26}$$

and at high fields takes the form

$$\begin{aligned} X_s = {} & \frac{1}{2\pi}\omega\mu d\frac{H_s}{J_c d}\left(\theta - \frac{7}{4}\sin\theta + \frac{1}{2}\sin 2\theta - \frac{1}{12}\sin 3\theta\right) \\ & + \frac{1}{\pi}\omega\mu d\left(\pi - \theta + 2\sin\theta - \frac{1}{2}\sin 2\theta\right) - \frac{2}{\pi}\omega\mu d\frac{J_c d}{H_s}\sin\theta \end{aligned} \tag{27}$$

with $\cos\theta = 1 - 2J_c d/H_s$. At fields $H_s < J_c d$, θ is fixed at π and Eq. (27) leads to Eq. (26). At higher fields $H_s > J_c d$, θ approaches 0 and X_s saturates at $\omega\mu d$.

Kim-Anderson critical state

Anderson and Kim[3] have reported evidence that J_c is not independent of flux density but drops with increasing field as

$$J_c(H) = J_o/(1 + B/B_0) \tag{28}$$

The work performed around a cycle in fields that exceed B_0 is fortuitiously unaffected by the specific Kim-Anderson modification of the critical current.[4]

Surface barrier

A barrier to the entry and exit of vortices may be represented by a high critical current density J_c at the surface. We assume for simplicity a critical current J_b from $z = 0$ to $z = b$ and a critical current J_c for $z > b$ with $J_b >> J_c$. The magnetic field $B(z)$ as a function of increasing H_{ext} is sketched in Fig. 2 (a) and for decreasing H_{ext} in Fig. 2 (b). There is a threshold field for flux to penetrate the barrier

$$H_b^* = b\, J_b \tag{29}$$

We examine a situation in which the external field has been increased to $H_0 >> H_b$ and then reduced. Vortices leave the barrier region, but the vortex density in the bulk is initially unaffected. As is apparent from Fig. 2 (b), to remove vortices from the bulk the external field must be reduced by at least $2H_b$. This is the field increment required for the establishment of a full reverse critical state.

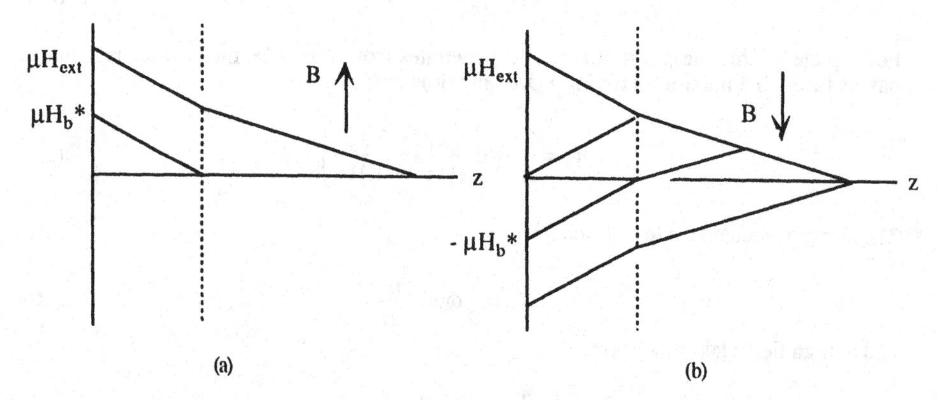

Figure 2. Representation of a surface barrier by a region in which the critical current density J_b is greater than the bulk critical current density J_c. a) The flux density $B(z)$ for increasing external fields. b) The flux density $B(z)$ for decreasing external fields.

---

[3] P. W. Anderson and Y. B. Kim, *Revs. Mod. Phys.* **36**, 39 (1964).

[4] A. M. Portis in *Earlier and recent aspects of superconductivity*, eds. G. Bednorz and K. A. Müller, Vol. 90 of the Springer Series in Solid State Sciences, (Springer, Berlin, 1990) pp. 278-303.

6.1.3. Elastic response

Thus far we have assumed with most investigators that a critical state with positive or negative J_c is formed immediately and expands into the volume. Such behavior leads to hysteresis for arbitrarily small changes in external field. Careful investigation has shown that there is a threshold field, which we designate as H', for hysteretic behavior. For field changes less than H', magnetization processes are reversible. For field changes greater than H' a full critical state develops with hysteresis.

We examine a superconductor cooled below T_c in a field $H_0 > H_{c1}$ We increase or decrease the field slightly and assume that instead of immediately developing a critical state, there is a range in which the incremental field penetrates exponentially and reversibly. For small incremental fields, the flux density within the superconductor is[5]

$$B(z) = \mu H_0 + \mu(H_{ext} - H_0) \exp(- x/\lambda') \tag{30}$$

The relaxation distance λ' may be a function of H, but is otherwise assumed to be independent of position. The field gradient

$$dB/dx = (B - \mu H_0)/\lambda' \tag{31}$$

is largest at the surface, where it is equal to $(H_{ext} - H_0)/\lambda'$. As described by Eq. (7), a flux gradient exerts a force on vortices. There is a maximum force and thus a maximum field gradient μJ_c that vortices can absorb without flow. Thus we expect exponential behavior only for field increments

$$|H_{ext} - H_0| < \lambda' J_c = H' \tag{32}$$

For field increments larger than H' the incremental field is assumed to drop linearly until it reaches H' and then exponentially. In Figure 3 (a) we sketch B(z) for increasing fields and in Fig. 3 (b) for decreasing fields.

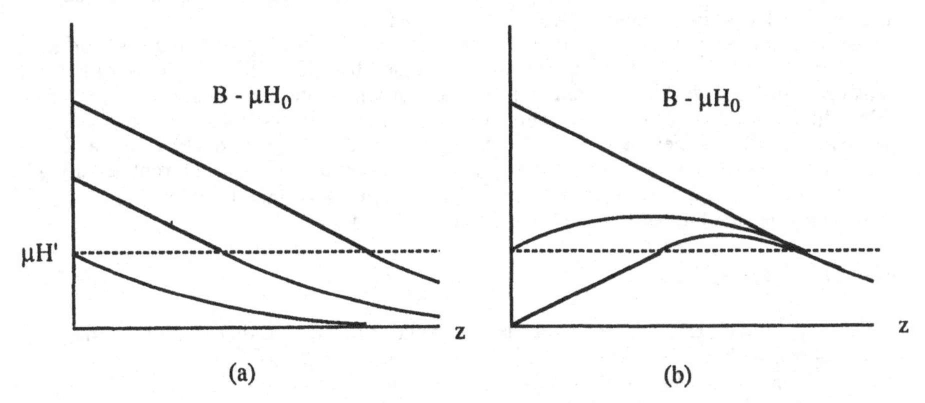

Figure 3. Behavior of incremental fields less that the field required for the full development of a critical state. (a) Increasing external fields. (b) Decreasing external fields.

[5] A. M. Campbell and J. E. Evetts, *Adv. Phys.* 21, 199 (1972). Reprinted as *Critical Currents in Superconductors* (Taylor and Francis, London, 1972) sec. 8.2.1.

We now generalize our treatment to J_c and λ' functions of position. We have in mind a surface barrier with J_c much larger at the surface than within the bulk. If there is little variation in H', λ' will vary inversely with J_c. For small field increments we write

$$B(x) = \mu H_0 + \mu(H_{ext} - H_0) \exp(-\int dx/\lambda') \tag{33}$$

with a field gradient

$$dB/dx = -(B - \mu H_0)/\lambda' \tag{34}$$

As the incremental field is increased, a critical state will first develop within the superconductor—not necessarily at the surface. In the region of the critical state, the field will drop linearly while it continues to drop exponentially elsewhere. With increasing fields we ultimately expect the field to drop everywhere at a local rate given by J_c except well within the superconductor where the incremental field has dropped below H' and the field approaches H_0 exponentially. If we now reduce the magnitude of the surface field, we expect a reverse exponential until a reverse critical state is developed somewhere. With further field reduction the reverse state should grow until it extends over most of the superconductor.

6.1.4. Field measurements

Lead-alloy films

Uchida *et al.*[6] have made a direct measurement of the critical state of an 0.52 mm-wide lead-alloy film with a magnetic field normal to the plane of the film. An array of Josephson junctions was deposited directly on the film with the increased quasi-particle current providing a measure of B. The basic film was Pb: 29% In. Adding 4% Au produced precipitates that pinned flux. Adding 29% Bi reduced the pinning, presumably because of the increased λ_L, which reduced the pinning potential.

In one series of measurements the films were cooled through T_c in zero field after which fields up to 155 Oe were applied normal to the film. The field at the edge of the film was up to 20 Oe lower than the external field and dropped linearly toward the center of the film. In a second series the films were cooled through T_c in 155 Oe. The field was then removed and B was measured across the film. The field at the edge of the film was 20 Oe and increased to about 120 Oe at the film center. The computed critical current density was $J_c = (10/4\pi) \times 100/.026 = 3 \times 10^3$ A/cm^2 and, except for surface barrier behavior at the film edge, appeared to be uniform for fields up to 155 Oe.

Ceramic $YBa_2Cu_3O_{7-\delta}$

Kisu *et al.*[7] have performed a direct measurement of the spatial distribution of magnetic flux trapped within a sample of ceramic $YBa_2Cu_3O_{7-\delta}$ 10 mm in diameter and 2 mm thick. The disk was cooled below T_c in a magnetic field normal to the plane of the disk. As the magnetizing field was reduced, the induction was measured as a function of radius with a GaAs Hall-sensor 1 mm on a side. The magnetization varied linearly with radius near the edge but was almost flat in the central region, indicating that a critical state was never fully

[6] N. Uchida, K. Enpuku, K. Yoshida and F. Irie, *J. Appl. Phys.* **56**, 2558 (1984).

[7] T. Kisu, K. Enpuku, K. Yoshida, M. Takeo and K. Yamafuji, *Jpn. J. Appl. Phys.* **26**, L1348 (1987).

developed and that the relaxation distance was of the order of the disk radius. A relaxation distance this large may be an indication of weak coupling between grains.

Thin films

Scharen *et al.*[8] have determined the critical current flowing in a laser-deposited film of $YBa_2Cu_3O_{7-\delta}$ from a direct measurement of the field produced by critical current flowing in the film. A field is applied normal to the film, increased above H^*, and then reduced to zero, generating a uniform screening current density J_c in the film. An axial Hall-probe is use to make a single measurement of the on-axis field, which at large distances is

$$H(z) \approx 2m/z^3 \tag{35}$$

For a rectangular film of sides $a < b$ the moment is given by $m = (1/4)J_c t a^2 b(1 - a/3b)$ where t is the film thickness. For a circular film of radius a, the moment is given by $m = (\pi/3)a^3 J_c t$.

Ceramic oxides

Tjukanov *et al.*[9] have measured the temperature dependence of flux trapped by ceramic tubes of $YBa_2Cu_3O_{7-\delta}$. The distinguishing characteristic of tube measurements in granular superconductors is that such measurements are sensitive only to *inter*granular currents and do *not* detect *intra*granular currents. This is because the total *intra*granular current flowing around the tube must be zero. On the other hand, there may be a net *intra*granular current. Similar observations have been made by Gyorgy *et al.*[10] on toroids 0.86 cm inside diameter and 0.188 cm high with a wall thickness 0.180 cm.

Song *et al.*[11] have measured the magnetization and critical current of a ceramic cube $2w = 0.5$ mm on an edge, forged and sintered to 95% of theoretical density. Diffractometer scans for directions parallel and perpendicular to the forging axis indicate that the ceramic is highly textured with the *c*-axis parallel to the forging axis. The magnetization M *vs.* H was fit using the critical current expression and a procedure developed by Fietz *et al.*[12] for flat plate geometry

$$(M\uparrow + M\downarrow) \approx 2M_{equil} - (\alpha w)^2/[3(H_{ext} + M_{equil} + H_0)^3] \tag{36}$$

$$(M\downarrow - M\uparrow) \approx \alpha w/(B + B_0) + (\alpha w)^3/[4(H_{ext} + M_{equil} + H_0)^5] \tag{37}$$

The parameters α, B_0 and M_{equil} were determined by a least-squares procedure.

Civale *et al.*[13] have made careful measurements of the magnetization of $La_{1.85}Sr_{0.15}CuO_4$ rods in fields as small as 10 mOe and find substantial hysteresis below a temperature of 10K. The field dependence of the magnetization is in general agreement

[8] M. J. Scharen, A. C. Cardona, J. Z. Sun, L. C. Bourne and J. R. Schrieffer, "A simple magnetization technique for determining critical currents in superconducting thin films," submitted to *Jpn. J. Appl. Phys.*

[9] E. Tjukanov, R. W. Cline, R. Krahn, M. Hayden, M. W. Reynolds, W. N. Hardy, J. F. Carolan and R. C. Thompson, *Phys. Rev. B* **36**, 7244 (1987).

[10] E. M. Gyorgy, G. S. Grader, D. W. Johnson, Jr., L. C. Feldman, D. W. Murphy, W. W. Rhodes, R. E. Howard, P. M. Mankiewich and W. J. Skocpol, *Appl. Phys. Lett.* **52**, 328 (1988).

[11] K. J. Song, D. Heskett, H. L. Dai, A. Liebsch and E. W. Plummer, *Phys. Rev. Lett.* **61**, 1380 (1988).

[12] W. A. Fietz, M. R. Beasley, J. Silcox and W. W. Webb, *Phys. Rev.* **136**, A335 (1964).

[13] L. Civale, H. Safar, F. de la Cruz, D. A. Esparza and C. A. D'Ovidio, *Solid State Commun.* **65**, 129 (1988).

with Bean's critical state model with critical current densities of the order of 100 A/cm^2 at the lowest temperatures and dropping to half that value at 20K. Such currents must be associated with intergranular processes.

Oussena *et al.*[14] have measured the magnetization of $La_{1.85}Sr_{0.15}CuO_4$ at fields up to 35 kOe and also find evidence of critical behavior but with critical currents of the order of 10^6 A/cm^2, which must be intragranular.

6.2. Flux Creep

In this section we first review experimental and theoretical studies of flux creep carried out over twenty years ago in a period of very active research on superconductivity, stimulated by the development of high current superconducting magnets. We then discuss recent studies of flux creep in the high-temperature superconductors and theories of relaxation.

6.2.1. Anderson dynamical theory

ac loss

H. London[15] analyzed the the resistance that is developed in a wire carrying alternating current densities above J_c. He assumed as did Bean a critical current density J_c largely independent of magnetic field. Kamper[16] measured the ac loss in rings of a Pb-Bi eutectic and obtained good agreement with London's analysis.

Flux creep in alloys

We begin with a discussion of the relaxation of flux inhomogeneities in type II superconductors. We review the experimental work of Kim, Hempstead and Strnad[17] and the related theoretical work of Anderson,[18,19] who analyzed the thermally activated motion of vortices past pinning centers. Anderson has emphasized that vortices interact with each other strongly and that creep is a collective process of *bundles* of moving vortices. Following the discussion of creep in high-temperature superconductors we discuss an alternative theoretical formulation that is modeled after lattice relaxation.

Persistent current decay

Kim *et al.*[16] measured the magnetic field within a flux tube as a function of time for durations up to 5000 seconds and obtained logarithmic relaxation of the field, which implies logarithmic relaxation of the persistent current flowing outside the flux tube

$$J(t) \approx - S \ln pt + const. \tag{38}$$

Anderson theory of flux creep

---

[14]M. Oussena, S. Senoussi and G. Collin, *Europhys. Lett.* **4**, 625 (1987); S. Senoussi, M. Oussena, M. Ribault and G. Collin, *Phys. Rev. B* **36**, 4003 (1987).

[15]H. London, *Phys. Lett.* **6**, 162 (1963).

[16]R. A. Kamper, *Phys. Lett.* **2**, 290 (1962).

[17]Y. B. Kim, C. F. Hempstead and A. R. Strnad, *Phys. Rev. Lett.* **9**, 306 (1962); *Phys. Rev.* **129**, 528 (1963); *Phys. Rev.* **131**, 2486 (1963); *Rev. Mod. Phys.* **36**, 43 (1964).

[18]P. W. Anderson, *Phys. Rev. Lett.* **9**, 309 (1963).

[19]P. W. Anderson and Y. B. Kim, *Rev. Mod. Phys.* **36**, 39 (1964).

To account for Eq. (38) Anderson[17,18] introduced a model of thermally activated flux creep driven by a magnetic pressure gradient. He took the field gradient to be microscopically inhomogeneous with creep taking place through the independent motion of flux bundles. He wrote for the relaxation of the associated macroscopic circulating current

$$\frac{dJ}{dt} = -\frac{1}{\tau} J \tag{39}$$

with $1/\tau$ a thermally activated relaxation rate

$$\frac{1}{\tau} = r_0 e^{-\beta u_0 [1 - J(t)/J_c]} \tag{40}$$

where r_0 is the attempt frequency with $\beta = 1/k_B T$. The force on the flux bundle reduces the barrier height u_0 by a fractional amount $J(t)/J_c$ to lowest order in $J(t)/J_c$ and increases the relaxation rate as indicated by Eq. (40). As the flux bundles relax, the associated circulating current $J(t)$ becomes smaller, the force is reduced, increasing the barrier height, and the relaxation is slowed. Substituting Eq. (40) into Eq. (39) we may write

$$e^{-\beta J(t) u_0 / J_c} dJ = -r_0 J(t) e^{-\beta u_0} dt \tag{41}$$

For high current $J(t) >> J_c/\beta u_0$ or equivalently, low temperature $k_B T/u_0 << J(t)/J_c$ the exponential on the left is rapidly changing and we may regard $J(t)$ on the right side of Eq. (41) as relatively fixed while treating the exponential as the important variable. We note that the left side of Eq. (41) is particularly sensitive to higher order terms in the dependence of the barrier height on current. Integrating Eq. (41) with the neglect of such terms gives

$$e^{-\beta u_0 J(t)/J_c} - e^{-\beta u_0 J(0)/J_c} \approx r_0 t \beta u_0 \frac{J(0)}{J_c} e^{-\beta u_0} \tag{42}$$

Solving Eq. (42) for $J(t)$, we obtain

$$J(t) \approx -\frac{J_c}{\beta u_0} \ln \left[e^{-\beta u_0 J(0)/J_c} + r_0 t \beta u_0 \frac{J(0)}{J_c} e^{-\beta u_0} \right] \tag{43}$$

For times sufficiently long that we can take

$$\frac{J(0)}{J_c} r_0 t >> \frac{1}{\beta u_0} e^{\beta u_0 [1 - J(0)/J_c]} \tag{44}$$

the first exponential may be dropped and we obtain logarithmic relaxation as observed by Kim *et al.* [16] and given by Eq. (39) with $S = J_c/\beta u_0$.

For an extended analysis of the Anderson theory of flux creep, see Beasley *et al.*[20] Campbell[21] has discussed modifications to the Anderson theory on the basis of weak-

---

[20] M. R. Beasley, R. Labusch and W. W. Webb, *Phys. Rev.* **181**, 682 (1969).

[21] A. M. Campbell, *Phil. Mag. B* **37**, 149, 169 (1978).

pinning models. Recent contributions include the work of Hagen and Giessen,[22] Feigel'man *et al.*[23] and Fisher *et al.*[24]

Manuel *et al.*[25] have used dynamical creep theory to interpret the exponential temperature dependence of the critical current $J_c(T)$

$$J_c(T) = J_c(0)e^{-T/T_0} \tag{45}$$

observed both in magnetization measurements and in transport.[26] It is found that a pinning potential that is flat out to radius a and then increases logarithmically for $r > a$

$$u(r) = u_0\left(1 + \ln\frac{r}{a}\right) \tag{46}$$

leads to the observed temperature dependence of $J_c(T)$.

Lottis *et al.*[27] have recently shown that a planar array of ferromagnetically coupled spins exhibits dynamical relaxation in anisotropy and demagnetizing fields that is similar to that in the Anderson model. The theory has been applied to a description of the decay of the remanent magnetization of CoCr films.[28]

6.2.2. *Kinetic theories*

Experimental survey

Müller, Takashige and Bednorz[29] first reported nonexponential relaxation toward the stable state of the magnetization of $La_{2-x}Ba_xCuO_{4-y}$. It was reported that the relaxation appeared exponential at short times and considerably slower at longer times. Mota *et al.*[30] have observed relaxation at 4.2 K of the magnetization of $La_{1.8}Sr_{0.2}CuO_4$ powder that is accurately logarithmic from 10 to 10^5 seconds. The slope on a semilogarithmic plot

$$S(t) = -\frac{dJ(t)}{d\ln t} \tag{47}$$

is found to increase linearly with T. In studies of the so-called *history effect,* S(t) in zero field is found to increase as the cube of the magnetizing field H_i.

---

[22] C. W. Hagen and R. Giessen, *Phys. Rev. Lett.* **62**, 2857 (1989).

[23] V. M. Feigel'man, V. B. Geshkenbein and A. I. Larkin, *Physica C* **167**, 177 (1990).

[24] D. S. Fisher, M. P. A. Fisher and D. Huse, *Phys. Rev. B* **43**, 130 (1991).

[25] P. Manuel, C. Aguillon and S. Senoussi, *Physica C* **177**, 281 (1991).

[26] E. Zeldov, N. M. Amer, G. Koren *et al. Appl. Phys. Lett.* **56**, 680, 1700 (1990).

[27] D. K. Lottis, R. M. White and E. D. Dahlberg, *Phys. Rev. Lett.* **67**, 362 (1991).

[28] D. K. Lottis, E. D. Dahlberg, J. A. Christner, J. I. Lee, R. L. Peterson and R. M. White, *J. Phys. (Paris), Colloq.* **49**, C8-1989 (1988); D. K. Lottis, R. M. White and E. D. Dahlberg (to be published).

[29] K. A. Müller, M. Takashige and J. G. Bednorz, *Phys. Rev. Lett.* **58**, 1143 (1987); M. Takashige, K. A. Müller and J. G. Bednorz, *Jpn. J. Appl. Phys.* **26**, (1987); K. A. Müller, K. W. Blazey, J. G. Bednorz and M. Takashige, *Physica B* **148**, 149 (1987).

[30] A. C. Mota, A. Pollini, P. Visani, K. A. Müller and J. G. Bednorz, *Phys. Rev. B* **36**, 4011 (1987).

Tjukanov *et al.*[10] report relaxation of the axial field from 1 to 10^4 seconds that is well described by the heuristic extended Kohlrausch form familiar in spin-glass relaxation[31]

$$B(t) = B(0)\left(\frac{t}{t_0}\right)^{-\alpha} \tag{48}$$

The form given by Eq. (48) is suggested by mean-field theory. The changes were quite small, however, and could as well have been fitted by a logarithmic time dependence. Hierarchical dynamics suggest the Kohlrausch stretched exponential[26] with $\beta < 1$

$$M(t) = M(0)e^{-(t/\tau)^{\beta}} \tag{49}$$

Giovannella *et al.*[32] have measured flux creep by torque measurements. Following an angular displacement of the magnetic field the torque is observed to decay in a strongly nonexponential manner, which is *not* a stretched exponential, but may be logarithmic in the time.

Malozemoff[33] has carefully reviewed studies of the macroscopic magnetic properties of the high-temperature superconductors with particular attention to the *irreversibility line* obtained when data are plotted in the H-T plane.

Kinetic theory of magnetic viscosity

The Anderson theory gives logarithmic relaxation of J(t) as a consequence of the reduced relaxation rate because of the reduction in J. An alternative model might involve the evolution of a distribution of relaxation rates.

Just such a theory has been developed by Street and Wooley[34] to describe magnetic relaxation in Alnico (a high coercivity alloy of Al, Ni, Co, Cu, and Fe). The approach, which is modeled after Cyril Smith's theory of transient creep in metals,[35] has also been applied to magnetic relaxation in Mn-Zn ferrite.[36]

The theory takes magnetization components M_i relaxing as

$$M_i(t) = M_i(0)e^{-r_i t} \tag{50}$$

The magnetization density is assumed to relax as a superposition of components

$$\frac{dM}{dt} = -\sum_i r_i M_i(0)e^{-r_i t} \tag{51}$$

with thermally activated (Arrhenius) relaxation rates given by

---

[31] K. Binder and A. P. Young, *Rev. Mod. Phys.* **58**, 101 (1986).

[32] C. Giovannella, G. Collin, P. Rouault and I. A. Campbell, *Europhys. Lett.* **4**, 109 (1987); C. Giovannella, G. Collin and I. A. Campbell, *J. Physique* **48**, 1835 (1987).

[33] A. P. Malozemoff, "Macroscopic magnetic properties of high temperature superconductors," Ch. 3 in *Physical Properties of High Temperature Superconductors I*, ed. D. M. Ginsberg (World, Singapore, 1989).

[34] R. Street and J. C. Woolley, *Proc. Phys. Soc. (London) A* **62**, 562 (1949).

[35] C. L. Smith, *Proc. Phys. Soc. (London)* **61**, 14, 201 (1948).

[36] R. Street and J. C. Woolley, *Proc. Phys. Soc. (London) A* **62**, 743 (1949).

$$r_i = r_0 e^{-\beta u_i} \tag{52}$$

with $\beta = 1/k_BT$ and the u_i fixed. We convert Eq. (51) to an integral over all u by taking $M_i(t) = m(u,t)\,du$ to obtain for the total magnetization

$$\frac{dM(t)}{dt} = -\int du\, r(u)\, m(u,0)\, e^{-r(u)t} \tag{53}$$

Substituting the differential of Eq. (52) which is $dr = -\beta r(u)\,du$ into Eq. (53) gives

$$\frac{dM(t)}{dt} = \frac{1}{\beta}\int dr\, m(u,0)\, e^{-r(u)t} \tag{54}$$

We now integrate Eq. (54) by parts. Taking $m(\infty, 0) = 0$ at the upper limit of barrier energy and $r(0) = \infty$ at the lower limit, we have

$$\frac{dM(t)}{dt} = \frac{1}{\beta t}\int dm(u,0)\, e^{-r(u)t} \tag{55}$$

6.2.3. Minimal glassy model

Waldner[37] has developed a kinetic theory that can be solved analytically and includes parameters that, depending on their values, lead to exponential, power-law or logarithmic relaxation. The model appears to account for observed "memory effects" as well as simulating the results of dynamical theories.

Waldner assumes the $M(u_i)$ to be initially distributed in a Poisson distribution, which for u continuous leads to the exponential distribution

$$m(u,0) = \frac{M(0)}{\langle u \rangle} e^{-u/\langle u \rangle} \tag{56}$$

The total magnetization takes the form

$$M(t) = \int du\, m(u,0) e^{-r(u)t} = \frac{M(0)}{\langle u \rangle}\int du \exp\left[-\frac{u}{\langle u \rangle} - r_0 t e^{-\beta u}\right] \tag{57}$$

For $\beta\langle u \rangle = 1$, this integral is simply

$$M(t) = \frac{1 - e^{-r_0 t}}{r_0 t} M(0) \tag{58}$$

More generally M(t) may be expressed in terms of tabulated integrals

$$M(t) = \frac{k_BT}{\langle u \rangle}\Gamma\left(\frac{k_BT}{\langle u \rangle}\right)\gamma^*\left(\frac{k_BT}{\langle u \rangle}, r_0 t\right)M(0) \tag{59}$$

[37] P. Erhart, A. M. Portis, B. Senning and F. Waldner, *J. Phys.: Condens. Matter* (in press).

where $\Gamma(x)$ is the Gamma function and $\gamma^*(x, y)$ is the incomplete gamma function.[38]

At high temperature $k_BT >> \langle u \rangle$, the rate $r(u)$ approximates r_0 at all but the highest energies, where there is very little magnetization density, and M(t) decays exponentially with rate r_0. In this connection we may think of $T = \langle u \rangle / k_B$ as a melting temperature. There is, of course, no phase transition but simply a growing insensitivity to energy barriers. Brandt[39] has suggested that the apparent melting of the flux-line lattice has this origin.

In the opposite limit, $k_BT << \langle u \rangle$, Eq. (59) gives logarithmic behavior. For $k_BT \approx \langle u \rangle$, power-law behavior is obtained with

$$M(t) \propto t^{-k_BT/\langle u \rangle} \tag{60}$$

Memory effect

Rossel *et al.*[40,41] have found "memory effects" in high-T_c superconductors very similar to those reported in spin-glasses.[42] In their experiments with the sample in a magnetic field, the temperature was abruptly changed at $t = 0$. After a waiting time t_w the applied magnetic field was changed by an amount ΔH. Rossel *et al* observe at time $2\ t_w$ an "echo" of the field change. They interpret the observed echo as an interference between the relaxation of the sample in the altered temperature and relaxation in the changed magnetic field. Erhart *et al.*[33] suggest that their model is capable of modeling such interference effects.

Comparison with dynamical theories

Dynamical theories like the Anderson flux-creep theory contain a single activation energy u that increases in magnitude as the system relaxes. Although the initial relaxation behavior of such theories is complex, the assymptotic relaxation is clearly logarithmic as given by Eq. (48). From the discussion of the minimal glassy model it seems clear that the relaxation is moving to higher energy as well. This is because magnetization at lower energy has largely relaxed, leaving the higher-energy magnetization to relax. Magnetization at still higher energy relaxes too slowly to make a significant contribution. This can only happen for a distribution of energies and at times $r_0t >> 1$ so that the actively relaxing components are Arrhenius. This argument suggests that the active energy at time t is given by $r_0t \exp(-\beta u) \approx 1$ or $u \approx k_BT \ln r_0t$ and the energy grows logarithmically with time.

6.3. Flux Flow

6.3.1. Bardeen-Stephen theory

In a study primarily concerned with surface superconductivity above H_{c2}, Hempstead and Kim[43] first reported a study of resistive behavior arising from viscous flow of vortices.

A charge q flowing through a magnetic field experiences a Lorentz force

---

[38] M. Abramowitz and A. I. Stegun, eds. *Handbook of Mathematical Functions* (Dover, New York, 1968) pp. 360-363.

[39] E. H. Brandt, *Z. Phys. B* **80**, 167 (1990).

[40] C. Rossel, Y. Maeno and I. Morgenstern, *Phys. Rev. Lett.* **62**, 681 (1989).

[41] C. Rossel in *Relaxation in Complex Systems and Related Topics* , eds. I. A. Campbell and C. Giovannella (Plenum, New York, 1990) pp. 105-112.

[42] L. Lundgren, *J. de Physique* **49**, C8-1001 (1988).

[43] C. F. Hempstead and Y. B. Kim, *Phys. Rev. Lett.* **12**, 145 (1964).

$$\mathbf{F} = q\, \mathbf{v} \times \mathbf{B} \tag{61}$$

Correspondingly, the force per unit length acting on a vortex line, across which a current flows is

$$\mathbf{F} = \Phi_0\, \mathbf{J} \times \hat{\mathbf{z}} \tag{62}$$

Under the action of such a force the vortex quickly reaches a terminal velocity

$$\mathbf{V} = (1/\eta)\, \mathbf{F} \tag{63}$$

where η is the vortex viscosity. Considerable theoretical attention[44,45,46,47,48] has been given to the origin of the viscosity but the matter is still not entirely resolved.[49,50,51]

The rate of energy dissipation per unit volume is

$$P = (B/\Phi_0)\, \mathbf{F} \cdot \mathbf{V} = (B/\eta\Phi_0)\, F^2 = (1/\eta)\, \Phi_0 B J^2 \tag{64}$$

which indicates an electrical resistivity

$$\rho = (1/\eta)\, \Phi_0\, B \tag{65}$$

Strnad *et al.*[17] found for temperatures well below T_c the empirical relation

$$\rho = \rho_n\, B/\mu H_{c2} \tag{66}$$

where ρ_n is the normal resistivity obtained for $H > H_{c2}$. Equating Eqs. (65) and (66) the vortex viscosity may be expressed empirically in terms of the normal resistivity as

$$\eta = \Phi_0\, \mu H_{c2}/\rho_n \tag{67}$$

This relation has the following simple interpretation. The moving vortex generates an electric field

$$\mathbf{E} = \mathbf{V} \times \mathbf{B} \tag{68}$$

This field drives a current through the normal core of the vortex with

---

[44] J. Bardeen, *Phys. Rev. Lett.* **13**, 747 (1964).

[45] A. R. Strnad, C. F. Hempstead and Y. B. Kim, *Phys. Rev. Lett.* **13**, 794 (1964).

[46] M. Tinkham, *Phys. Rev. Lett.* **13**, 804 (1964).

[47] M. J. Stephen and J. Bardeen, *Phys. Rev. Lett.* **14**, 112 (1965); John Bardeen and M. J. Stephen, *Phys. Rev.* **140**, A1197 (1965).

[48] M. Tinkham, *Introduction to Superconductivity*, (McGraw-Hill, New York, 1975), pp.162-171.

[49] Y. B. Kim and M. J. Stephen, "Flux flow and irreversible effects" Ch. 19 in *Superconductivity Volume II*, ed. R. D. Parks, (Dekker,, New York, 1969).

[50] J. C. Levet, M. Potel, P. Gougeon, H. Noel, M. Guilot and J. L. Tholence, *Nature* **331**, 307 (1988); J. L. Tholence, H. Noel, J. C. Levet, M. Potel, P. Gougeon, C. Chouteau and M. Guillot, *Physica C*, (1988), in press. X. Cai, R. Joynt and D. C. Larbalestier, *Phys. Rev. Lett.* **58**, 2798 (1987) have attributed jumps in the magnetoresistance of $YBa_2Cu_3O_{7-\delta}$ to superconductivity above this temperature.

[51] J. R. Clem, "The Bardeen-Stephen Theory," *J. Supercond.* **4**, 337 (1991).

$$J = (1/\rho_n)\,\mathbf{E} \tag{69}$$

producing a dissipation per unit length

$$P = \eta V^2 = \int dS\,\mathbf{J}\cdot\mathbf{E} \approx (\Phi_0 B_{core}/\rho_n)\,V^2 \tag{70}$$

which agrees with Eq. (67) so long as we take $B_{core} = \mu H_{c2}$.

6.3.2. Flux diffusion

The equation of motion of a damped vortex driven by a force per unit length F (x, t) is

$$m\frac{d^2u}{dt^2} + \eta\frac{du}{dt} + \kappa u = F(x,t) \tag{71}$$

where m is the mass per unit length,[52] η is the viscosity,[33,34,53,54] k is the force constant associated with pinning and u is the displacement. If the force is produced by a static current or magnetic field gradient with

$$F = \Phi_0 J = -\Phi_0\, dH/dx \tag{72}$$

Eq, (71) becomes, neglecting m

$$\eta\frac{du}{dt} + \kappa u = \Phi_0 J \tag{73}$$

As J is increased there is a critical current J_c at which vortices flow with velocity

$$v = \frac{du}{dt} = \frac{\Phi_0}{\eta}(J - J_c) \tag{74}$$

The condition for conservation of magnetic flux

$$\frac{\partial H}{\partial t} + \frac{\partial vH}{\partial x} = 0 \tag{75}$$

gives finally with Eqs. (74) and (72) the nonlinear diffusion equation[55]

$$\eta\frac{\partial H}{\partial t} = \Phi_0\frac{\partial}{\partial x}\left[H\left(\frac{\partial H}{\partial x} + J_c\right)\right] \tag{76}$$

---

[52] H. Suhl, *Phys. Rev. Lett.* **14**, 226 (1965).

[53] C. Caroli, P. G. de Gennes and J. Matricon, *Phys. Lett.* **9**, 307 (1964).

[54] J. Gittleman and B. Rosenblum, *Phys. Lett.* **20**, 453 (1966).

[55] C. P. Bean in *Superconductivity and Applications*, eds. H. S. Kwok, Y.-H. Kao and D. T. Shaw (Plenum, New York, 1989) pp. 767-772.

With fields at the surface periodic at frequency ω, Eq. (76) leads to two regimes:

(i) *Pinning-limited flow.* For $\omega H < \Phi_0 J_c^2/\eta$, flux penetrates a distance H/J_c as given by the Bean model.

(ii) *Diffusion-limited flow.* For $\omega H > \Phi_0 J_c^2/\eta$, flux entry is limited by diffusion to a penetration distance of the order of $\sqrt{\Phi_0 H/\omega\eta}$.

7
FILM TRANSMISSION LINES AND RESONATORS

7.1. Transmission Lines

We begin with a review of the theory of the propagation of electromagnetic waves on transmission lines. The theory is applied in this chapter to patterned lines and resonators. In the following chapter we treat coaxial cables and waveguides by distributed-element transmission lines and the cavity resonator by a lumped-element circuit.

7.1.1. Telegrapher's equations

The Telegrapher's equations[1] describe the relation between the voltage V across a transmission line and the current I that flows through the line shown in Fig. 1.

$$\frac{\partial V}{\partial z} = -L' \frac{\partial I}{\partial t} \qquad (1)$$

$$\frac{\partial I}{\partial z} = -C' \frac{\partial V}{\partial t} \qquad (2)$$

where L' and C' are the series inductance and shunt capacitance per unit length of transmission line. Combining Eqs. (1) and (2) leads to the wave equation

$$\frac{\partial^2 V}{\partial z^2} = L'C' \frac{\partial^2 V}{\partial t^2} \qquad (3)$$

Assuming a traveling wave of the form

[1] S. Ramo, J. R. Whinnery and T. Van Duzer, *Fields and Waves in Communication Electronics* (Wiley, New York, 1984) sec. 5.2.

$$V(z, t) = V\, e^{i(kz - \omega t)} \tag{4}$$

substitution of Eqn. (4) into Eqn. (3) yields the dispersion relation

$$\omega = k/\sqrt{L'C'} \tag{5}$$

7.1.2. Characteristic impedance

The characteristic impedance of the transmission line $Z_0 = V/I$ is obtained by substitution of Eq. (4) into Eq. (1) together with Eq. (5)

$$Z_0 = \omega L'/k = \sqrt{L'/C'} \tag{6}$$

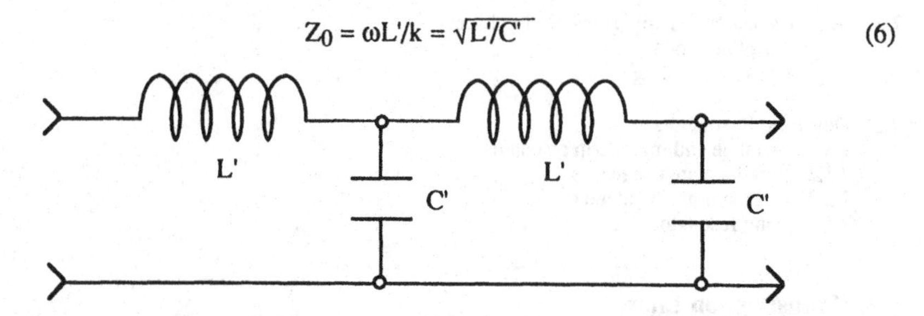

Figure 1. Classical distributed-element transmission line. In series with the line is an inductance L' per unit length. Across the line is shunt capacitance C' per unit length.

7.1.3. Reflection coefficient

Terminating the line at z = 0 leads to voltage and current on the line

$$V(z, t) = \tfrac{1}{2}\, V_0\, [e^{ikz} + \rho\, e^{-ik\text{-}z}]\, e^{-i\omega t} \tag{7}$$

$$I(z, t) = \tfrac{1}{2}\, (V_0/Z_0)\, [e^{i\beta z} - \rho\, e^{-ikz}]\, e^{-i\omega t} \tag{8}$$

where ρ is the complex voltage reflection coefficient and V_0 is the voltage of a source that is matched to the transmission line. At the termination of the line, the voltage is the product of the current and the teminating impedance Z_L

(9)

which leads to the expression for the complex voltage reflection coefficient

$$\rho = 2\frac{V_L}{V_0} - 1 = \frac{Z_L - Z_0}{Z_L + Z_0} \tag{10}$$

The voltage at the load is

$$V_L = \tfrac{1}{2}(1+\rho)V_0 e^{-i\omega t} = \frac{Z_L}{Z_L + Z_0} V_0 e^{-i\omega t} \tag{11}$$

and may be represented by a voltage generator V_0 with internal impedance Z_0 and working into a load Z_L.

7.2. Striplines and Microstrip Transmission Lines

7.2.1. *Stripline modes*

Swihart[2] has analyzed the transmission of an electromagnetic wave on a superconducting strip transmission line. Van Duzer and Turner[3] obtain the Swihart modes from the equations for the Josephson transmission line by setting the superconducting current to zero. A number of investigators have used striplines to study the characteristics of patterned films.[4,5]

Fig. 2 shows sections through three film transmission line geometries, the stripline, the microstrip line and the coplanar transmission line. The stripline has the advantage of no modal dispersion but does require three films. The coplanar line is patterned from a single film and because of the very small gap between the signal line conductor and the ground planes, provides the best isolation. Losses are lowest for the stripline geometry and highest for the coplanar line because of current-crowding at the edges of both the signal line and the ground planes.

We discuss propagation on a microstrip line in terms of the electromagnetic fields between the signal line conductor and the ground plane. We assume that the separation b_j between the signal conductor and the ground plane is small compared with the width of the signal conductor. The region between the two conductors is characterized by a conductivity σ_j and a permittivity ε_j. The electric and magnetic fields at the surfaces of the conductors are E_g and H_g.

The Maxwell equations give for the relation between the electric and magnetic fields in the region between the conductors

$$\frac{\partial E_z}{\partial y} - \frac{\partial E_y}{\partial z} = -\frac{\partial B_x}{\partial t} \tag{12}$$

$$\frac{\partial H_x}{\partial z} = J_y + \frac{\partial D_y}{\partial t} \tag{13}$$

We take $B_x = \mu_j H_g$ and $\partial E_z/\partial y = 2E_g/b_j = (2Z_g/b_j)H_g$ where $Z_g = E_g/H_g$ is the surface impedance at the conducting surfaces. We write for the microstripline equations with E_j the transverse electric field in the region between the conductors,

$$\frac{\partial E_j}{\partial z} = \frac{2Z_g}{b_j} H_g + \mu_j \frac{\partial H_g}{\partial t} \tag{14}$$

---

[2] J. C. Swihart, "Field solution for a thin-film superconducting strip transmission line," *J. Appl. Phys.* **32**, 461 (1961).

[3] T. Van Duzer and C. W. Turner, *Principles of Superconductive Devices and Circuits* (Elsevier, New York, 1981) sec. 4.04.

[4] S. M. Anlage, H. Sze, H. J. Snortland, S. Tahara, B. Langley, C.-B. Eom and M. R. Beasley, *Appl. Phys. Lett.* **54**, 2710 (1989).

[5] L. C. Bourne, R. B. Hammond, McD. Robinson, M. M. Eddy, W. L. Olson and T. W. James, *Appl. Phys. Lett.* **56**, 2333 (1990).

$$\frac{\partial H_g}{\partial z} = J + \varepsilon_j \frac{\partial E_j}{\partial t} \tag{15}$$

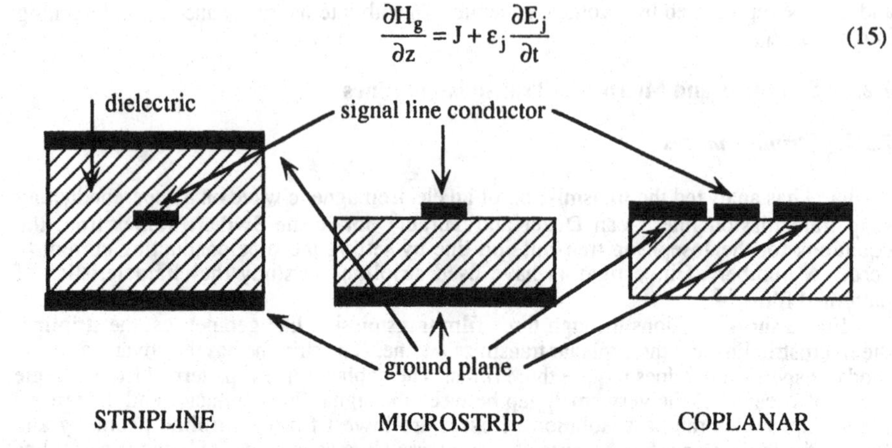

Figure. 2. Sections through three film transmission lines, the stripline, the microstrip line and the coplanar transmission line. The stripline offers the best shielding but requires three films. The coplanar line is patterned from a single film but provides little shielding. The microstrip line, which is half a stripline, is intermediate.

To obtain equations that resemble the transmission line equations we write $V = -E_j b_j$ and the current per unit width $I' = H_g$ with $L_j = \mu_j b_j$ and $C''_j = \varepsilon_j / b_j$

$$\frac{\partial V}{\partial z} = -2Z_g I' - L_j \frac{\partial I'}{\partial t} \tag{16}$$

$$\frac{\partial I'}{\partial z} = J_j - C''_j \frac{\partial V}{\partial t} \tag{17}$$

Comparing Eqs. (16) and (17) with Eqs. (1) and (2) we note that Eq. (16) includes a voltage drop across the surface impedance of the upper and lower guiding surfaces and that Eq. (17) includes the shunt conduction current J_j in addition to the displacement current.

Assuming a wave of the form

$$V(z,t) = Ve^{i(kz - \omega t)} \tag{18}$$

leads with the simplification $Z_g = -i\omega L_g$ and $J_j = 0$ to the dispersion relation

$$k^2 = \omega^2 \mu\varepsilon(1 + 2L_g/L_j) \tag{19}$$

For $L_g \gg L_j$ the phase velocity ω/k is substantially reduced, as pointed out by Swihart.[2]

7.2.2. *Current-crowding*

Van Duzer and Turner[6] discuss the distribution of current flowing on an isolated current-

[6] T. Van Duzer and C. W. Turner, *Principles of Superconductive Devices and Circuits* (Elsevier, New York, 1981)

carrying thin strip of thickness b and width a. For $b << \lambda_L$ the current is uniform through the thickness of the film. Near the center of the strip the current increases as[7]

$$J(x) = J(0)[1 - (2x/a)^2]^{-1/2} \tag{20}$$

Near the edge of the film the current takes the form

$$J(x) = J(a/2)\exp[-(a-2|x|)\, t/2a\lambda_L] \tag{21}$$

Connecting Eqs. (20) and (21) gives for the ratio of the current at the edge to that at the center

$$J(w/2)/J(0) = 1.165\,(wt/a\lambda_L)^{1/2} \tag{22}$$

The problem of a thin strip over a ground plane may be analyzed by the method of images.[8,9,10]

Charged conducting strip of ellipsoidal cross-section.

That the current density on a flat strip of width a increases as $[1 - (2x/a)^2]^{-1/2}$, where x is the distance from the midline of the strip out to the edge, may also be shown with the use of the ellipsoidal homeoid[11] with axes a, b and c. A thin strip is simulated by taking c >> a >> b. Placing the ellipsoidal surface at constant electrostatic potential gives for the surface charge density

$$\sigma_s = 4\rho_0/(x^2/a^4 + y^2/b^4)^{1/2} \tag{23}$$

where

$$4x^2/a^2 + 4y^2/b^2 = 1 \tag{24}$$

is the equation of the elliptical cross-section. The charge *per unit length* along x is then

$$\sigma_x = \rho_0 b^2/4y = \rho_0 b/2(1 - 4x^2/a^2)^{1/2} \tag{25}$$

in agreement with the result of Rhoderick and Wilson.[7]

The magnetic problem and conjugate functions

The electric field lines, which must be normal to the surface, are conjugate to the magnetic field lines as shown in Fig. 3 so long as the magnetic field lines are parallel to the surface as is the case for b large compared with the penetration or skin depth, just opposite to the

---

secs. 3.08 and 3.09.

[7] E. H. Rhoderick and E. M. Wilson, "Current distribution in thin superconducting films," *Nature* **194**, 1167 (1962).

[8] V. L. Newhouse, *Applied Superconductivity* (Wiley, New York, 1964) pp. 103-105.

[9] W. H. Chang, "The inductance of a superconducting strip transmission line," *J. Appl. Phys.* **50**, 8129 (1979).

[10] E. Muchowski and A. Schmidt, "On the current distribution in a shielded superconducting film," *Z. Physik* **255**, 187 (1972).

[11] A. M. Portis, *Electromagnetic Fields: Sources and Media* (Wiley, New York, 1978), sec. 1-24 and probs. 1-27 and 5-19.

limit considered by Van Duzer and Turner.[5] For a discussion of conjugate functions see the classic electromagnetism text of Smythe.[12] Smythe's argument indicates that the current density on the surface of an ellipsoid should vary as

$$I_s = 4J_0\,(x^2/a^4 + y^2/b^4)^{-1/2} \tag{26}$$

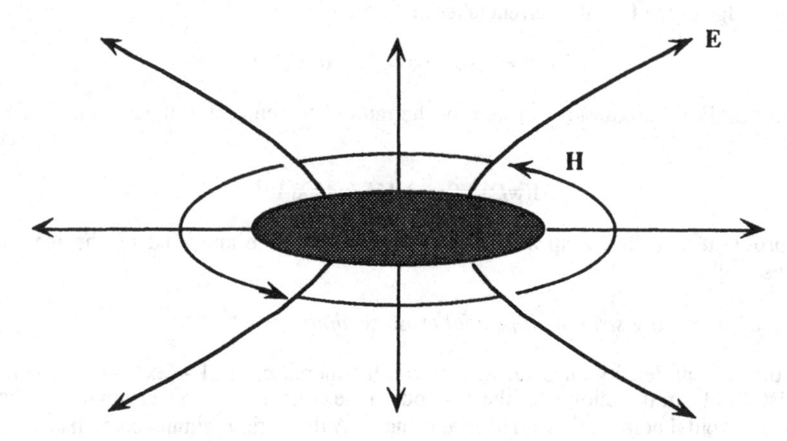

Figure 3. The electric field lines, which must be normal to the surface, are conjugate to the magnetic field lines.

The current density per unit interval of x is then

$$J_x = J_0 b^2/4y \tag{27}$$

The charge density on a *flat* strip should be pretty much equal to the surface charge density σ_x calculated on the elliptical surface. This would suggest that the current density along a flat strip should increase as J_x. There is presumably a cutoff at a/b times the current density at the center of the strip, but this ratio is sufficiently large for a thin film that the current density at the edge may be taken as singular, the approximation of Rhoderick and Wilson.[7]

Sheen *et al.*[13] have performed an explicit calculation of the magnetic field distribution over the entire surface of a patterned thin film stripline. The calculation is complicated by the fact that the penetration depth is comparable to the film thickness. As the penetration depth may well vary with the surface magnetic field, it must be obtained self-consistently. Unlike previous work, this paper considers the resistivity as well as the inductivity in obtaining the current distribution, utilizing the approach of Weeks *et al.*[14]

7.3. Thin Film Resonators

7.3.1. Stripline and microstrip resonators

---

[12] W. R. Smythe, *Static and Dynamic Electricity* (McGraw-Hill, New York, 1939) secs. 4.09 and 4.10.

[13] D. M. Sheen, S. M. Ali, D. E. Oates, R. S. Withers and J. A. Kong, *IEEE Trans. Appl. Superconduct.* **1**, 108 (1991).

[14] W. T. Weeks, L. L. Wu, M. F. McAllister and A. Singh, *IBM J. Res. Develop.* **23**, 652 (1979).

DiIorio *et al.*[15] have studied Y-Ba-Cu-O thin films with a stripline technique used earlier[16] in the study of Nb, NbN and Nb_3Sn. A stripline has a narrow conducting strip centered between upper and lower ground planes and separated by symmetric dielectrics. The resonator is a length of transmission line one-half wavelength long at its fundamental frequency. Higher harmonics occur at all multiples of the fundamental. The resonator is capacitively coupled to the external circuit across gaps at the ends of the stripline. Such a stripline concentrates the losses in the center conductor, which supports a much higher current density than do the ground planes. This high current density results in microwave magnetic fields at the surface of the conductor substantially higher than can be obtained conveniently at the walls of a resonant cavity. For example, an input power of -20 dBm leads to a current density of 10^7 A/cm^2 and a surface field of 100 Oe in a 50 Ω resonator with a quality factor $Q = 10^6$.

A superconducting stripline structure is constructed by depositing superconducting films on three separate substrates and then mechanically clamping the assembly with the substrate of the bottom film outside the stripline so that it does not contribute to the loss. DiIorio *et al.*[13] used thin-film Nb for both the top ground plane and the central conductor while the HTS thin film was used only for the bottom ground plane. The Nb films were deposited on low-loss ($\tan\delta < 10^{-6}$) sapphire substrates.

In a stripline resonator, the surface resistance of a superconducting film is calculated from the quality factor Q, which is obtained from the separation between half-power points for each resonance. The ultimate sensitivity of the stripline is limited by dielectric losses. Operating at a fundamental frequency of 0.5 GHz the sensitivity to the surface resistance of a ground plane is about 5 μΩ. The sensitivity is nearly 50 times greater (nearly 0.1 μΩ) at this frequency for a patterned center conductor with the reason for the increased sensitivity the much greater current density in the narrow central conducting strip.

The microwave surface resistance R_s of various HTS thin-film materials have been compared in this way using a stripline resonator structure.[17,18]

As the width of the central strip is reduced, the Q of the line is found to decrease nearly linearly with the with, indicating that the resistance of the strip increases faster than the inductance and points to edge-losses. Tahara *et al.*[19] find that the dc critical current density increases with reduced strip width, suggesting the possibility of edge pinning that inhibits flux entry. Although the dc and rf results appear contradictory, so long as grain boundaries do not cross the strip, dc currents can flow around these boundaries and the critical current is unaffected. Microwave currents, on the other hand, flow more uniformly and can be expected to cross grain boundaries.

Anlage *et al.*[20] in addition to measuring the penetration depth[21] have used the microstrip resonator technique to determine the surface resistance R_s of $YBa_2Cu_3O_{7-\delta}$ thin

---

[15] M. S. DiIorio, A. C. Anderson and B. Y. Tsaur, *Phys. Rev. B* **38**, 7019 (1988).

[16] A. C. Anderson, R. S. Withers, S. A. Reible and R. W. Ralston, *IEEE Trans. Mag.* **19**, 485 (1983).

[17] D. E. Oates and A. C. Anderson, "Stripline measurements of surface resistance: relation to HTSC film Properties and deposition methods" in *SPIE Proceedings* **1187**, 326 (1989), *Processing of Films for High T_c Superconducting Electronics*, Santa Clara, CA, 10-12 October 1989, ed. T. Venkatesan (SPIE, Bellingham, WA, 1989).

[18] D. E. Oates, A. C. Anderson and P. M. Mankiewich, "Measurement of the surface resistance of $YBa_2Cu_3O_{7-x}$ thin films using stripline resonators," *J. Supercond.* **3**, 251 (1990).

[19] S. Tahara, S. M. Anlage, C.-B. Eom, D. K. Fork, T. H. Geballe and M. R. Beasley, *Physica C* **162-164**, (1989).

[20] S. M. Anlage, H. Sze, H. J. Snortland, S. Tahara, B. Langley, C.-B. Eom, M. R. Beasley and R. Taber, *Appl. Phys. Lett.* **54**, 2710 (1989).

[21] S. M. Anlage, B. W. Langley, H. J. Snortland, C. B. Eom, T. H. Geballe and M. R. Beasley, "Magnetic penetration depth measurements with the microstrip resonator technique," *J. Supercond.* **3**, 311 (1990).

films. They have also made power measurements[22] and find that the surface resistance increases linearly with the surface rf field.

Thin-film transmission lines and resonators are prepared by standard photolithographic patterning techniques. These processes appear not to deteriorate the microwave properties of high quality laser-deposited YBaCuO films.[23] For films of lesser quality, on the other hand, substantial deterioration in microwave properties can occur for a variety of reasons—even exposure to room air for several days can produce substantial changes. The environmental sensitivity of these films may result from poor stoichiometry and poor grain boundary contacts, while high quality epitaxial films, free of grain boundaries, are found to be much more stable.[12]

Anlage *et al.*[18] have used the microstrip resonator technique to determine the superconducting penetration depth of $YBa_2Cu_3O_{7-\delta}$ thin films. The microstrip resonator differs from the stripline resonator in having only a single ground plane. Ordinarily two films are evaporated and one is patterned. The two films are clamped together separated by a very thin (10 μm) mylar dielectric sheet with the substrates on the outside. One advantage of this arrangement is that both substrates are outside the resonator. The fact that the thickness of the dielectric is comparable to the penetration depth into the films increases the sensitivity of microstrip resonator to penetration depth. The ultimate precision with which the penetration depth λ can be determined is limited by the width of the resonance, which is the operating frequency divided by quality factor Q.

7.3.2. *Parallel plate resonators*

A technique developed by Taber[24] that utilizes square resonant films has been used to measure the surface resistance of high quality films.[25,26,27] The equations that describe the surface current density I'(x, z) and the voltage V(x, z) on a plate follow from Eqs. (16) and (17)

$$\frac{\partial V}{\partial z} = -ZI'_z \qquad \frac{\partial V}{\partial y} = -ZI'_y \qquad \frac{\partial I'_z}{\partial z} + \frac{\partial I'_y}{\partial y} = -Y''V \tag{28}$$

The admittance per unit area Y" and the impedance per square Z are

$$Y'' = -\frac{i\omega\varepsilon}{b} \qquad Z = -i\omega(L + 2L_s) \tag{29}$$

with $L = \mu b$. Combining these equations leads to the wave equation

[22] S. M. Anlage, B. Langley, J. Halbritter, S. Tahara, H. Sze, N. Switz, C.-B. Eom, H. J. Snortland, R. Taber, T. H. Geballe and M. R. Beasley, *Physica C* **162-164**, 1645 (1989).

[23] L. Schultz, B. Roas, P. Schmitt and G. Endres, "Preparation and characterization of pulsed laser deposited HTSC films" in *SPIE Proceedings* **1187**, 204 (1989), *Processing of Films for High T_c Superconducting Electronics*, Santa Clara, CA, 10-12 October 1989, ed. T. Venkatesan (SPIE, Bellingham, 1989).

[24] R. C. Taber, "A parallel pl;ate resonator techniuque for microwave loss measurements on superconductors," *Rev. Sci. Instrum.* **61**, 2200 (1990).

[25] N. Newman, K. Char, S. M. Garrison, R. W. Barton, R. C. Taber, C. B. Eom, T. H. Geballe and B. Wilkens, *Appl. Phys. Lett.* **57**, 520 (1990).

[26] K. Char, N. Newman, S. M. Garrison, R. W. Barton, R. C. Taber, S. S. Laderman and R. D. Jacowitz, *Appl. Phys. Lett.* **57**, 409 (1990).

[27] S. S. Laderman, R. C. Taber, R. D. Jacowitz, J. L. Moll, C. B. Eom, T. L. Hylton, A. F. Marshall, T. H. Geballe and M. R. Beasley, *Phys. Rev. B.* **43**, 2922 (1991).

$$\frac{d^2V}{dx^2} + \frac{d^2V}{dz^2} = Y''ZV \tag{30}$$

with

$$k^2 = k_x^2 + k_z^2 = -Y''Z = \omega^2\mu\varepsilon(1 + 2L_s/L) \tag{31}$$

With the plates open along the edges, open boundary conditions yield

$$k_x = m\pi/a \qquad\qquad k_z = p\pi/d \tag{32}$$

leading for a square resonator to

$$ka = \alpha_{mp} = \pi\sqrt{m^2 + p^2} \tag{33}$$

The values of α_{mp} for the first twenty modes of a square resonator are given in Table 1.

Table 1. Values of $ka = \alpha_{mp}$ for a square resonator with open boundary conditions.

	m = 0	m = 1	m = 2	m = 3	m = 4	m = 5
p = 0	0.00000	3.14159	6.28310	9.42478	12.56637	15.70796
p = 1		4.44288	7.02481	9.93459	11.75476	16.01904
p = 2			8.88577	11.32717	14.04963	16.91799
p = 3				13.32865	15.70796	18.31848
p = 4					17.77153	20.11601
p = 5						22.21441

Disk modes

In cylindrical coordinates Eq. (30) takes the form

$$\frac{1}{r}\frac{\partial}{\partial r}\left(r\frac{\partial V}{\partial r}\right) + \frac{1}{r^2}\frac{\partial^2 V}{d\phi^2} = Y''ZV \tag{34}$$

with Y" and Z given by Eq. (29). The solutions are of the form

$$V(r, \phi) = J_n(kr)\cos n\phi \tag{35}$$

The boundary condition is $V'(r, \phi) = 0$ at the edge of the disk with $r = R$ with

$$k^2 = -Y''Z = \omega^2(\mu\varepsilon)(1 + 2L_s/L) \tag{36}$$

where the parameters are the same as for the linear and rectangular resonators. The modes are designated by the indices n and m, where m is order of the zero of $J'(R, \phi)$. The values of kR that meet this condition are given in Table 2.

The lowest mode of the disk resonator is the (1,1) mode, which resembles the (1,0) mode of a square resonator. The next mode of the disk resonator is the (2,1) mode, which resembles the (1,1) mode of a square resonator. The third disk mode is (0,1), which resembles the (2,2) mode of the square resonator.

Table 2. Values of $kR = \alpha_{nm}$ for a disk resonator with open boundary conditions.

	m = 1	m = 2	m = 3
n = 0	3.83171	7.01559	10.17347
n = 1	1.84118	5.33144	8.53632
n = 2	3.05424	6.70613	9.96947
n = 3	4.20119	8.01524	
n = 4	5.31755	9.28240	
n = 5	6.41562		
n = 6	7.50127		
n = 7	8.57784		
n = 8	9.64742		

7.3.3. Resonant patch antenna

Chaloupka *et al* [28,29] have developed the electrically small planar HTS patch antenna patterned from $YBa_2Cu_3O_{7-\delta}$, deposited onto $LaAlO_3$ that is shown in Fig. 4. The antenna dimensions are b = 6 mm, d = 1.5 mm and w = 0.15 mm. The antenna is separated from a ground plane of normal-conducting Cu by a $LaAlO_3$ substrate of thickness h = 1mm.

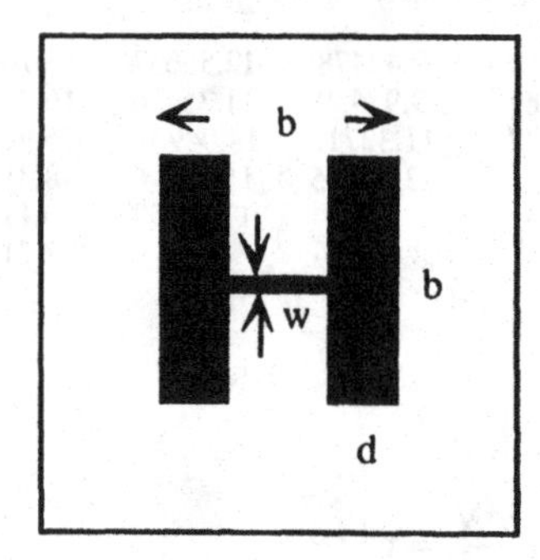

Figure 4. Outline of a planar HTS patch antenna patterned from $YBa_2Cu_3O_{7-\delta}$ deposited onto $LaAlO_3$. The antenna dimensions are given in the text.

We analyze the antenna response by considering the lumped-element representation of a single-mode antenna shown in Fig. 5. The elements L, C and R characterize the antenna, which is excited capacitively by the voltage V acting through the coupling capacitance c. The antenna radiates inductively through the mutual inductance m. The element Z_0 is the impedance presented by the radiation field.

The circuit equations are

$$-\frac{I_0}{i\omega c} - \frac{I_0}{i\omega C} + \frac{I_1}{i\omega C} = V \tag{37}$$

$$-\frac{I_0}{i\omega C} + \frac{I_1}{i\omega C} - I_1 R + i\omega L I_1 + i\omega m I + \frac{I_1}{i\omega C} = 0 \tag{38}$$

$$i\omega \ell I - Z_0 I + i\omega m I_1 = 0 \tag{39}$$

---

[28] H. Chaloupka, "High-temperature superconductors—a material for miniaturized or high-performance microwave components," *Frequenz* **44**, 141 (1990).

[29] H. Chaloupka, N. Klein, M. Peiniger, H. Piel, A. Pischke and G. Splitt, "Miniaturized HTS microstrip patch antenna," *IEEE Trans. Microwave Theory Tech.* **39**, 1513 (1991).

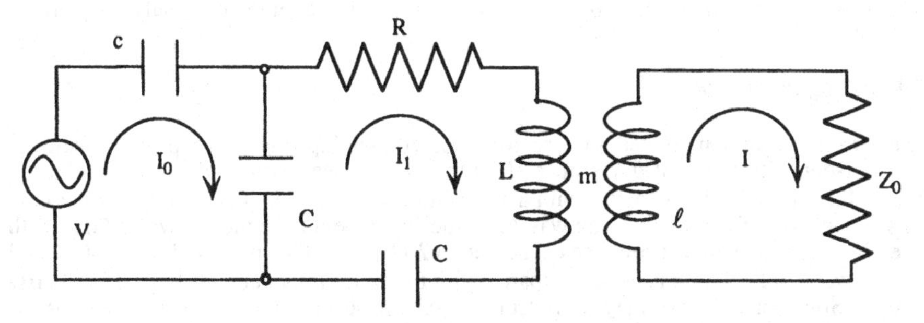

Figure 5. Lumped-element representation of a single-mode antenna. The antenna is excited capacitively through c and radiates inductively through the mutual inductance m. The element Z_0 is the impedance presented by the radiation field.

Writing $\gamma = 1 + C/c$, the radiation current is

$$I = \frac{i\omega mV/\gamma}{\omega^2 m^2 + (i\omega\ell - Z_0)\left[(2 - 1/\gamma)(1/i\omega C) + i\omega L - R\right]} \tag{40}$$

with the current circulating in the antenna

$$I_1 = -\frac{i\omega\ell - Z_0}{i\omega m} I \tag{41}$$

At resonance with $\omega_0 L = (2 - 1/\gamma)/\omega_0 C$ and assuming $\omega l << Z_0$ we obtain for the maximum circulating current

$$I_{max} = \frac{i\omega mV/\gamma}{\omega^2 m^2 + Z_0 R} \tag{42}$$

For maximum power transfer we have from Eq. (42) $\omega m = \sqrt{Z_0 R}$ with $m = \sqrt{\ell L}$ which leads to $\omega\ell/Z_0 = R/\omega L << 1$ and justifies the above assumption $\omega\ell << Z_0$. Writing for the radiation current

$$I = \frac{1}{2 - (1 - \omega_0^2/\omega^2) i\omega L/R} \frac{iV}{\gamma Z_0 R} \tag{43}$$

the power radiated is

$$I^2 Z_0 = (V^2/\gamma^2 R)\{4 + (\omega L/R)^2 (1 - \omega_0^2/\omega^2)^2\}^{-1} \tag{44}$$

with a full width at half maximum $\Delta\omega = 2L/R$. On the other hand, for an undercoupled antenna we have in place of Eq. (8)

$$I^2 Z_0 = (V^2/\gamma^2 R)(\omega^2 m^2 / Z_0 \{1 + (\omega L/R)^2 (1 - \omega_0^2/\omega^2)^2\}^{-1} \tag{45}$$

which has a full width at half-maximum $\Delta\omega=L/R$, just half that of the optimally coupled antenna.

7.3.4. Ring resonator

Pyee *et al.*[30] have constructed a microstrip superconducting ring (or loop-gap) resonator by laser-depositing $YBa_2Cu_3O_{7-x}$ onto MgO. The ring was patterned by laser-ablation. The radius of the loop was 4 mm with a gap 80 to 200 μm wide. The dielectric thickness was 0.5 mm and the ground plane was Ag. The fundamental frequency was 5 GHz with linearly spaced modes that could be excited up to 20 MHz. The measured R_S at 30 K and 5 GHz was 1 mΩ, which is to be compared with the best film values, which are in the μΩ range. Some of this loss may arise from the silver substrate, but much of the loss is presumably associated with junctions at the ring edge and current-crowding. R_s was found to increase as ω^2 as expected for intrinsic material as well as for grain boundaries. The penetration at 30 K was determined to be 210 nm, indicating some grain boundary penetration, presumably along the edge of the ring.

---

[30] M. Pyee, P. Meisse, M. Chaubet, D. Chambonnet and D. Sinobad in *High T_c Superconductor Thin Films*, ed. L. Correra, (North-Holland, Amsterdam, 1992) pp. 201-206.

8
WAVEGUIDES AND CAVITY RESONATORS*

8.1. Waveguides

8.1.1. Introduction

Present-day microwave technology can be traced to the research and development effort that led to the perfection of radio detection and ranging (radar) during the Second World War. Much of this work took place at the M. I. T. Radiation Laboratory and has been exhaustively reported in a series of 28 volumes.[1] The material presented in this series has

* This chapter is largely based on A. M. Portis, D. W. Cooke and E. R. Gray, "rf properties of high-temperature superconductors: Cavity methods," *J. Supercond.* **3**, 297 (1990).

[1] *Massachusetts Institute of Technology Radiation Laboratory Series*, editor-in-chief, L. N. Ridenour (McGraw-Hill,

by now found its way into a number of guides and texts.[2,3,4,5] Much of the background required for this section has been drawn from the intermediate-level text on electromagnetic fields and waves by Ramo, Whinnery and Van Duzer.[6]

8.1.2. Coaxial lines

The simplest enclosed transmission line is based on coaxial cylinders. In the lowest-order mode of the coaxial line, both the electric and magnetic fields are transverse to the cylinder axis with this mode designated as transverse-electromagnetic (TEM). The magnetic field is azimuthal and the electric field is radial. Higher order modes of the coaxial line do exist[7,8] but they are ordinarily non-propagating at the operating frequency of the line.

The capacitance per unit length of a coaxial transmission line with dielectric is

$$C = 2\pi\varepsilon/\ln(b/a) \tag{1}$$

and the inductance per unit length is

$$L = (\mu/2\pi)\ln(b/a) \tag{2}$$

where *a* is the radius of the inner conductor and *b* is the radius of the outer conductor. A wave at frequency ω propagates down the line with a wavevector

$$k = \omega(LC)^{1/2} \tag{3}$$

The characteristic impedance of the line is

$$Z_0 = V/I = \omega L/k = k/\omega C = (L/C)^{1/2} \tag{4}$$

Using Eqs. (1) and (2) leads for the coaxial transmission line to

$$Z_0 = (1/2\pi)(\mu/\varepsilon)^{1/2}\ln(b/a) \tag{5}$$

with the impedance of free space $\eta = (\mu_0/\varepsilon_0)^{1/2} = 376.730\ \Omega$.

Losses originate in the surfaces of the conducting cylinders and within the dielectric that separates the inner and outer conductors. Such losses may be treated by replacing the inductance L by the series combination of L and a resistance R per unit length, leading to a series impedance $R - i\omega L$, and by replacing the capacitance C by the parallel combination of C and a conductance G per unit length, leading to the shunt admittance $G - i\omega C$.

The wavevector of the line becomes complex with k replaced by

$$k + i\kappa = [(\omega L + iR)(\omega C + iG)]^{1/2} \tag{6}$$

New York, 1948-1951).

[2]E. L. Ginzton, *Microwave Measurements* (McGraw-Hill, New York, 1957).

[3]R. N. Ghose, *Micxrowave Circuit Theory and Analysis* (McGraw-Hill, New York, 1963).

[4]J. L. Altman, *Microwave Circuits* (Van Nostrand, Princeton, 1964).

[5]R. E. Collins, *Foundations for Microwave Engineering* (McGraw-Hill, New York, 1966).

[6]S. Ramo, J. R. Whinnery and T. Van Duzer, *Fields and Waves in Communication Electronics*, 2nd. ed. (Wiley, New York, 1984).

[7]Ramo, Whinnery and Van Duzer, Section 8.10.

[8]*Waveguide Handbook*, ed. N. Marcuvitz (McGraw-Hill, New York, 1951).

The characteristic impedance of the line is also complex with

$$Z_0 = [(\omega L + iR)/(\omega C + iG)]^{1/2} \tag{7}$$

For $\kappa << k$ the attenuation factor is approximately

$$\kappa \approx \frac{1}{2}(R/Z_0 + G\,Z_0) \tag{8}$$

8.1.3. Rectangular waveguides

The electromagnetic fields **E** and **H** of a guided wave are described by the equations

$$\nabla^2 \mathbf{E} = -k^2\mathbf{E} \qquad \nabla^2 \mathbf{H} = -k^2\mathbf{H} \tag{9}$$

with $k^2 = \omega^2\mu\varepsilon$, where μ is the permeability of the medium and ε is the permittivity. For a waveguide of uniform cross-section (a cylindrical waveguide), the phase of the fields advances uniformly with wavevector

$$\beta = (k^2 - k_c^2)^{1/2} \tag{10}$$

where the cutoff wavevector k_c increases with the transverse curvature of the propagating mode. The longest wavelength that will propagate without attenuation is $\lambda_c = 2\pi/k_c$.

Waves may propagate with either the electric field transverse (TE) or the magnetic field transverse (TM) to the axis of the waveguide. For a TM wave in a rectangular waveguide, the longitudinal electric field is

$$E_z = E \sin k_x x \sin k_y y\, e^{i\beta z} \tag{11}$$

The boundary condition $E_z = 0$ leads for a rectangular waveguide of width *a* and height *b* to the cutoff condition

$$k_c^2 = k_x^2 + k_y^2 = (\pi m/a)^2 + (\pi n/b)^2 \tag{12}$$

Note from Eq. (11) that TM modes with m = 0 or n = 0 lead to $E_z = 0$ and are excluded. For a TE wave in a rectangular waveguide the longitudinal magnetic field is

$$H_z = H \cos k_x x \cos k_y y\, e^{i\beta z} \tag{13}$$

The boundary condition $E_z = 0$ leads to the same cutoff condition Eq. (11) as for a TM wave.

8.1.4. Circular waveguide

Circular waveguides are of interest because of their reduced attenuation for propagating waves. The longitudinal electric field of a TM mode is of the form

$$E_z(r, \phi) = E\, J_m(k_c r) \cos m\phi \tag{14}$$

The boundary condition $E_z(a, \phi) = 0$ allows modes with $k_c a = u_{mn}$ where u_{mn} is the n'th zero of the Bessel function $J_m(u)$. The longitudinal magnetic field of a TE wave is

$$H_z(r, \phi) = H\, J_m(k_c r) \cos m\phi \tag{15}$$

The boundary condition $E_z(a, \phi) = 0$ allows modes with $k_c a = u'_{mn}$ where u'_{mn} is the n'th zero of $(d/du)\, J_m(u)$. The recurrence relation[9]

$$J_{m+1}(u) = -\, u^m\, (d/du)[u^{-m}\, J_m(u)] \tag{16}$$

gives for $m = 0$, $J_1(u) = (d/du)J_0(u)$ which leads to $u_{1n} = u'_{0n}$ and a cutoff wavelength λ_c of the circular TM_{1n} wave that is the same as that of the circular TE_{0n} wave.

8.2. Cavity Resonators

Microwave cavity resonators are enclosed structures that support a resonant electromagnetic mode at microwave frequencies . The simplest cavities are resonant sections of transmission lines. The earliest use of microwave cavities to study the electrodynamics of superconductors was the work of Pippard.[10]

8.2.1. Circular cylindrical resonators

Modes of a cylindrical resonator of length d are designated as TE_{mnp} or TM_{mnp} with

$$(\omega/c)^2 = k_{mn}{}^2 + (p\pi/d)^2 \tag{17}$$

where $k_{mn} = 2\pi/\lambda_{mn}$ is the appropriate cutoff wavevector and p is the number of half wavelengths along the axis of the cavity. The modes of a circular cylindrical cavity are characterized by the radial and azimuthal dependence of the longitudinal component of field. The degeneracy of the circular TM_{1np} and TE_{0np} modes must be lifted by modification of an end wall .

Bohn *et al.*[11,12] have constructed six circular Cu and Nb cavities resonant in TE_{011} and TE_{012} modes at frequencies from 1.5 to 40 GHz. Centered in the top plate is a hole that lifts the degeneracy of the TE_{01p} and TM_{11p} modes. The sample replaces the end wall of the cavity as is discussed in Sec. 8.3.1. An advantage of the TE_{01p} modes is that the electric field is entirely azimuthal with no radial component. So long as circular symmetry is preserved no currents cross joints normal to the cylinder axis.

Rubin *et al.*[13] have built a superconducting niobium 6 GHz circular TE_{011} cavity with removable endplates. A groove is placed in one of the endplates to lift the degeneracy with the TM_{111} mode. A pair of coupling loops are located in the upper endplate. The input loop is variable while the output loop, which monitors the transmitted power and the transient decay, is fixed. The opposite end plate contains a niobium tube that is beyond cutoff. The sample is mounted on a small-diameter sapphire rod on the axis of the tube. The extension of the sample into the cavity is varied externally.

---

[9] G. Arfken, *Mathematical Methods for Physicists*, 2nd. ed. (Academic, New York, 1970), sec. 11.1.

[10] A. B. Pippard, *Proc. Roy. Soc. (London)* **A216**, 547 (1953).

[11] C. L. Bohn, J. R. Delayen and M. T. Lanagan, *Proceedings of the Workshop on High-T_c Superconductivity*, Huntsville, AL, 23-25 May 1989. GACIC Publication PR-89-02, IIT Research Institute, Chicago.

[12] C. L. Bohn, J. R. Delayen, U. Balachandran and M. T. Lanagan, *Appl. Phys. Lett.* **55**, 304 (1989).

[13] D. L. Rubin, J. Gruschus, J. Kirchgessner, D. Moffat, H. Padamsee, J. Sears, Q. S. Shu, S. Tholen, E. Wilkins, R. Buhrman, S. Russek and T. W. Noh, *Proceedings of the Third Workshop on rf Superconductivity*, Argonne, IL, September 1987, edited by K. W. Shepard, Argonne National Laboratory Publication ANL-PHY-88-1. Pages 211-227.

Sridhar and Kennedy[14] have used a circular TE_{011} cavity constructed of oxygen-free high-conductivity (OFHC) copper. The cavity is held at 4.2 K with the sample, similarly mounted on a sapphire rod, at elevated temperature. Microwave radiation is coupled into and out of the resonator through two coaxial lines, each terminated by a loop within circular cutoff tubes. Coupling to the resonator is varied by moving the loops in and out of their cutoff tubes.

Müller *et al.*[15] have developed an OFHC copper cavity with a circular TE_{021} mode at 86 GHz and a circular TE_{013} mode at 87 GHz. Carini *et al.*[16,17] have constructed millimeter-wave copper transmission cavities resonant in a circular TE_{011} mode at 102 and 148 GHz.

Button and Alford[18] have described the construction of cylindrical cavities of polycrystalline yttria-stabilized zirconia (YSZ), coated with $YBa_2Cu_3O_x$. YSZ tubes 70 mm in diameter and 75 mm long were coated with YBCO thick-film ink, which was then dried and sintered. The end-plates were coated disks. Soft copper shims prevented leakage between the barrel and end-plates. A Q in excess of 7×10^5 was obtained at 77 K and 5.66 GHz.

8.2.2. Rectangular resonators

The modes of a resonator of rectangular cross-section are designated as rectangular TE_{mnp} or TM_{mnp} where m and n indicate the number of nodes in the longitudinal field along the rectangular axes and p indicates the number of half wavelengths contained by the cavity. TE and TM modes with the same indices are degenerate except for m or n equal to 0 for which there can be no TE mode.

Microwave spectrometers for the observation of electron spin resonance have commonly used cavities resonant in TE_{10p} modes.[19,20,21] Fuller *et al.*[22] and Rachford *et al.*[23] have placed samples on the walls of their rectangular TE_{103} copper cavity resonant at 9.2 GHz with microwave power coupled into the cavity by means of a Gordon coupler,[24] which is adjusted externally. Friedberg and Strandberg[25] have used a TE_{101} cavity for the measurement of magnetoresistance at microwave frequencies.

---

[14] S. Sridhar and W. L. Kennedy, *Rev. Sci. Instrum.* **59**, 531 (1988).

[15] G. Müller, D. J. Brauer, R. Eujen, M. Hein, N. Klein, H. Piel, L. Ponto, U. Klein and M. Peiniger, *IEEE Trans. Magn.* **25**, 2402 (1989).

[16] J. P. Carini, A. N. Awasthi, W. Beyermann, G. Grüner, T. Hylton, K. Char, M. R. Beasley and A. Kapitulnik, *Phys. Rev. B* **37**, 9726 (1988).

[17] J. Carini, L. Drabeck and G. Grüner, *Mod. Phys. Lett.* **3**, 5 (1989).

[18] T. W. Button and N. McN. Alford, *Appl. Phys. Lett.* **60**, 1378 (1992).

[19] D. J. E. Ingram, *Spectroscopy at Radio and Microwave Frequencies, Second Edition*, Butterworths, London, 1967.

[20] C. P. Poole, Jr., *Electron Spin Resonance, A Comprehensive Treatise on Experimental Techniques*, Interscience Publishers, New York, 1967.

[21] R. S. Alger, *Electron Paramagnetic Resonance: Techniques and Applications*, Interscience Publishers, New York, 1968.

[22] W. W. Fuller, F. J. Rachford, W. L. Lechter, P. R. Broussard, L. H. Allen and J. H. Classen, *IEEE Trans. Magn.* **25**, 2394 (1989).

[23] F. J. Rachford, W. W. Fuller and M. S. Osofsky, *J. Supercond.* **1**, 165 (1988).

[24] J. P. Gordon, *Rev. Sci. Instrum.* **32**, 658 (1961).

[25] C. B. Friedberg and M. W. P. Strandberg, "Microwave magnetoresistance measurements," *J. Appl. Phys.* **40**, 2475 (1969).

8.2.3. Coaxial TEM resonators

Measurements on bulk superconductors at frequencies below 1.7 GHz may usefully be made with a half-wave coaxial resonator. The outer conductor can be copper with cavity losses dominated by the sample, a superconducting axial rod. The entire apparatus may be filled with either liquid nitrogen or liquid helium. An important advantage of this arrangement is that the considerable heat generated by the sample in high-power critical rf field measurements is readily transferred to the cryogen.

Argonne investigators[2,26] have measured thin cylindrical rods of $YBa_2Cu_3O_{7-\delta}$ contained by a quartz tube on the axis of a half-wave resonant coaxial line. The line is shielded by a copper outer conductor. The outer cylinder extends beyond the ends of the inner conductor and in these regions acts as a circular TM_{01} transmission line beyond cutoff. Two cavities have been constructed, one for operation in the frequency range 150-600 MHz and the second in the range 600-1500 MHz.

The advantages of the coaxial cavity are (i) its transverse dimensions may be quite small, (ii) high surface fields can be achieved at the inner conductor with moderate power input leading to high sensitivity and (iii) it is not necessary for the shield to be superconducting.

8.2.4. Special structures

Spheroidal and ellipsoidal resonators

The modes of a spherical resonator[27] are designated TE_{mnp} or TM_{mnp} where m designates the r dependence, n the ϕ variation and p the θ dependence. The resonant wavelength of the TM_{101} mode in a sphere of radius r is $\lambda = 2.29$ times r and that of the TE_{101} mode is $\lambda = 1.395$ times r. Spheroidal and ellipsoidal resonators are used in part because of their reduced degeneracy. Examples of these resonators are given in Table III.

Piel's group at Wuppertal[28] have used a 2.93 GHz oblate spheroidal superconducting cavity of niobium, anodized to form with a highly insulating layer of Nb_2O_5. Coupling to the cavity is through loops contained in circular beyond-cutoff wavegudes along the resonator axis. The rf input loop is at the top and an rf monitor loop is at the bottom.

Rubin *et al.*[29] have used an ellipsoidal 8.6 GHz superconducting niobium cavity to investigate the microwave losses in a single crystal flake of surface area about 3 mm^2. The crystal was placed in a high magnetic field region of the cavity and the cavity was cooled to 1.5 K. The surface resistance of the crystal flake was determined by comparison with the losses of a titanium alloy foil scaled in wavelength and placed in a 1.5 GHz copper cavity.

Cooke *et al.*[30] have constructed a 3 GHz niobium superconducting cavity in its fundamental TM_{010} mode.

---

[26] J. R. Delayen, K. C. Goretta, R. B. Poeppel and K. W. Shepard, *Appl. Phys. Lett.* **52**, 930 (1988).

[27] J. A. Stratton, *Electromagnetic Theory* (McGraw-Hill Book Company, Inc., New York, 1941), sec. 9.24.

[28] M. Hagen, M. Hein, N. Klein, A. Michalke, G. Müller, H. Piel, R. W. Röth, F. M. Mueller, H. Sheinberg and J. L. Smith, *J. Mag. Mag. Mat.* **68**, L1 (1987).

[29] D. L. Rubin, J. Gruschus, J. Kirchgessner, D. Moffat, H. Padamsee, J. Sears, Q. S. Shu, S. Tholen, E. Wilkins, R. Buhrman, S. Russek and T. W. Noh, in *Third Workshop on rf Superconductivity*, Argonne, IL, September 1987 (Argonne National Laboratory Publication ANL-PHY-88-1, 1988).

[30] D. W. Cooke, B. Bennett, E. R. Gray, R. J. Houlton, W. L. Hults, M. A. Maez, A. Mayer, J. L. Smith and M. S. Jahan, *Appl. Phys. Lett.* **55**, 1038 (1989).

Reentrant resonators

The Argonne group has constructed a reentrant circular cavity[31] resonant at 850 MHz. A cavity constructed in this way has an over-all length that is about one-eighth the free-space wavelength. This is because the gap at the end acts much like the capacitive termination of a coaxial line. The small cavity size gives the advantage of larger sample filling factors and higher rf fields.

A reentrant cavity may be modeled as a shorted segment of transmission line that is capacitively shunted.[20] The input impedance of a line of length ℓ line is

$$Z = - iZ_0 \tan 2\pi\lambda/\ell \tag{18}$$

For $\ell < \lambda/4$ the line is inductive and may be resonated by a capacitor. In particular, for a line $\lambda/8$ long, a capacitor $C = 1/\omega_0 Z_0$ resonates the line. Such a resonator is one-quarter the volume of a half-wavelength coaxial cavity. The volume filling factor is then four times as great. The Q is proportional to the volume/surface ratio and is about half for the reentrant cavity. The energy density at a given incident power is doubled and the surface rf field H_s is about $\sqrt{2}$ greater for the reentrant cavity.

Split-ring resonators

Becks *et al.*[32] have reported the construction of a compact split-ring resonator for use in an atomic hydrogen maser that is to go into earth-orbit. The frequency and field distribution of the cavity are determined by two adjacent half-cylindrical electrodes separated by a gap. Similar resonators have been used in magnetic resonance spectroscopy.[33,34] The cavity is 5 cm in diameter and 5 cm high with a gap of approximately 0.5 cm and is resonant at the hydrogen hyperefine transition frequency 1.420405 GHz.[35] The electrodes are silver that has been coated electrophoretically[36] with $YBa_2Cu_3O_{7-\delta}$. To eliminate radiation loss, the resonator is enclosed in a cylindrical shield.

Bonn *et al.*[37] have developed a split-ring resonator for studying small samples of the oxide superconductors in the frequency range 0.3 to 5 GHz. An unloaded Q of 1.2×10^6 has been achieved at 1.78 GHz, making such cavities useful for samples with surface areas as small as 0.1 mm^2 and surface resistances as low as microhms.

Hughey *et al.*[38] have described the construction of a 35-GHz fully split high-Q superconducting cavity operating in the TM_{010} mode. The cavity is tuned by varying the spacing between halves. A quality factor $Q = 4 \times 10^7$ has been achieved at $T = 2$ K in both

---

[31] C. L. Bohn, J. R. Delayen and C. T. Roche, "Apparatus for measurement of surface resistance *vs.* rf magnetic field of high-T_c superconductors, *Rev. Sci. Instrum.* **61**, 2207 (1990).

[32] M. Becks, A. Brust, H. Chaloupka, S. Haindl, M. Hein, H. Heinrichs, M. Jansen, M. Jeck, N. Klein, U. Klein, S. Lauterjung, E. Mahner, A. Michalke, N. Minatti, B. Mönter, G. Müller, D. Opie, S. Orbach, M. Peiniger, H. Piel, A. Pischke, L. Ponto, D. Reschke, R. W. Röth, J. Schurr, N. Tellmann, D. Wehler and J. Zander, *Fourth Workshop on rf Superconductivity*, KEK Tsukaba-shi, Japan, August 1989.

[33] W. N. Hardy and L. A. Whitehead, *Rev. Sci. Instrum.* **52**, 213 (1981).

[34] W. Froncisz and J. S. Hyde, *J. Mag. Res.* **47**, 515 (1982).

[35] D. Opie, H. Schone, M. Hein, G. Müller, H. Piel, D. Wehler, V. Folen and S. Wolf, *IEEE Trans. Magn.* **27**, 2944 (1991).

[36] M. Hein, E. Mahner, G. Müller, H. Piel, L. Ponto, M. Becks, U. Klein and M. Peiniger, *J. Appl. Phys.* **66**, 5940 (1989).

[37] D. A. Bonn, D. C. Morgan and W. N. Hardy, *Rev. Sci. Instrum.* **62**, 1819 (1991).

[38] B. J. Hughey, T. R. Gentile, D. Kleppner and T. W. Ducas, *Rev. Sci. Instrum.* **61**, 1940 (1990).

lead-plated and solid niobium cavities. Coating with high-temperature superconductors is feasible.

8.2.5. Optical resonators

Fabry-Perot resonators

High reflectivity in the far infrared has been demonstrated for high quality films[39,40,41] and for single crystals[42,43] so that low loss resonators appear feasible. At higher frequency the reflectivity drops either as the result of excitation across the gap[44] or absorption by defects.[45]

Renk *et al.*[46] have reported the operation of a far-infrared Fabry-Perot resonator with reflectors of high T_c superconducting material. For normally incident radiation, the transmissivity is given by Airy's formula[47]

$$\tau = \tau_{max}\,[1 + (2F/\pi)^2 \sin^2\delta/2]^{-1} \tag{19}$$

where τ_{max} is the maximum transmissivity and F is the finesse, the ratio of fringe separation to half-width. The quantity $\delta = 2(kd + \theta)$ is the phase shift for a round trip between plates of separation d with k the wavevector and θ the phase shift on reflection from either of the plates. The characteristic properties of the resonator depend on the absorptivity A, the transmissivity D and the reflectivity R of the plates

$$\tau_{max} = (1 + A/D)^{-2} \qquad F = \pi R^{1/2}/(1 - R) \tag{20}$$

Resonances occur for $\delta = 2n\pi$ where n = 1, 2, . . is the order of the resonance. The half-width of the n'th resonance is

$$\Gamma_n = k_n/nF(k_n) \tag{21}$$

with k_n/n the fringe separation.

---

[39] J. Schützmann , W. Ose, J. Keller, K. F. Renk, B. Roas, L. Schultz and G. Saemann-Ishenko, *Europohys. Lett.* **8**, 679 (1989).

[40] K. F. Renk, J. Schützmann, A. Prückl, B. Roas, L. Schultz and G. Saemann-Ischenko, *Physica B* **165-166**, 1253 (1990).

[41] K. Kamarás, S.L. Herr, C. D. Porter, N. Tache, D. B. RTanner, S. Etemad, T. Venkatesan, E. Chase, A. Inam, X. D. Wu, M. S. Hegde and B. Dutta, *Phys. Rev. Lett.* **64**, 84 &1962 (1990).

[42] G. A. Thomas, J. Orenstein, D. H. Rapkine, M. Capizzi, A. J. Millis, R. N. Bhatt, L. F. Schneemeyer and J. V. Waszczak, *Phys. Rev. Lett.* **61**, 1313 (1988).

[43] J. Orenstein, G. A. Thomas, A. J. Millis, S. L. Cooper, D. H. Rapkine, T. Timusk, L. F. Schneemeyer and J. V. Waszczak, "Frequency- and temperature-dependent conductivity in $YBa_2Cu_3O_{6+x}$ crystals," *Phys. Rev. B* **42**, 6342 (1990).

[44] J. Orenstein, G. A. Thomas, A. J. Millis, S. L. Cooper, D. H. Rapkine, T. Timusk, L. F. Schneemeyer and J. V. Waszczak, "Frequency- and temperature-dependent conductivity in $YBa_2Cu_3O_{6+x}$ crystals," *Phys. Rev. B* **42**, 6342 (1990).

[45] K. Kamarás, S.L. Herr, C. D. Porter, N. Tache, D. B. Tanner, S. Etemad, T. Venkatesan, E. Chase, A. Inam, X. D. Wu, M. S. Hegde and B. Dutta, *Phys. Rev. Lett.* **64**, 84 &1962 (1990).

[46] K. F. Renk, J. Betz, J. Schützmann, A. Prückl, B. Brunner and H. Lengfellner, "Use of high T_c superconductors for far-infrared Fabry-Perot resonators," *Appl. Phys. Lett.* **57**, 2148 (1990).

[47] M. Born and E. Wolf, *Principles of Optics* (Pergamon, Oxford, 1980) 6th ed, pp. 323-326.

Confocal resonators

Martens *et al.*[48] have developed a half-confocal resonator for the measurement of superconducting surface resistance. Measurements have been performed over the frequency range 29 to 39 GHz. Because the fields are distributed over the spherical surface but concentrated on the planar sample, the bulk of loss arises in the sample. For the radius of curvature r large compared with the wavelength λ, the spot size is $\sqrt{r\lambda / 2\pi}$. Radiation loss ($Q \approx 10^7$) is only a few percent of sample loss and can generally be neglected.

8.3. Coupling of Transmissions Lines and Resonators

We begin with a summary of the results of transmission line theory, where waveguides and coaxial cables are represented by distributed-element transmission lines and the microwave resonator by a lumped-element circuit. The coupling of the transmission lines to the resonator is represented by ideal transformers.

8.3.1. Transmission lines

The Telegrapher's equations[49] describe the relation between the voltage V across a transmission line and the current I that flows through the line and leads to the expression for the characteristic impedance of the line

$$Z_0 = V/I = \omega L/k = \sqrt{L/C} \tag{22}$$

where L is the series inductance per unit length, C is the shunt capacitance per unit length and $k = \omega\sqrt{LC}$ is the wavevector on the line.

Terminating the line with an impedance Z_L leads to a terminal voltage V_L and terminal current I_L satisfying

$$V_L = I_L Z_L = [(1 + \rho)/(1 - \rho)] I_L Z_0 \tag{23}$$

where ρ is the complex reflection coefficient. Assuming an incident wave of amplitude one-half V_0, the reflection coefficient is

$$\rho = 2\frac{V_L}{V_0} - 1 = \frac{Z_L - Z_0}{Z_L + Z_0} \tag{24}$$

The voltage at the load is that generated by a voltage source V_0 with internal impedance Z_0 working directly into a lumped-element load Z_L.

8.3.2. Equivalent circuit

The equivalent circuit of a resonator coupled to input and output transmission lines is shown in Fig. 1. The magnetic energy stored in the resonator is associated with the

---

[48] J. S. Martens, V. M. Hietala, D. S. Ginley, T. E. Zipperian and G. K. G. Hohenwarter, *Appl. Phys. Lett.* **58**, 2543 (1991).

[49] S. Ramo, J. R. Whinnery and T. Van Duzer, *Fields and Waves in Communication Electronics, 2nd Ed.* (John Wiley & Sons, New York, 1984). Secs. 8.11, 10.10 and 11.14.

inductor L. The electrical energy is associated with the capacitor C, and dissipation is through the resistor R. An equivalent circuit is taken with the three elements in parallel. Coupling to the input transmission line of characteristic impedance Z_0 is through an ideal transformer with turns-ratio m:1. The resonator is loaded by an output line of impedance Z_0' through an ideal transformer of turns-ratio m':1. The microwave source is matched to the input line and represented by a voltage generator V_0. The detector is matched to the output line.

The reciprocal impedance of the unloaded resonator is

$$1/Z = 1/R + i\ (1/\omega L - \omega C) \tag{25}$$

The complex resonant frequency is at the pole of Z

$$\omega = [\omega_0^2 - 1/(2\tau)^2]^{1/2} - i/2\tau \tag{26}$$

with $\omega_0 = 1/\sqrt{LC}$ and energy relaxation time $\tau = RC$. The unloaded quality factor of the resonator is defined as

$$Q_u = \omega_0\tau = R/\omega_0 L \tag{27}$$

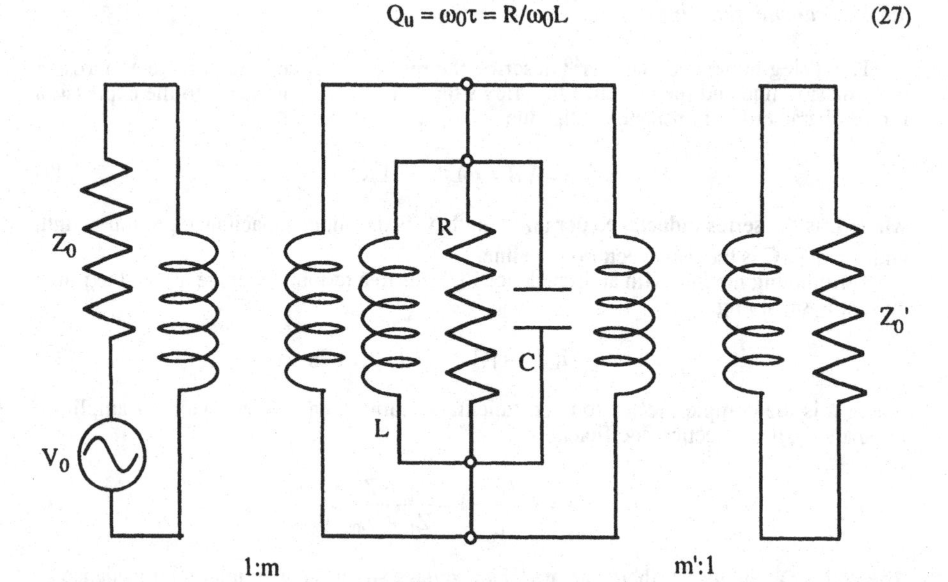

Figure 1. Equivalent circuit of a two-port resonator coupled to input and output transmission lines. The magnetic energy stored in the resonator is associated with the inductor L. The electrical energy is associated with the capacitor C, and dissipation is through the resistor R. The resonator is driven from an input line of impedance Z_0 through an ideal transformer of turns ratio 1:m and loaded by an output line of impedance Z_0' through an ideal transformer of turns ratio m':1. The microwave source is matched to the input line and represented by a voltage generator V_0. The detector is matched to the output line.

8.3.3. Resonator response

The expected response of a resonator is best studied by transforming the driving voltage and loads into shunt with the resonator as shown in Fig. 2. In parallel with R are the load resistances m^2Z_0 and m'^2Z_0'. We characterize these loads by external quality factors

$$Q_e = m^2Z_0/\omega_0L = Q_u/\beta \qquad Q_e' = m'^2Z_0'/\omega_0L = Q_u/\beta' \tag{28}$$

The total cavity loss is represented by the loaded quality factor Q_ℓ with

$$1/Q_\ell = 1/Q_u + 1/Q_e + 1/Q_e' \tag{29}$$

The voltage V across the resonator terminals

$$V = \frac{1/Q_e}{1/Q_l + i(\omega_0/\omega - \omega/\omega_0)} V_0 \tag{30}$$

is a measure of the response of the resonator.

8.3.4. Reflection

The voltage reflection coefficient ρ is obtained by treating the resonator as a load. From Eq. (24), the voltage at a load is $V_L = {}^1/_2\,(1 + \rho)\,V_0$ which gives

$$\rho = \frac{(2/Q_e - 1/Q_\ell) - i(\omega_0/\omega - \omega/\omega_0)}{1/Q_\ell + i(\omega_0/\omega - \omega/\omega_0)} \tag{31}$$

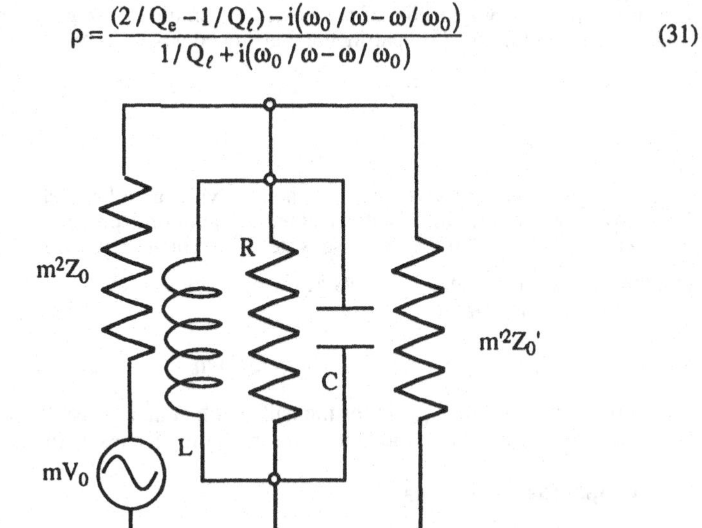

Figure 2. Transformation of the driving voltage and loads into shunt with the resonator. In parallel with R are the load resistances m^2Z_0 and m'^2Z_0'.

Both Q_e and Q_ℓ can be determined from the power reflection coefficient

$$R = |\rho|^2 = \frac{(2/Q_e - 1/Q_\ell)^2 + (\omega_0/\omega - \omega/\omega_0)^2}{(1/Q_\ell)^2 + (\omega_0/\omega - \omega/\omega_0)^2} \tag{32}$$

8.3.5. Transmission

The voltage transmission coefficient τ is the fractional voltage transmitted through an inserted resonator to a matched load. The power transmission coefficient $T = |\tau|^2$ is the ratio of the power delivered through the resonator to the incident power

$$T = P'/P_0 = (|V|^2/m'^2 Z_0')/(V_0^2/4m^2 Z_0) \tag{33}$$

Using Eq. (30) we obtain the power transmission coefficient

$$T = \frac{4/Q_e Q'_e}{1/Q_\ell^2 + (\omega_0/\omega - \omega/\omega_0)^2} \tag{34}$$

The bandwidth $2\delta\omega$ of the power transmission coefficient is the full frequency interval between half-power points. The half-power condition from Eq. (34) is $\omega/\omega_0 - \omega_0/\omega = \pm 1/Q_\ell$. With $\omega = \omega_0 \pm \delta\omega$ we obtain from Eq. (34) for the half-width at half-power

$$2\delta\omega = \omega_0/Q_\ell \tag{35}$$

Measuring the bandwidth $2\delta\omega$ of the power transmitted through the resonator gives the loaded quality factor Q_ℓ. Comparing the transmitted power from Eq. (34) gives

$$\Delta T/T = \Delta Q_\ell^{\,2}/Q_\ell^{\,2} \approx -2Q_\ell \, \Delta (1/Q_\ell) \tag{36}$$

8.3.6. Relaxation

Measurement of the power relaxation time provides a further method of determining Q_ℓ. The power relaxation time is determined by applying power to the resonator for a sufficiently long time that the response is steady and then terminating the drive power. The voltage across the resonator decays at a frequency $\omega = \omega_0\sqrt{1 - 1/4Q_\ell^2}$ with a power relaxation time given by

$$\tau = Q_\ell/\omega_0 \tag{37}$$

Comparison with Eq. (35) gives for the half-width at half-power $2\delta\omega = 1/\tau$. Sridhar and Kennedy[5] have described the electronics for measuring the resonator decay time τ.

8.4. Sample Configurations

8.4.1. Replacement

In the replacement configuration, the end wall, usually of a circular TE_{011} cavity, is replaced by a thick superconducting film or a film deposited on a metallic conductor. In this geometry no currents flow across the junction between the end wall and the body of the cylindrical cavity and there are no additional losses. Delayen *et al.* [15] have replaced the copper center rod of their coaxial cavity by ceramic superconducting rods.

Bohn *et al.*[2,3] have used samples ranging from 1.25 to 15 cm in diameter to form the bottom end wall of an appropriate size cavity. Cooke *et al.*[50,51] obtain the temperature dependence of the surface resistance R_s of the end wall of a copper cavity by cooling to T = 15 K with a closed-cycle refrigerator and slowly warming to room temperature while measuring Q values. A computer-controlled network analyzer automatically determines the resonance peak and half-power points from which the cavity quality factor is calculated.

Radcliffe *et al.*[52] have used a clamped ceramic superconducting disk as a replacement for the end wall of a circular brass transmission cavity. A number of the approximately sixty detectable modes in the frequency range 7 to 20 GHz were selected. Under computer control, the power transmission coefficient, bandwidth and frequency were measured between 4.2 K and room temperature.

Carini *et al.*[7,8] have measured surface impedance in TE_{011} transmission cavities at the millimeter-wave frequencies 102 and 148 GHz. The sample was mounted as the end wall of the cavity and the power transmitted through the cavity was measured as a function of microwave frequency, giving the central frequency ω_0 and the bandwidth $2\delta\omega$.

8.4.2. Cavity perturbation

Specimens may be measured in a cavity by the perturbation method with the electromagnetic field penetrating either one or both surfaces of the sample depending on sample placement.

Rubin *et al.*[4] have measured the temperature dependence of the surface resistance of ceramic pellets in a superconducting niobium 6 GHz circular TE_{011} resonant cavity. As described in Sec. 1.2, samples are mounted on a sapphire rod enclosed by a niobium cutoff tube. The sapphire rod makes contact with the helium bath through a thermal resistor. Magnetic fields in the TE_{011} mode are radial at the end wall and vanish on axis. For this reason the contribution of a small-diameter sample to the increase in cavity losses is substantially reduced. By measuring the rise in the temperature of a thermal resistor half-way up the sapphire rod, Rubin *et al.* [4] detect resistances 100 times smaller than can be measured from the reduction in cavity Q. Rubin *et al.*[53] have examined the microwave properties of a number of small single crystals that were attached to the end of the sapphire rod and the cavity Q was measured as a function of crystal temperature.

Awasthi *et al.*[54] and Carini *et al.*[8] have measured the temperature dependence at millimeter-wave frequencies of the surface impedance of ceramic superconductors. The surface resistance is determined from the change in $1/Q_\ell$, as discussed in Sec. 8.4.2. The penetration depth is computed from the surface reactance, which produces a shift in cavity frequency that is discussed in Sec. 8.4.3.

---

[50] D. W. Cooke, E. R. Gray, R. J. Houlton, B. Rusnak, E. A. Meyer, J. G. Beery, D. R. Brown, F. H. Garzon, I. D. Raistrick, A. D. Rollett and R. Bolmaro, *Appl. Phys. Lett.* **55**, 914 (1989).

[51] D. W. Cooke, E. R. Gray, R. J. Houlton, H. H. S. Javadi, M. A. Maez, B. L. Bennett, B. Rusnak, E. A. Meyer, P. N. Arendt, J. G. Beery, D. R. Brown, F. H. Garzon, I. D. Raistrick, A. D. Rollett, B. Bolmaro, N. E. Elliott, N. Klein, G. Müller, S. Orbach, H. Piel, J. Y. Josefowicz, D. B. Rensch, L. Drabeck and G. Grüner, *Physics C* **162- 164**, 1537 (1989).

[52] W. J. Radcliffe, J. C. Gallop, C. D. Langham, M. Gee and M. Stewart, *Physica C* **153-155**, 635 (1988).

[53] D. L. Rubin, K. Green, J. Gruschus, J. Kirchgessner, D. Moffat, H. Padamsee, J. Sears, Q. S. Shu, L. F. Schneemeyer and J. V. Waszczak, *Phys. Rev. B* **38**, 6538 (1988).

[54] A. M. Aswasthi, J. P. Carini, B. Alavi and G. Grüner, *Solid State Commun.* **67**, 373 (1988).

8.5. Analysis

8.5.1. Surface impedance

The surface impedance of a conductor is the ratio of the electric to the magnetic field at the surface

$$Z_S = R_S - i\,X_S = E_S/H_S \tag{38}$$

The penetration depth λ and the classical skin depth δ may be determined from the surface impedance of a superconductor thick compared with the penetration depth

$$Z_S = -\,i\omega\lambda\mu/[1 - 2i(\lambda/\delta)^2]^{1/2} \tag{39}$$

This expression may require correction for microwave penetration of the substrate.

8.5.2. Geometry factor

The energy dissipated at the cavity walls is

$$P = \frac{1}{2}\int_S dS\; R_s H_s{}^2 \tag{40}$$

while the energy stored in the cavity may be written as the integral

$$U = \frac{1}{2}\,\mu_0 \int_V dV\, H^2 \tag{41}$$

The reciprocal of the unloaded cavity Q is

$$\frac{1}{Q_u} = \frac{P}{\omega_0 U} = \frac{1}{\omega_0\mu_0}\frac{\int_S dS H_s^2 R_s}{\int_V dV H^2} \tag{42}$$

A change in cavity losses as the result of wall replacement or the introduction of additional absorbing surface leads to

$$\Delta\frac{1}{Q_u} = \frac{\Delta P}{\omega_0 U} = \frac{1}{\omega_0\mu_0}\frac{\int_S dS H_s^2 \Delta R_s}{\int_V dV H^2} \tag{43}$$

The sample geometry factor is defined as

$$G = \omega_0\mu_0 \frac{\int_V dV H^2}{\int_{\Delta S} dS H_s^2} \tag{44}$$

where ΔS is the area over which a change in R_S has been made, either by replacement of wall or axial material or by the insertion of material into the cavity. So long as the coupling is not affected by ΔR_S, Eq. (39) gives

$$\Delta(1/Q_\ell) = \Delta(1/Q_u) \tag{45}$$

and the change in surface resistance is

$$\Delta R_s = G\,\Delta(1/Q_\ell) \tag{46}$$

8.5.3. Frequency shift

The connection between the change in surface reactance ΔX_s and the frequency shift $\Delta\omega_0$ is obtained from the following argument given by Pippard.[11] We should be able to relate $\Delta Z_s = \Delta R_s - i\,\Delta X_s$ to the change in complex frequency $\Delta(\omega_0 - i/2\tau)$. From Eqs. (37) and (46) we write

$$\Delta R_s = (G/\omega_0)\,\Delta\,(1/\tau) \tag{47}$$

The required complex relation is then

$$\Delta Z_s = \Delta R_s - i\,\Delta X_s = (2iG/\omega_0)\,\Delta(\omega_0 - i/2\tau) \tag{48}$$

which gives for the change in surface reactance

$$\Delta X_s = -\,2G\,\Delta\omega_0/\omega_0 \tag{49}$$

Slater[55,56] has obtained this result from an energy argument.

8.6. Measurement

8.6.1. Reflection

Rachford *et al.*[13] have determined the Q_u of circular TE_{011} and rectangular TE_{103} cavities from the frequency dependence of the reflected power by Eq. (41) with $1/Q_e' = 0$ for a reflection cavity. During an experimental run the cavity and sample temperature were allowed to rise slowly while the unloaded Q and frequency shift were computed and recorded at regular intervals.

8.6.2. Transmission

Cooke *et al.*[6,17] obtain the total surface resistance R_s rather than ΔR_s as in Eq. (46) from the fact that a known fraction f of the losses in their cavity occur at an end wall. Their argument is that if the end wall were lossless, the unloaded quality factor of the copper cavity Q_c would be increased by a factor $1/(1 - f)$ which gives

$$R_s = G\Delta\frac{1}{Q_\ell} + \frac{fG}{Q_c} \tag{50}$$

The sample geometry factor G, discussed in Sec. 4.2, is determined from Eq. (50) when a

---

[55] J. C. Slater, *Microwave Electronics* (D. Van Nostrand Co., Inc., New York, 1950). See also J. C. Slater, *Rev. Mod. Phys.* **18**, 441 (1946).

[56] J. C. Slater, *Microwave Transmission* (McGraw-Hill Book Company, Inc., New York, 1942; reprinted by Dover Publications, New York, 1959).

material such as stainless steel with known R_s is used as the end wall. Bohn *et al.*[13] have similarly determined R_s from the unloaded quality factor of a circular cavity operating in a TE_{011} or TE_{012} mode with the sample forming the bottom surface.

Klein *et al.*[57,58] have determined the surface resistance and reactance of several c-axis oriented epitaxial thin films of $YBa_2Cu_3O_7$ at 87 GHz with a circular TE_{013} cavity through the use of Eqs. (46) and (49). Partial penetration of microwave radiation into the substrate leads to an enhancement of Z_s and must be corrected for the determination of λ and δ through Eq. (39).

Carini *et al.*[7,8] have measured the surface impedance at millimeter-wave frequencies from the power transmitted through the cavity as a function of frequency, giving the central frequency ω_o and the bandwidth $2\delta\omega$. The procedure was repeated with a polished OFHC copper end wall. The difference between the surface resistances of the sample and copper end wall is given by Eq. (46). The difference in surface reactance is related to the shift in cavity frequency by Eq. (49).

Minehara *et al.*[25] have determined the unloaded Q of an all $YBa_2Cu_3O_{7-\delta}$ cavity from the power transmission coefficient and the bandwidth of the transmitted power. The surface resistance was obtained from Eq. (46) with $R_s = G/Q_u$, where G is the sample geometry factor computed from Eq. (44) for a cavity in which the sample covers the entire surface.

Delayen *et al.*[15] have used coaxial cavities from 150 to 1500 MHz[3] with surface rf magnetic fields up to 640 Oe at 190 MHz at an input power of 2.2 kW with the resonant line immersed in liquid nitrogen. The surface rf magnetic field at the sample is determined from the voltage at a calbrated pickup probe located midway between the end-plates of the cavity.

8.6.3. Relaxation

The surface resistance as a function of temperature, frequency and power may be determined from a measurement of the decay time τ of pulsed cavity excitation. Sridhar and Kennedy[5] have fed pulsed microwave power into one port of a circular TE_{011} cavity and observed the transmitted signal, following detection, on a fast oscilloscope. The loaded quality factor of the resonator was determined from Eq. (37). The resonant frequency $\omega_0/2\pi$ was directly obtained from the synthesizer output. The surface resistance and reactance were determined by the sample perturbation relations, Eqs. (46) and (49).

Rubin *et al.*[4] take the power transmission coefficient T as a measure of the cavity unloaded Q. The rise in temperature of a niobium pellet at room temperature, where the microwave surface resistance of niobium is accurately known, gives the magnetic field strength at the pellet surface. A sintered ceramic pellet then replaces the niobium standard and the decay time is measured as a function of pellet temperature. Coupling to the cavity is adjusted to fix the rf field at the sample.

Kato *et al.*[59] determine Q_ℓ from τ on reflection from a circular TM_{011} cavity resonant at 2.86 GHz. A measurement of the the power reflection coefficient R through Eq. (32) leads to the unloaded cavity Q and $\Delta(1/Q_u)$ when the cavity is perturbed.

Delayen and Bohn[60] determined the differential decay rate $\Delta(1/\tau)$ associated with sample losses. From the known surface resistance R_s of Pb at room temperature, the

---

[57] N. Klein, G. Müller, H. Piel, B. Roas, L. Schultz, U. Klein and M. Peiniger, *Appl. Phys. Lett.* **54**, 757 (1989).

[58] N. Klein, G. Müller, S. Orbach, H. Piel, H. Chaloupka, B. Roas, L. Schultz, U. Klein and M. Peiniger, *Physica C* **162-164**, 1549 (1989).

[59] K. Kato, K. Takahashi, H. Mitera and K. Minami, *Jpn. J. Appl. Phys.* **27**, 1641 (1988).

[60] J. R. Delayen and C. L. Bohn, *Phys. Rev.* **40**, 5151 (1989).

sample geometry factor $G\,\Delta(1/\tau) = \omega_0\,\Delta R_s$ was determined as given by Eq. (47). The geometry factor is independent of the material properties of the central conductor, but does depend on its diameter as may be seen from Eq. (44). Replacing the sample by a copper rod yields a shorter decay time. This procedure gives the difference between R_s of the superconductor and R_s of copper.[61] The surface resistance of the superconductor may be obtained by analogy with Eq. (50)

$$R_s = G[\omega_0\,\Delta(1/\tau) + f/Q_c] \tag{51}$$

where Q_c is the unloaded Q of the cavity with copper as the center conductor and $f \approx 1$ is the fraction of unloaded loss that arises from the center conductor.

At high rf levels with the central conductor a superconducting ceramic, the pulse decay is nonexponential with a long tail. This behavior indicates that the surface resistance R_s of the ceramic increases with the strength of the rf magnetic field. Under these conditions the initial decay rate is taken as a measure of the surface resistance at peak power. The decay time may be determined as a function of rf magnetic field from the local logarithmic derivative.

---

[61] J. R. Delayen, K. W. Shepard, K. C. Goretta and R. B. Poeppel, *Proceedings of the Third Workshop on rf Superconductivity*, Argonne, IL, September 1987, edited by K. W. Shepard, Argonne National Laboratory Publication ANL-PHY-88-1.

9
ELECTRODYNAMICS OF TYPE II SUPERCONDUCTIVITY*

9.1. Carrier Dynamics

For a plane electromagnetic wave of frequency ω and wavevector k, propagating in a homogeneous medium, the displacement and electric field are related by the permittivity $\varepsilon(\omega, k)$

$$\mathbf{D}(\omega, k) = \varepsilon(\omega, k)\, \mathbf{E}(\omega, k) \tag{1}$$

Similarly, the magnetic induction and the magnetic field are related by the permeability $\mu(\omega, k)$

$$\mathbf{B}(\omega, k) = \mu(\omega, k)\, \mathbf{H}(\omega, k) \tag{2}$$

The phase velocity and the wavevector of the wave are

$$v = \omega/k = 1/\sqrt{\mu\varepsilon} \tag{3}$$

$$k = \omega\sqrt{\mu\varepsilon} \tag{4}$$

The admittivity $y(\omega, k)$ relates the displacement current to the electric field

$$\partial\mathbf{D}/\partial t = -i\omega\, \varepsilon(\omega, k)\, \mathbf{E} = y(\omega, k)\, \mathbf{E} \tag{5}$$

and yields for the admittivity

*This chapter draws on A. M. Portis and D. W. Cooke, "Effect of magnetic fields and variable power on the microwave properties of granular superconductors," in *High Temperature Superconductors*, J. J. Pouch, S. A. Alterovitz and R. R. Romanofsky, eds., Materials Science Forum (Trans Tech Publications, Aedermannsdorf, Switzerland, 1992).

$$y(\omega, k) = -i\omega\varepsilon(\omega, k) \tag{6}$$

The admittivity (and its reciprocal, the specific impedance) suggest the representation of a dielectric or conducting medium by lumped electrical elements—resistors, capacitors and inductors in series and parallel combinations. Such representations are helpful not only for granular superconductors, where intergranular regions may dominate the electrodynamics of the medium, but also for bulk superconductors, where the conductivity is commonly represented by resistive elements in *parallel* with inductive elements. Flux-flow processes, on the other hand, are represented by resistive elements in *series* with inductive elements. The main rationale for this representation is the visualization of physical processes.

9.1.1. Surface Impedance

At an interface, the characteristic impedance of the medium is called the surface impedance

$$Z_S = E/H = \omega\mu/k = (\mu/\varepsilon)^{1/2} \tag{7}$$

In terms of the specific impedance $z(\omega, k) = 1/y(\omega, k)$ [Eq. (6)] the surface impedance is

$$Z_S = R_S - iX_S = (-i\omega\mu z)^{1/2} \tag{8}$$

The real part R_S of the surface impedance is called the surface resistance and the imaginary part X_S is called the surface reactance. The specific admittance of a medium may be written as

$$y = \frac{2}{\omega\mu\delta^2} + \frac{i}{\omega\mu\lambda^2} = \frac{1 - i\tan\phi}{-i\omega\mu\lambda^2} \tag{9}$$

with the loss-tangent given by $\tan\phi = 2(\lambda/\delta)^2$. The lengths λ and δ are introduced simply to characterize the real and imaginary parts of the admittivity. The surface impedance is

$$Z_s = R_s - iX_s = -i\omega\mu\lambda\sqrt{\cos\phi}\, e^{i\phi/2} \tag{10}$$

which gives for the components of the surface impedance

$$R_s = \omega\mu\lambda\sqrt{\cos\phi}\,\sin\phi/2 \qquad X_s = \omega\mu\lambda\sqrt{\cos\phi}\,\cos\phi/2 \tag{11}$$

9.1.2. Penetration depth

From Eqs. (4), (6) and (9) the complex wavevector is

$$k = \sqrt{i\omega\mu/z} = (i/\lambda\sqrt{\cos\phi})\, e^{-i\phi/2} \tag{12}$$

The attenuation constant is the reciprocal of the penetration depth

$$1/\lambda_s = (\cos\phi/2)/\lambda\sqrt{\cos\phi} \tag{13}$$

Comparing Eqs. (11) and (13) leads to the expression for the surface reactance

$$X_S = \omega\mu\lambda_s\cos^2\phi/2 \tag{14}$$

Values of the penetration depth for rf currents in the conducting planes of $YBa_2Cu_3O_7$ have been measured in thin films with values ranging between 140 and 170 nm.[1] The highest quality films exhibit the shortest penetration depths.

The penetration depth λ_L has been measured by muon spin rotation (μSR),[2] mutual inductance[3] and self-inductance.[4] Sridhar *et al.*[5] have measured the temperature dependence of the penetration depth λ_{ab} at an rf frequency of 6 MHz for flux parallel to the *ab*-plane of a single crystal of $YBa_2Cu_3O_x$. The crystal was placed in the resonant coil of a highly stable tunnel-diode oscillator and fractional shift in oscillator frequency was taken to be proportional to the shift in λ_{ab}. The variation in λ_{ab} can be fit near T_c to weak-coupling theory with $\Delta(0) = 1.76\ k_BT_c$. By assuming instead $\Delta(0) = 2.15\ k_BT_c$, the variation with temperature of λ_{ab} can be fit taking $\lambda_{ab}(0) = 1400$ Å .

The admittivity, which is the reciprocal of the specific impedance is

$$y = 1/z = -i\omega\varepsilon = -i\omega\varepsilon_{core} + inq^2/\omega m = -i\omega\varepsilon_{core} + i/\omega\ell \tag{15}$$

where

$$\ell = m/nq^2 = \mu\lambda_L^2 \tag{16}$$

is the specific inductance of the superconducting electrons. The specific impedance z is represented in Fig. 1. The IS unit of ℓ is the Henry-meter. This inductance arises from the inertia of the carriers and is often called kinetic inductance. From Eqs. (8), (15) and (16) the surface impedance, neglecting ε_{core}, is given by the familiar expression that connects the surface reactance with the penetration depth

$$Z_S = \sqrt{-i\omega\mu z} = -i\omega\sqrt{\mu\ell} = -i\omega\mu\lambda_L \tag{17}$$

A number of investigators have measured the reflectivity and transmissivity of thin high-T_c superconducting films in space. Ho *et al.*[6,7] have measured the transmission and reflection coefficients at 60 GHz. Volkov *et al.*[8] have measured transmission in the submillimeter band from 8 to 32 cm^{-1}.

Nichols *et al.*[9] have measured transmission at 9 GHz through a film placed across a

---

[1] N. Klein, H. Chaloupka, G. Müller, S. Orbach, H. Piel, B. Roas, L. Schultz, U. Klein, and ? Peiniger, *J. Appl. Phys.* **67**, 6940 (1990).

[2] B. Pümpin, H. Keller, W. Kündig, W. Odermatt, I. M. Savic, J. W. Schneider, H. Simmler, P. Zimmermann, J. G. Bednorz, Y. Maeno, K. A. Müller, C. Rossel, E. Kaldis, S. Rusiecki, W. Assmus and J. Kowalewski, *Physica C* **162-164**, 151 (1989).

[3] Ph. Flueckiger, J.-L. Galivano, Ch. Leemann, P. Martinoli, B. Dam, G. M. Stollmann, P. K. Srivastava, P. Debely and H. E. Hintermann, *Physica C* **162-164**, 1563 (1989).

[4] P. L. Gammel, A. F. Hebard, C. E. Rice and A. F. J. Levi, *Physica C* **162-164**, 1565 (1989).

[5] S. Sridhar, D.-H. Wu and W. Kennedy, *Phys. Rev. Lett.* **63**, 1873 (1989).

[6] W. Ho, P. J. Hood, W. F. Hall, P. Kobrin, A. B. Harker and R. E. DeWarmes, *Phys. Rev. B* **38**, 7029 (1988).

[7] P. H. Kobrin, W. Ho, W. F. Paul, P. J. Hood, I. S. Gergis and A. B. Harker, *Physica C* **162-164**, 877 (1989).

[8] A. A. Volkov, B. P. Gorshunov, G. V. Kozlov, S. I. Krasnosvobodtsev, E. V. Pechen', O. I. Sirotinskii and Ya. Pettselt, *Sov. Phys. JETP* **68**, 148 (1989).

[9] C. S. Nichols, N. S. Shiren, R. B. Laibowitz and T. G. Kazaka, *Phys. Rev. B* **38**, 11970 (1988).

waveguide. Miranda *et al.*[10,11] have made similar measurements in the 26.5 to 40.0 GHz frequency range. Wijeratne *et al.*[12] have measured transmission at 80 GHz through a bulk sample placed in a waveguide and Tateno and Masaki[13] have measured the reflection at 9 GHz from a bulk sample terminating a waveguide.

Because of the high temperature-dependent permittivity of $SrTiO_3$, the surface resistance of moderately absorbing films is sensitive to resonant modes in the substrate.[14,15]

9.1.3. Two-fluid surface impedance

The two-fluid model, originated by Gorter and Casimir,[16] provides a phenomenological description of the electrodynamics of a superconductor over a wide frequency and temperature range. The transport current is taken to be the sum of contributions from two groups of carriers, those that behave normally and those that satisfy London electrodynamics[17] with

$$\ell_s \mathbf{J}_s = -\mathbf{A} \tag{18}$$

where ℓ_s is the kinetic inductivity of the superconducting carriers, $\mathbf{J}_s$ is the superconducting current density and $\mathbf{A}$ is the vector potential in the London gauge. Taking the time-derivative of Eq. (18) yields the acceleration equation

$$-\,d\mathbf{A}/dt = \mathbf{E} = \ell_s\, d\mathbf{J}_s/dt \tag{19}$$

It is customary to decribe the acceleration of normal carriers with the Drude equation[18]

$$\ell_n\, d\mathbf{J}_n/dt + \rho_n \mathbf{J}_n = \mathbf{E} \tag{20}$$

where ℓ_n is the kinetic inductivity of the normal carriers, ρ_n is their resistivity and $\mathbf{J}_n$ is the normal current density.

The specific impedance of the superconductor is represented by the circuit shown in Fig. 1. Normal current flows through the branch with ℓ_n and ρ_n in series and displacement current through the capacitance c_d. Superconducting current flows through ℓ_s. The specific impedance, neglecting c_d, is

[10]F. A. Miranda, W. L. Gordon, K. B. Bhasin, V. O. Heinen, J. D. Warner and G. J. Valco, NASA TM-102345 (1989). To be published in *Superconductivity and Applications* (Plenum, New York, 1990).

[11]F. A. Miranda, K. B. Bhasin, V. O. Heinen, R. Kwor and T. S. Kalkur, *Physica C* **168**, 91 (1990).

[12]A. T. Wijeratne, G. L. Dunifer, J. T. Chen, L. E. Wenger and E. M. Logothetis, *Phys. Rev. B* **37**, 615 (1988).

[13]J. Tateno and N. Masaki, *Jpn. J. Appl. Phys.* **26**, L1654 (1987).

[14]N. Klein, G. Müller, S. Orbach, H. Piel, H. Chaloupka, B. Roas, L. Schultz, U. Klein and M. Peiniger, "Millimeter wave surface resistance and London penetration depth of epitaxially grown $YBa_2Cu_3O_{7-x}$ thin films," *Physica C* **162-164**, 1549 (1989).

[15]L. Drabeck, K. Holczer, G. Grüner and D. J. Scalapino, "Ohmic and radiation losses in superconducting films," *J. Appl. Phys.* **68**, 892 (1990).

[16] C. J. Gorter and H. B. G. Casimir, *Phys. Z.* **35**, 963 (1934).

[17]F. London, F. and H. London, H.: *Proc. Roy. Soc. A* **149**, 71 (1935).

[18]P. Drude, *Ann. Phys.* **1**, 566 and **3**, 369 (1900).

$$z = \frac{-i\omega\ell_s(\rho_n - i\omega\ell_n)}{\rho_n - i\omega(\ell_s + \ell_n)} \tag{21}$$

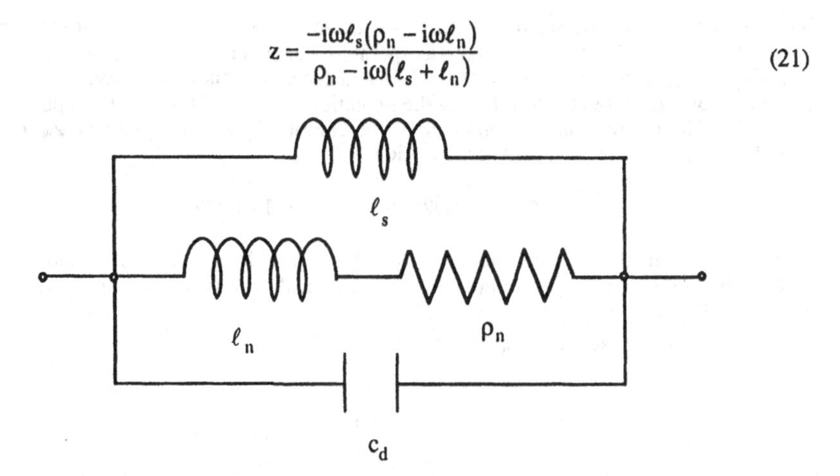

Figure 1. The two-fluid model is represented with normal transport characterized by a resistivity ρ_n and an inductivity ℓ_n and superconducting transport by an inductivity ℓ_s. Displacement current is represented by the capacitance c_d.

At sufficiently low frequency the specific impedance is approximately

$$z \approx -i\omega\ell_s + \omega^2\ell_s^2 / \rho_n \tag{22}$$

and is independent of ℓ_n, which enters only in higher order.[19] To the same order of approximation the surface impedance from Eq. 8 is

$$Z_s \approx -i\omega\sqrt{\mu\ell_s} + \frac{\omega^2\sqrt{\mu\ell_s^3}}{2\rho_n} \tag{23}$$

The surface reactance is seen to increase linearly with frequency while the surface resistance is a quadratic function of frequency.

Mattis and Bardeen[20] have analyzed the temperature and frequency dependence of the surface resistance for the nonlocal case. Halbritter[21] has obtained from BCS theory a surface resistance

$$R_s = (C/T)\omega^{3/2} \exp[-\Delta(T)/k_B T] \tag{24}$$

for $\hbar\omega$ less than the gap frequency $\Delta(T)$ and at temperatures less than $\frac{1}{2}T_c$. This theory

[19]K. F. Renk, B. Gorshunov, J. Schützmann, A. Prückl, B. Brunner, J. Betz, S. Orbach, N. Klein, G. Müller and H. Piel, *Europhys. Lett.* **15**, 661 (1991).

[20]D. C. Mattis and J. Bardeen, *Phys. Rev.* **111**, 412 (1958).

[21]J. Halbritter, Z. *Physik* **266**, 209 (1974).

has been modified by Blaschke[22,23] for cases where the gap is anisotropic or smeared.

Sridhar *et al.*[5,6] have measured the temperature dependence of the surface resistance R_s of a single crystal of $YBa_2Cu_3O_x$ at 10 GHz in a Nb superconducting cavity. The surface resistance was reported to drop below the detection limit of 400 $\mu\Omega$ for temperatures 5 K below T_c. The temperature dependence of the complex surface impedance Z_s may be fit near T_c by using the local-limit approximation

$$Z_s = R_n\{(1/2i)[1 + i\,\delta_n^2/2\lambda(T)^2]\}^{-1/2} \tag{25}$$

with the experimentally determined penetration depth for flux in the plane $\lambda_{ab}(0) = 1400$ Å and a skin depth $\delta_n = 2.5$ μm calculated in the classical limit from the normal-state surface resistance R_n.

For $\rho << \omega\ell$ the surface impedance is

$$Z_s \approx (1 - i)\sqrt{\frac{\omega\mu\rho}{2}} \tag{26}$$

which is the classical expression for the surface impedance of a normal conductor with

$$R_s = X_s = \frac{1}{2}\omega\mu\delta \tag{27}$$

for the resistive part of the surface impedance and a skin depth

$$\delta = \sqrt{\frac{2\rho}{\omega\mu}} \tag{28}$$

9.1.4. Quasiparticle conductivity

Classical BCS superconductors show evidence in both their acoustic and electromagnetic properties of a "coherence peak" just below T_c. There has been considerable interest in whether the high-temperature superconductors exhibit such a coherence peak, particularly since (see Sec. 5.2) there is no evidence of such a peak in longitudinal nuclear relaxation.

Experimental

Recent experiments find little evidence for a coherence peak in the conductivity *just* below T_c but do find a broad maximum in conductivity *well* below T_c. The observed maximum is believed to arise from a gradual suppression of quasiparticle scattering with decreasing temperature.

Nuss *et al.*[24] have determined both the real and imaginary parts of the sub band-gap conductivity in $YBa_2Cu_3O_7$ in the 0.5 to 2.5 THz range as shown in Fig. 2. This data was obtained from a measurement with coherent time-domain spectroscopy of the power transmitted through thin high-quality films of $YBa_2Cu_3O_7$. Electromagnetic far-infrared radiation is generated and detected by time-resolved optoelectronic techniques. A 50 μm

[22] R. Blaschke, J. Ashkenazi, O. Pictet, D. D. Koelling, A. T. v. Kessell and F. M. Müller, *J. Phys. F* **14**, 175 (1984).

[23] R. Blaschke and R. Blocksdorf, *Z. Phys. B* **49**, 99 (1982).

[24] M. C. Nuss, P. M. Mankiewich, M. L. O'Malley, E. H. Westerwick and P. B. Littlewood, *Phys. Rev. Lett.* **66**, 3305 (1991).

photoconductively switched dipole transmitting antenna emits a short pulse with a broad frequency spectrum from dc to almost 2.5 THz when a fast-rising 100 fs optical pulse from a colliding-pulse mode-locking dye laser switches the dipole antenna, which is dc-biased. A photoconductive dipole receiving antenna is switched by a delayed pulse, sampling the temporal shape of the electromagnetic pulse transmitted through the superconducting film. The frequency response of the superconductor is obtained by Fourier transformation of the time-domain data. The reactive component of the conductivity σ_2 rises below T_c from a small value at the transition. Below 50 K, σ_2 decreases with frequency as $1/\omega$. The absorptive component σ_1, starting from the transition, rises with decreasing temperature with a maximum around 70 K and decreases at lower temperatures. The magnitude of the maximum decreases with increasing frequency. The authors suggest that the maximum observed in σ_1 results from a competition between an increasing mean free path and a decreasing density of states in decreasing temperature. Smearing of the density of states around the gap energy 2Δ wipes out the coherence peak observed in conventional superconductors.

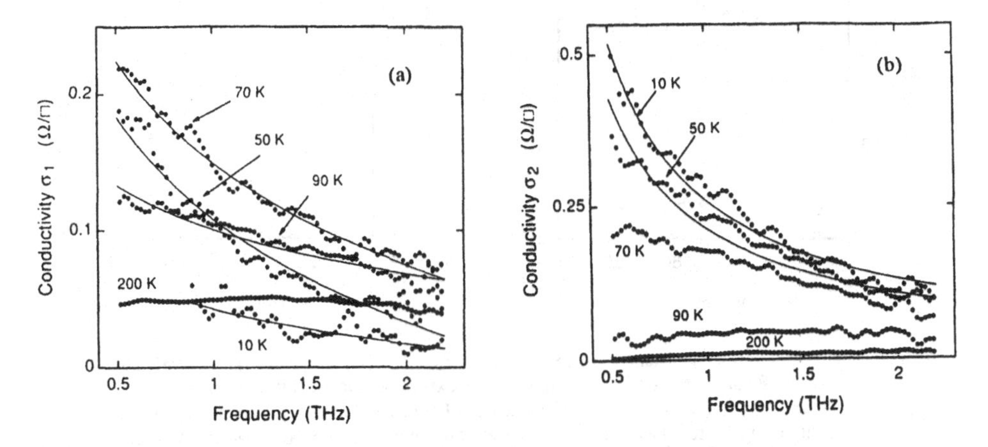

Figure 2. Conductivity of YBaCuO from time-domain terahertz spectroscopy [24]. (a) Real part σ_1 of the conductivity. Solid lines are guides to the eye. (b) Imaginary part σ_2 of the conductivity. The solid lines at the two lowest temperatures are fits to a $1/\omega$ frequency dependence.

Holczer *et al.*[25] have observed at 60 GHz and just below T_c a sharp peak in σ_1 of $Bi_2Sr_2CaCu_2O_8$ single crystals. The peak is attributed to BCS coherence effects. Olsson and Koch[26] have been able to fit the data of Holzer *et al.* as well as their own rf data with a broadened transition. Kobrin *et al.*[27] have also found in the σ_1 of $YBa_2Cu_3O_7$ films just below T_c a peak-like structure, which has been interpreted by Glass and Hall[28] as arising from two effects: (i) an enhancement in conductivity arising from sample inhomogeneity followed by (ii) supression of the expected coherence because of the increase in the normal carrier mean free path l over the superconducting carrier coherence length ξ_0.

[25]K. Holczer, L. Forro, L. Mihály and G. Grüner, *Phys. Rev. Lett.* **67**, 152 (1991).

[26]H. K. Olsson and R. H. Koch, *Phys. Rev. Lett.* **68**, 2406 (1992).

[27]P. H. Kobrin, W. Ho, W. F. Hall, P. J. Hood, I. S. Gergis and A. B. Harker, *Phys. Rev.* **42**, 6259 (1990).

[28]N. E. Glass and W. F. Hall, *Phys. Rev. B* **44**, 4495 (1991).

Romero *et al.*[29] have determined transmittance of $Bi_2Sr_2CaCu_2O_8$ ($T_c = 82$ K) to far infrared synchrotron radiation and fit their data to a conductivity similar to that obtained by Nuss *et al.*[27]

Bonn *et al.*[30] have measured the microwave surface resistance at 2.95 GHz of a small high quality single crystal of $YBa_2Cu_3O_7$ in the superconducting split-ring resonator described in Ch. 8 with the results shown in Fig. 3.. After a sharp drop in R_s by a factor of 5×10^3 at T_c, the surface resistance rises to a maximum at 35 K and then falls at lower temperature. This peak is believed to arise from suppression of the scattering of quasiparticles at reduced temperature in agreement with Nuss *et al.* and Romero *et al.*

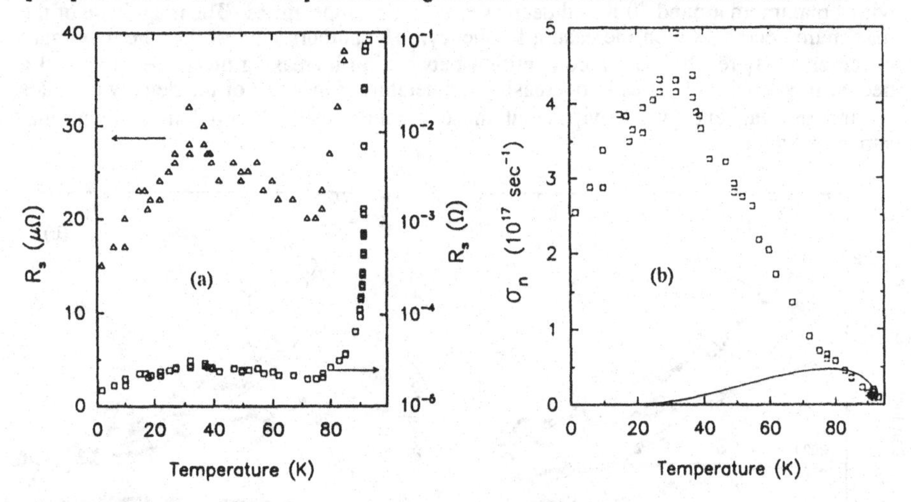

Figure 3. Surface resistance at 2.95 GHz of a sub-millimeter single crystal of YBaCuO [30]. (a) R_s drops below T_c more rapidly than expected from BCS theory—nearly four orders of magnitude by 80 K. (b) The real part of the conductivity σ_1 extracted from the measured surface resistance. The solid curve is the expected BCS coherence peak in the absence of damping.

Theoretical

The absence of an NMR coherence peak, as discussed in Sec. 5.2.2 has been attributed to strong pair-breaking, which smears the singular BCS density of states.[31,32,33] Many of the unusual normal-state properties of the high-temperature superconductors have been attributed to strong scattering from low-energy spin and charge fluctuations with a flat

---

[29] D. B. Romero, C. D. Porter, D. B. Tanner, L. Forro, D. Mandrus, L. Mihály, G. L. Carr and G. P. Williams, *Phys. Rev. Lett.* **68**, 1590 (1992).

[30] D. A. Bonn, P. Dosanjh, R. Liang and W. N. Hardy, *Phys. Rev. Lett.* **68**, 2390 (1992).

[31] Y. Kuroda and C. M. Varma, *Phys. Rev. B* **42**, 8619 (1990).

[32] L. Coffey, *Phys. Rev. Lett.* **64**, 1071 (1990).

[33] F. Marsiglio, *Phys. Rev. B* **44**, 5373 (1991).

spectrum extending to energies of the order of k_BT.[34,35] In the superconducting state these fluctuations are expected to develop a gap of order 2Δ, suppressing the low-energy scattering of quasi-particles. As a result, the quasiparticle mean free path should rapidly increase below T_c.

9.2. Flux Dynamics

We begin this section by reviewing studies of flux penetration at low frequencies. The following sections consider the contributions of flux flow and dynamic creep to the high-frequency surface impedance. In the final section we review microwave surface impedance experiments in high magnetic fields.

9.2.1 Bean critical state

Hysteresis

Kamper,[36] Bean[37,38] and H. London[39] have observed and interpreted hysteretic magnetic loss in type II superconductors for cyclic magnetic fields exceeding the critical field H_{c1}. The basic idea is that at magnetic fields above H_{c1} flux enters a type II superconductor in quanta that take the form of current vortices. These vortices are pinned by defects and are depinned by the magnetic pressure that results from the current associated with a gradient in the flux density. For a pinning potential independent of induction, the flux density should decrease linearly into the material in increasing fields and increase linearly into the material in decreasing fields. This difference leads to hysteresis in the magnetization of a type II superconductor and to work around the cycle of an alternating field.

Bean calculated that so long as flux does not entirely penetrate the sample, the magnetic work performed around a cycle of the magnetic field is

$$W = {}^2\!/_3\, \mu H_s^3/J_c \qquad (29)$$

It is usual at ac frequencies to relate the absorption to the imaginary part of a complex magnetic permeability $\mu(\omega)$. Because we are primarily concerned with high frequency studies, we choose instead to describe the surface power absorption at frequency ω in terms of a surface resistance

$$P = \frac{\omega}{2\pi} W = {}^1\!/_2\, H_s^2 R_s \qquad (30)$$

which leads to the expression for the critical surface resistance

$$R_S = \frac{2\omega}{3\pi} \frac{\mu H_s}{J_c} \qquad (31)$$

---

[34] C. M. Varma, P. B. Littlewood, S. Schmitt-Rank, E. Abrahams and A. E. Ruckinstein, *Phys. Rev. Lett.* **63**, 1996 (1989); **64**, 497(E) (1990).

[35] P. B. Littlewood and C. M. Varma, *J. Appl. Phys.* **69**, 4979 (1991).

[36] R. A. Kamper, *Phys. Lett.* **2**, 290 (1962).

[37] C. Bean, *Phys. Rev. Lett.* **8**, 250 (1962).

[38] C. Bean, *Rev. Mod. Phys.* **36**, 31 (1964).

[39] H. London, *Phys. Lett.* **6**, 162 (1963).

Extensive studies of the magnetization of the high-temperature superconductors establish the existence of bulk critical-state phenomena with critical current densities that may be increased by the controlled introduction of defects.

Field dependence

Anderson and Kim[40] have considered the effect of flux density on the critical current associated with the uniform motion of flux into or out of a superconductor. With increased flux density the critical current should decrease because of the averaging of pinning forces over a flux distribution. In particular, Anderson and Kim fit the critical current of a cylindrical shell with the form

$$J_c(B) = \frac{J_c(0)}{1 + B / B_0} \tag{32}$$

The effect of a dc magnetic field on ac critical behavior has been treated by Campbell and Evetts.[41] They find that coupling between flux vortices leads to a threshold field H' for hysteretic behavior. For H_{ac} less than H', magnetic processes are reversible and no work is done around a cycle. For H_{ac} greater than H', the work done around a cycle increases as H_{ac}^3. The field H' arises because forces between vortices permit limited increases or decreases in flux concentration without irreversible processes in the pinning potential. As the applied magnetic field increases, the forces between vortices becomes increasingly important and H' is expected to increase.

Transients

For the development of a critical state, flux vortices must be nucleated or denucleated and then flow respectively into or out of the sample. Bean[42] has recently discussed the development of a critical state in response to a step in magnetic field. He finds using values appropriate to $YBa_2Cu_3O_7$ that the relaxation time when a field is stepped from 0 to 100 Oe is of the order of 1 μsec. Since nucleation times are expected to be shorter than the relaxation time that Bean obtains, the development of a bulk critical state in the high-temperature superconductors is presumably limited by viscous damping of flux flow, as considered in Sec. 6.3.2.

9.2.2. Flux pinning and flow

We consider next the contribution of flux-flow to the surface impedance of superconductors. This problem was examined early for classical type II superconductors.

---

[40] P. W. Anderson and Y. B. Kim, *Rev. Mod. Phys.* **36**, 39 (1964).

[41] A. M. Campbell and J. E. Evetts, *Adv. Phys.* **21**, 199 (1972); reprinted as *Critical Currents in Superconductors*, (Taylor and Francis, London, 1972).

[42] C. P. Bean in *Superconductivity and Applications*, eds H. S. Kwok, Y.-H. Kao and D. T. Shaw (Plenum Press, New York, 1990) p 767.

[43,44,45] Based on the pioneering work of Martinoli *et al.*,[46] Coffey and Clem[47,48,49,50] have recently undertaken an exhaustive study of flux-flow loss, including as well a careful consideration of flux pinning and creep.

We take the surface of a superconductor to be in the xy-plane with z-positive directed into the superconductor. The static flux density B and an rf magnetic field H_s are along the y-direction. The rf current flows in the x-direction with flux driven into and out of the sample along the z-direction as sketched in Fig. 4.

The equation of motion of flux vortices at low temperatures is

$$m\frac{d^2u}{dt^2} + \eta\frac{du}{dt} + \kappa u = \Phi_0 J \tag{33}$$

where m is the mass per unit length of vortices, η is the viscosity per unit length and κ is

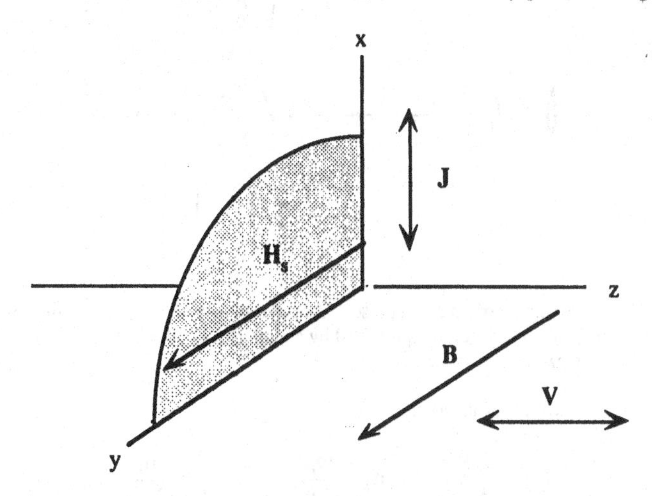

Figure 4. Principal directions for flux-flow. The surface of a superconductor is in the xy-plane with positive-z directed into the superconductor. The static flux density B and an rf magnetic field H_S are along the y-direction. The rf current flows along the x-direction with flux driven into and out of the sample along the z-direction.

the spring-constant that binds vortices to pinning centers. Operating frequencies ω are well below η/m and the inertial term is usually dropped. The electric field in the superconductor is the sum of two terms, one from the kinetic inductance and the other from flux-flow

---

[43]M. Cardona, G. Fischer and B. Rosenblum, *Phys. Rev. Lett.* **12**, 101 (1964).

[44]M. Cardona, J. I. Gittleman and B. Rosenblum, *Phys. Lett.* **17**, 92 (1965).

[45]J. I. Gittleman and B. Rosenblum, *Proc. IEEE* **52**, 1138 (1964); *Phys. Rev. Lett.* **16**, 7345 (1966); *J. Appl. Phys.* **39**, 2617 (1968).

[46]P. Martinoli, Ph. Flückiger, V. Marisco, P. K. Srivastava, Ch. Leemann and J. L. Gavilano, *Physica B* **165 & 166**, 1163 (1990).

[47]M. W. Coffey and J. R. Clem, *IEEE Trans. Magn.* **27**, 2136, 4396(E) (1991).

[48]J. R. Clem and M. W. Coffey, *Physica C*, **185-189**, 1915 (1991).

[49]M. W. Coffey and J. R. Clem, *Phys. Rev. Lett.* **67**, 386 (1991).

[50]M. W. Coffey and J. R. Clem, *Phys. Rev. B* **45**, 9872, 10 527 (1992).

$$E = \ell_s \frac{dJ}{dt} + B\frac{du}{dt} \tag{34}$$

where ℓ_s is the kinetic inductivity of the supercarriers and normal current is neglected. Eqs. (33) and (34) may be interpreted with the *ansatz* that Eq. (33) describes the decomposition of current through three circuit elements connected in parallel, leading to the specific impedance of the superconductor represented in Fig. 5. The supercarriers are represented by the inductivity ℓ_s. The flux-pinning contribution to the impedance is represented by an inductor ℓ_f with damping by a resistor ρ_f and inertia by a parallel capacitor c_f.

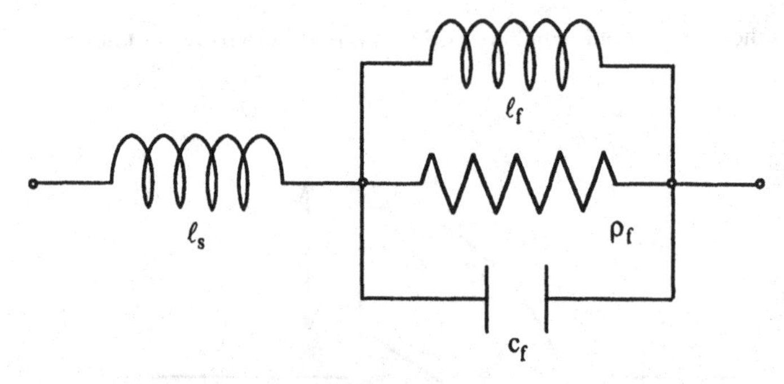

Figure 5. The supercarriers are represented by the inductivity ℓ_s. The flux-pinning contribution to the impedance is represented by an inductor ℓ_f, damping by a resistor ρ_f, and inertia by a parallel capacitor c_f.

The values of the vortex parameters are

$$\ell_f = \frac{\Phi_0 B}{\kappa} \qquad \rho_f = \frac{\Phi_0 B}{\eta} \qquad c_f = \frac{m}{\Phi_0 B} \tag{35}$$

and the specific impedance, neglecting c_f is

$$z_s = -i\omega\ell_s + \frac{\rho_f}{1 + i\omega_f / \omega} \tag{36}$$

where $\omega_f = \rho_f/\ell_f = \kappa/\eta$ is the flux crossover frequency. The circuit representation of the flux-flow impedance shown in Fig. 5 is analogous to the circuit representation of the two-fluid model shown in Fig. 1.

9.2.3. Dynamic flux creep

As the sample temperature approaches T_c, two changes take place in the impedance represented by Fig. 5. The first change is that quasi-normal excitations develop within the grains, leading to normal current flow in shunt with the supercurrent as described by the two-fluid model in Sec. 9.1.3. The second change may be unique to flux flow. At elevated temperature, flux diffusion becomes thermally activated with vortices free to creep under the driving force of the current. Flux creep is detected, however, only at frequencies low compared with a flux-creep crossover frequency ω_c. At frequencies above ω_c vortices are pinned by the inductivity ℓ_f in Fig. 5. This problem has been carefully considered by Coffey and Clem. A strictly phenomenological representation of their theory may be obtained by modifying Eq. (33) to

$$m\frac{d^2u}{dt^2}+\eta\frac{du}{dt}+(d/dt+\omega_c)^{-1}\kappa\frac{du}{dt}=\Phi_0 J \tag{37}$$

with both κ and ω_c functions of temperature. At frequencies $\omega >> \omega_c$, Eq. (37) reverts to Eq. (33) as required. At frequencies $\omega << \omega_c$, on the other hand, Eq. (37) becomes

$$m\frac{d^2u}{dt^2}+(\eta+\kappa/\omega_c)\frac{du}{dt}=\Phi_0 J \tag{38}$$

and the vortices move freely with viscosity $\eta + \kappa/\omega_c$. A circuit representation of Eq. (37) is given in Fig. 6 by inserting a creep resistivity

$$\rho_c = \frac{\Phi_0 B}{\kappa/\omega_c} \tag{39}$$

in series with the pinning inductivity ℓ_f.

9.2.4. *Flux-flow surface impedance*

The flux-flow contribution to the specific impedance is shown in Fig. 6 and (neglecting inertia) is of the form obtained by Coffey and Clem

$$z_f = \frac{\rho_f(\rho_c - i\omega\ell_f)}{\rho_f+\rho_c-i\omega\ell_f} = \frac{\varepsilon - i\omega\tau}{1-i\omega\tau}\rho_f \tag{40}$$

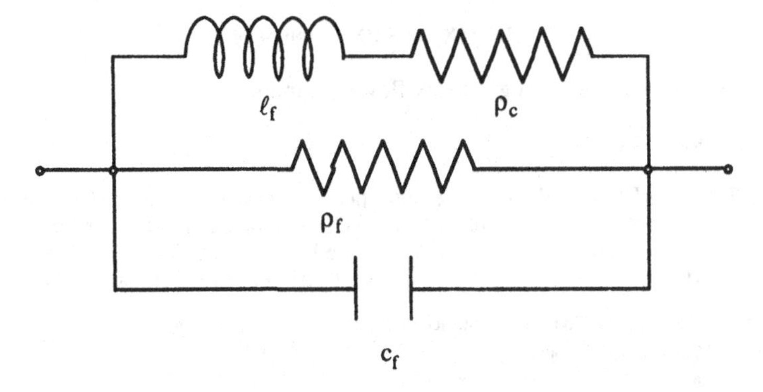

Figure 6. Dynamic flux creep is modeled by inserting a creep-resistivity ρ_c in series with the pinning inductivity ℓ_f.

with the phenomenological flux-creep factor $\varepsilon = \rho_c/(\rho_c + \rho_f) = (1 + \kappa/\eta\omega_c)^{-1}$ and a phenomenological relaxation time given by $\tau = \ell_f/(\rho_c + \rho_f) = \varepsilon/\omega_c$. At sufficiently low magnetic fields z_f is much smaller than z_s (the impedance of the carriers) and the surface impedance of a homogeneous superconductor is to lowest order in z_f

$$Z_s = \sqrt{-i\omega\mu(z_s + z_f)} = \sqrt{-i\omega\mu z_s} + \frac{1}{2}\sqrt{\frac{-i\omega\mu}{z_s}}\, z_f \qquad (41)$$

At sufficiently high magnetic field and frequency, flux flow dominates the specific impedance of the sample and the surface impedance is that of an inductivity and a resistivity in series

$$Z_S = \sqrt{-\omega\mu z} = \sqrt{-i\omega\mu(\rho - i\omega\ell)} \qquad (42)$$

In order to examine the response of a sample for flux-flow loss, it is convenient to introduce the loss-tangent $\rho/\omega\ell = \tan\phi$ and to write the surface impedance as

$$Z_S = -iR_c\sqrt{\frac{2}{\sin\phi}}\, e^{i\phi/2} \qquad (43)$$

with the surface reactance and resistance

$$X_S = R_c\sqrt{\cot\phi/2} \qquad (44)$$

$$R_S = R_c\sqrt{\tan\phi/2} \qquad (45)$$

with the classical surface resistance given by $R_c = (\omega\mu\rho/2)^{1/2}$. The components of the surface impedance R_S and X_S are plotted in Fig. 7 as functions of ϕ.

R_S is plotted against X_S in Fig. 8. Because there may be other magnetically induced contributions to Z_S, it is useful to plot as well the ratio of the change in R_s to that in X_s when the magnetic field is changed. So long as both η and κ are independent of magnetic field [Eq. (35)], field changes leave the loss-tangent unchanged with

$$dR_S/dX_S = R_s/X_s = \tan {}^1/_2\,\phi \qquad (46)$$

against X_S as a way of identifying the flux-flow contribution.

9.2.5. Microwave studies

Marcon, Giura *et al.*[51,52,53,54,55,56] have undertaken an extensive study of the microwave surface impedance of ceramic superconductors in high magnetic fields over a wide range of temperature. We limit this review to studies above H_{c1} (of the order of 100 Oe), where vortices are expected to be nucleated in bulk. Discussion of low-field effects is postponed

[51] R. Marcon, R. Fastampa, M. Giura and C. Matacotta, *Phys. Rev. B* **39**, 2796 (1989).

[52] M. Giura, R. Marcon and R. Fastampa, *Phys. Rev. B* **40**, 4437 (1989).

[53] M. Giura, R. Fastampa, R. Marcon and E. Silva, *Phys. Rev. B* **42**, 6228 (1990).

[54] R. Marcon, R. Fastampa and M. Giura, *Europhys. Lett.* **11**, 561 (1990).

[55] R. Marcon, R. Fastampa, M. Giura and E. Silva, *Phys. Rev. B* **43**, 2940 (1991).

[56] M. Giura, R. Marcon, R. Fastampa and E. Silva, *Phys. Rev. B* **45**, 7387 (1992).

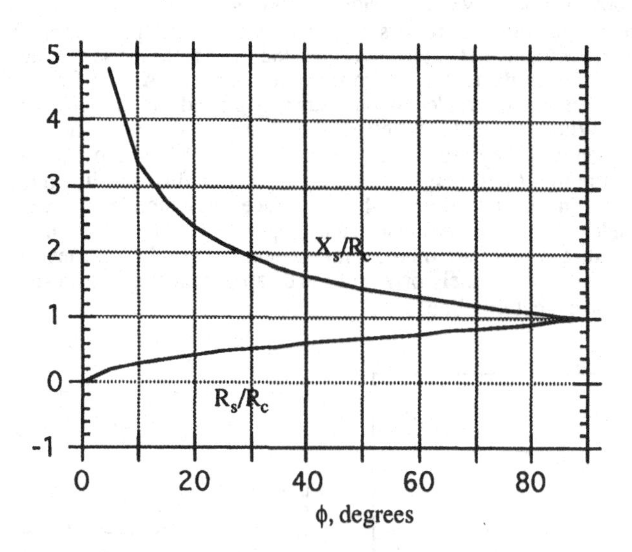

Figure 7. Components of the relative surface impedance Z_S/R_C as functions of ϕ.

to Ch. 12 because of the strong connection of these effects with granularity.[49,50] Marcon *et al.*[53] have convincingly demonstrated the presence of microwave loss from flux flow in samples of Y-Ba-Cu-O and Bi-Sr-Ca-Cu-O.

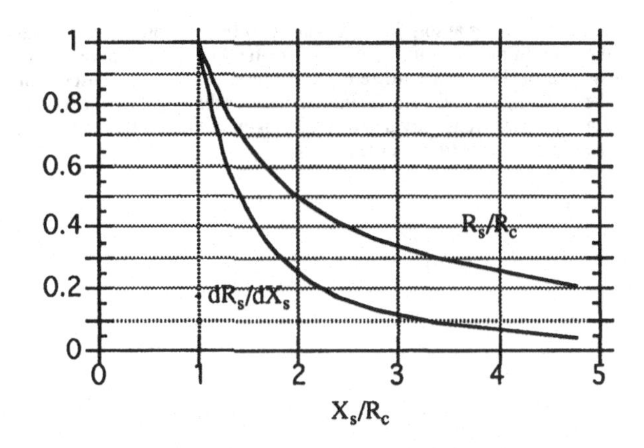

Figure 8. Plot of the surface resistance R_S against the surface reactance X_S.

Giura *et al.*[54] report hysteresis in microwave surface resistance at 48 GHz as shown in Fig. 9 when a magnetic field is cycled at temperatures up to 50 K, above which temperature the effect disappears. Although one might expect that the losses would be higher in decreasing fields because of the larger amount of contained flux, just the reverse is observed. Similar observations have been made by Pakulis *et al.*[57]

Giura *et al.*[54] propose that their results are a consequence of flux-creep. When the magnetic field is increased above H_{c1}, those vortices that first enter the sample are weakly pinned and contribute strongly to the microwave absorption. As the field is further increased, vortices that enter the sample are more strongly pinned and the surface resistance grows more slowly. When the field is reversed, those vortices that were the first to enter are now the first to leave. They are weakly pinned and contribute most to the surface resistance. Thus, although there may be little hysteresis in flux density, there can be pronounced hysteresis in flux-flow loss with the surface resistance lower in decreasing than in increasing fields. We can expect that as the temperature is increased, flux-creep will set in with the result that vortices are mixed with no memory of whether they entered the sample early or late. This is the likely origin of the disappearance of hysteresis above 50 K, a sort of *dynamic* irreversibility line.

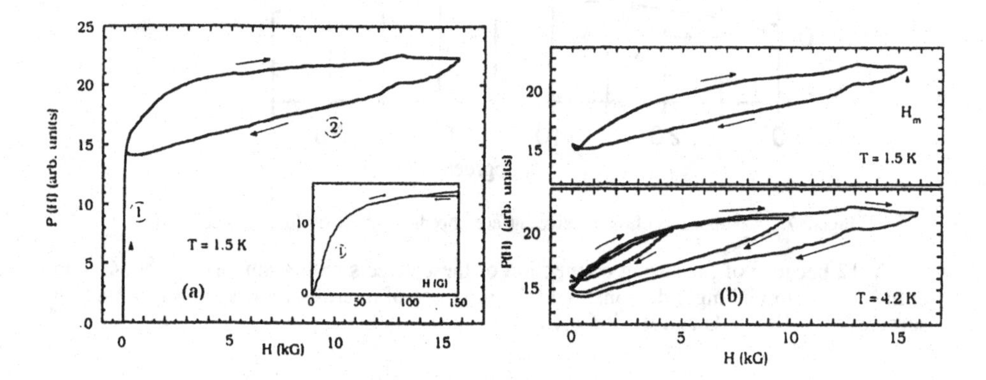

Figure 9. Microwave power absorption following zero-field cooling [54]. (a) Curve 1 is the power absorbed in the first run following zero-field cooling. Curve 2 is obtained when the magnetic field is reduced to zero. In the inset is shown the absorption at low fields. (b) The upper figure gives the absorption in runs subsequent to (a). The loops are reprodicible in subsequent runs provided that the maximum field is the same. The lower figure shows hysteresis loops for lower values of H_{max}.

[57] E. J. Pakulis and T. Osada, *Phys. Rev. B* **37**, 5940 (1988); E. J. Pakulis and G. V. Chandrashekhar, *Phys. Rev.* **39**, 808 (1989).

10
JOSEPHSON ELECTRODYNAMICS

10.1. Josephson Effects

The observation by Giaever[1] of electron-tunneling between superconductors and between a superconductor and a normal metal led to the discovery of the Josephson effects, which we consider in this chapter, and of macroscopic interference phenomena, which are treated in Ch. 15. In a remarkable series of theoretical contributions, B. D. Josephson[2,3,4,5] showed that the tunnelling current between superconductors should be of the form

$$J = J_{n1} + J_{n2} \cos \delta + J_0 \sin \delta \qquad (1)$$

with

$$\Phi_0 \frac{\partial \delta}{\partial t} = 2\pi V \qquad (2)$$

where δ is the difference in phase between superconductors and V is the voltage difference with the fluxoid $\Phi_0 = h/q$. The first two terms in Eq. (1) are the quasiparticle contributions

[1] I. Giaever, *Phys. Rev. Lett.* **5**, 147, 464 (1960).
[2] B. D. Josephson, *Phys. Lett.* **1**, 251 (1962) .
[3] B. D. Josephson, *Rev. Mod. Phys.* **34**, 216 (1964); *Advan. Phys.* **14**, 419 (1965).
[4] B. D. Josephson, "Weakly coupled superconductors" in *Superconductivity*, ed. R. D. Parks (Dekker, New York, 1969) pp. 423-447.
[5] B. D. Josephson, *Rev. Mod. Phys.* **46**, 251 (1974).

to the tunneling observed by Giaever. The third term was entirely new and unexpected, but within a year was observed by Anderson and Rowell.[6]

10.1.1. Current-phase relation

The Schrödinger wave function in the superconducting state, as we have seen in Ch. 3, is of the form

$$\Psi(\mathbf{r}) = \psi(\mathbf{r})\, e^{i\phi(\mathbf{r})} \tag{3}$$

with $\psi(\mathbf{r})$ the probability amplitude of a superconducting pair. As we have also seen, the phase of the wave function is

$$\phi(\mathbf{r}) = \int \mathbf{k} \cdot d\mathbf{r} \tag{4}$$

where $\mathbf{k}$ is the gradient of the phase and the total momentum $\hbar\, \mathbf{k}$ is given by

$$\hbar\, \mathbf{k}(\mathbf{r}) = m\, \mathbf{v}(\mathbf{r}) + q\, \mathbf{A}(\mathbf{r}) \tag{5}$$

with $q = -2e$ and $m = 2m_e$. To obtain the current-phase relation, we follow Feynman's treatment,[7] which uses the Schrödinger wavefunction and a transfer Hamiltonian.[8] We label the superconductors on the two sides of a junction as 1 and 2 with wavefunctions Ψ_1 and Ψ_2. The Schrödinger equations on the two sides are then

$$\mathcal{H}\, \Psi_1 = U_1 \Psi_1 + T\, \Psi_2 = i\hbar \frac{\partial \Psi_1}{\partial t} \tag{6}$$

$$\mathcal{H}\, \Psi_2 = U_2\, \Psi_2 + T\, \Psi_1 = i\hbar \frac{\partial \Psi_2}{\partial t} \tag{7}$$

The energies on opposite sides of the junction are U_1 and U_2. The parameter T is a transfer integral, which is a measure of coherent pair transfer through the junction. With a potential V across the junction, and the zero of energy taken at the middle of the junction, we have

$$i\hbar \frac{\partial \Psi_1}{\partial t} = \tfrac{1}{2} qV\, \Psi_1 + T\, \Psi_2 \tag{8}$$

$$i\hbar \frac{\partial \Psi_2}{\partial t} = -\tfrac{1}{2} qV\, \Psi_2 + T\, \Psi_1 \tag{9}$$

If we write $\Psi_1 = \psi_1 e^{i\phi_1}$ and $\Psi_2 = \psi_2 e^{i\phi_2}$ we obtain the equations with $\delta = \phi_2 - \phi_1$

$$\frac{\partial \Psi_1}{\partial t} = (T/\hbar)\, \psi_2 \sin \delta \tag{10}$$

---

[6] P. W. Anderson and J. Rowell, *Phys. Rev. Lett.* **10**, 230 (1963).

[7] Richard P. Feynman, Robert B. Leighton and Matthew Sands, *The Feynman Lectures on Physics, Quantum Mechanics*, Vol. III, (Addison-Wesley, Reading, MA, 1965.) secs. 21-1 through 21-9.

[8] J. Bardeen, *Phys. Rev. Lett.* **6**, 57 (1961); **9**, 147 (1962). For a discussion of the contributions of Bardeen and others to tunneling theory, see L. M. Falicov, *J. Supercond.* **4**, 331 (1991).

$$\frac{\partial \Psi_2}{\partial t} = -(T/\hbar)\, \psi_1 \sin \delta \tag{11}$$

$$\frac{\partial \phi_1}{\partial t} = (T/\hbar)(\psi_2/\psi_1) \cos \delta - \frac{\pi V}{\Phi_0} \tag{12}$$

$$\frac{\partial \phi_2}{\partial t} = (T/\hbar)(\psi_1/\psi_2) \cos \delta + \frac{\pi V}{\Phi_0} \tag{13}$$

Eqs. (10) and (11) lead by charge conservation arguments to the flow of pair current through the junction

$$J = J_0 \sin \delta \tag{14}$$

with the maximum pair current J_0 proportional to the transfer integral T.

10.1.2. ac Josephson effect

Eqs. (12) and (13) give Eq. (2) for the rate of change of the phase difference $\delta = \phi_2 - \phi_1$

$$\Phi_0 \frac{\partial \delta}{\partial t} = 2\pi V \tag{15}$$

which may be integrated to give for V constant

$$\delta(t) = \delta_0 + \frac{2\pi}{\Phi_0} Vt \tag{16}$$

Substituting Eq. (16) into Eq. (14) leads to an oscillating current at frequency $\omega = 2eV/\hbar$. For $V = 0$, there is a current $J = J_0 \sin \delta$ which can be anything between $-J_0$ and $+J_0$. We have the curious result that without a voltage the current is indeterminate[9] and applying a voltage leads to an oscillating current with zero average value![10]

Josephson inductance

From Eqs. (14) and (15), the rate of change of the total current I is

$$\frac{dI}{dt} = \frac{\cos \delta}{L_0} V \tag{17}$$

with $L_0 = \Phi_0/2\pi I_0$ where I_0 is the maximum Josephson current.. Eq. (17) has the form of an inductive response and suggests the expression for the Josephson inductance

$$L_J = \frac{L_0}{\cos \delta} \tag{18}$$

---

[9] J. E. Mercereau, "dc Josephson effect," Chapter 31 in *Tunneling Phenomena in Solids*, eds. E. Burstein and S. Lundqvist (Plenum, New York, 1969).

[10] D. N. Langenberg, "ac Josephson tunneling—experiment," Chapter 33 in *ibid.*

Inverse ac Josephson effect

The development of a dc voltage across an unbiased Josephson junction in the presence of rf radiation at frequency ν is called the inverse ac Josephson effect.[11,12] The voltage is a rational fraction of $\nu\Phi_0$ and provides a highly precise standard of voltage.[13]

10.1.3. Phase difference and vector potential

We now derive from Eq. (15) the relation between the phase *difference* and the vector potential in situations where V is established by changing magnetic flux. We begin by integrating the electric field under the assumption that there is no electrostatic component

$$\mathbf{E} = -\nabla\phi - \frac{\partial}{\partial t}\mathbf{A} \Rightarrow -\frac{\partial}{\partial t}\mathbf{A} \tag{19}$$

over a closed contour that crosses the junction twice at positions separated by Δx and is closed sufficiently far from the junction that the vector potential is zero. Under the assumption $\mathbf{E} = 0$ in the superconductor, we have the result

$$\Delta V = \Delta x \frac{dV}{dx} = \frac{\partial}{\partial t}\oint d\mathbf{r}\cdot\mathbf{A} = \Delta x \frac{d}{dx}\frac{\partial}{\partial t}\int_{-\infty}^{\infty} d\mathbf{r}\cdot\mathbf{A} \tag{20}$$

Substituting for δ from Eq. (15) and integrating gives for the phase difference across the junction with the contour integral in a plane orthogonal to the junction.

$$\Phi_0\delta = 2\pi\int_{-\infty}^{\infty}\mathbf{A}\cdot d\mathbf{r} \quad \text{or} \quad \Phi_0\frac{d\delta}{dx} = 2\pi\frac{d\Phi}{dx} \tag{21}$$

10.1.4. Josephson coupling

P. W. Anderson[14] has obtained for the maximum Josephson current at $T = 0$

$$I(0) = \frac{\pi\Delta(0)}{qR_n(0)} \tag{22}$$

where R_n is the normal-state tunneling resistance between conductors at 0 K. Ambegaokar and Baratoff[15] have obtained at elevated temperatures

$$I_0(T) = \frac{\pi\Delta(T)}{qR_n(T)}\tanh\frac{\Delta(T)}{2k_BT} \tag{23}$$

---

[11] D. N. Langenberg, D. J. Scalapino, B. N. Taylor and R. E. Eck, *Phys. Lett.* **20**, 563 (1966).

[12] J. T. Chen, R. J. Todd and Y. W. Kim, *Phys. Rev.* **B5**, 1843 (1972).

[13] J. Niemeyer, J. H. Hinken and R. L. Kautz, *IEEE Trans. Instrum. Meas.* **34**, 185 (1985).

[14] P. W. Anderson, Ravello Spring School, 1963.

[15] V. Ambegaokar and A. Baratoff, *Phys. Rev. Lett.* **10**, 486 (1963).

Expansion of Eq. (23) near T_c through Eq. (2-39) gives

$$I_0(T) \approx \frac{\pi}{2qk_BT_cR_n}\Delta^2(T) \approx 8.38\frac{\Delta(0)}{qR_n}(1-t) \tag{24}$$

The tunnelling of Cooper pairs gives rise to a coupling of the phases of the superconducting wave functions on the two sides of a junction as pointed out by P. W. Anderson and developed by Josephson. The coupling energy depends on the phase difference δ in the form

$$\mathcal{F}(\delta) = \hbar\frac{I_0}{q}(1-\cos\delta) = L_0I_0^2(1-\cos\delta) \tag{25}$$

The Josephson current is associated with the coupling through the relation

$$I = \frac{2\pi}{\Phi_0}\frac{\partial\mathcal{F}}{\partial\delta} = I_0\sin\delta \tag{26}$$

which is Eq. (14). (See the later discussion of Josephson systems in Sec. 15.5.5.)

The presence of normal-phase material within a junction suppresses the gap parameter $\Delta(T)$, leading near T_c to a reduced current density

$$J_0(T) \propto (1 - t)^2 \tag{27}$$

This reduction is accentuated by the short coherence length of the new superconductors as pointed out by Deutscher and Müller.[16]

10.2. Resistively Shunted Junction

10.2.1. Dynamic behavior

McCumber[17] and Stewart[18] have developed the theory of the resistively shunted Josephson junction shown in Fig. 1. This circuit is described by the equation

$$C_j\frac{dV}{dt} + \frac{V}{R_j} + I_0\sin\delta = I \tag{28}$$

Using Eq. (2), this equation may be rewritten as

$$L_0C_j\frac{d^2\delta}{dt^2} + \frac{L_0}{R_j}\frac{d\delta}{dt} + \sin\delta = \frac{I}{I_0} \tag{29}$$

with the small-current Josephson inductance from Eq. (18) $L_0 = \Phi_0/2\pi I_0$. (Note that C_j plays the role of a mass.) The complex oscillation frequency for small δ is

---

[16] G. Deutscher and K. A. Müller, *Phys. Rev. Lett.* **59**, 1745 (1987).

[17] D. E. McCumber, *J. Appl. Phys.* **39**, 2503, 3113 (1968).

[18] W. C. Stewart, *Appl. Phys. Lett.* **12**, 277 (1968).

$$\omega = \sqrt{\omega_J^2 - \frac{1}{4\tau^2}} - \frac{i}{2\tau} \tag{30}$$

with the Josephson plasma frequency $\omega_J = 1/\sqrt{L_0 C_j}$ and the energy relaxation time $\tau = R_j C_j$. Eqs. (2) and (22) lead to a maximum frequency for coherent oscillation

$$\omega_c = \frac{2\pi V}{\Phi_0} < \frac{\pi \Delta(0)}{\hbar} \tag{31}$$

For $Q = \omega_J \tau > 1/2$ the circuit executes oscillations whose energy decays with relaxation time τ. For $Q = \omega\tau < 1/2$ the circuit is overdamped and decays exponentially.

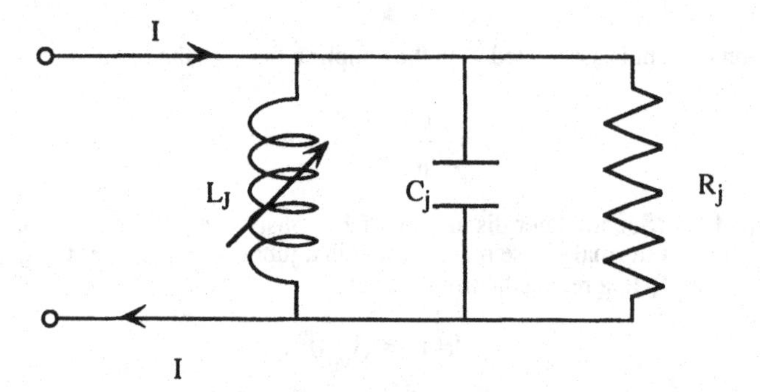

Figure 1. In the resistively shunted junction (RSJ) model the junction, represented by a variable inductance, is shunted by its capacitance and a normal resistance.

10.2.2. Current-voltage characteristic

The current-voltage characteristics of a high-Q junction are shown in Fig. 2 (a). When the input current exceeds I_0, the voltage jumps to $I_0 R$ and continues up the resistive branch with further increase in current. When the current is reduced, the circuit remains on the resistive branch down to zero voltage.

A low-Q junction is obtained by letting the shunt capacitance C_j (which plays the role of a mass) go to zero. The plasma frequency ω_J goes to infinity and the relaxation time τ goes to zero. The phase difference across the junction is described by the equation

$$\frac{L_0}{R_J}\frac{d\delta}{dt} + \sin\delta = \frac{I}{I_0} \tag{32}$$

When the current I exceeds I_0 the circuit undergoes relaxation oscillations at a frequency

$$\nu = \frac{\sqrt{(I/I_0)^2 - 1}}{2\pi}\frac{R_j}{L_0} = \sqrt{I^2 - I_0^2}\,\frac{R_j}{\Phi_0} \tag{33}$$

The *mean* voltage across a shunted junction carrying a constant current I

$$\langle V\rangle = v\Phi_0 = \sqrt{I^2 - I_0{}^2}\,R_j \tag{34}$$

is shown in Fig 2 (b).[19,20] In the low-Q limit there is no hysteresis and the current moves smoothly from the zero-voltage value onto the resistive branch.

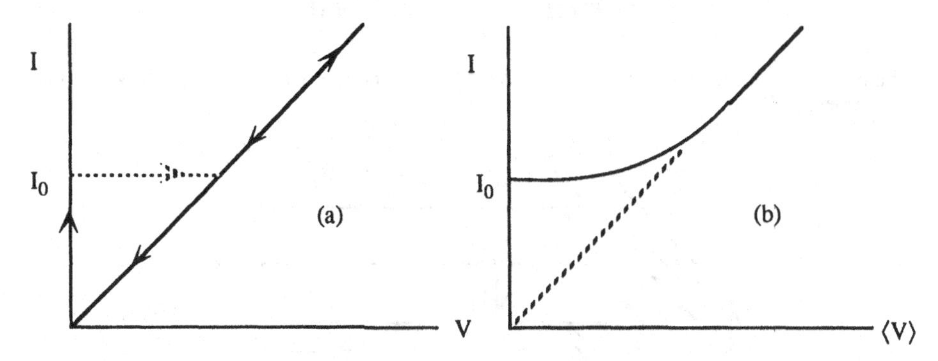

Figure 2. Current-voltage characteristics of a resistively shunted junction. (a) When the input current to a high-Q junction exceeds I_0, the voltage jumps to I_0R and continues up the resistive branch with further increase in current. When the current is reduced, the circuit remains on the resistive branch down to zero voltage. (b) The current in a low-Q junction as a function of the mean voltage. In this limit there is no hysteresis and the current moves smoothly from the zero-voltage value onto the resistive branch.

Junctions of intermediate Q show hysteresis between the limiting cases Figs. 2 (a) and (b).

10.3. Josephson Penetration

We consider the electrodynamics of the structure shown in Fig. 3 with two superconducting surfaces parallel to the yz-plane and separated by a distance b. A transverse electromagnetic wave propagates with magnetic field in the y-direction

$$\mathbf{H} = \hat{y}H_j(x,z,t) \tag{35}$$

and the Josephson current in the x-direction

$$\mathbf{J} = \hat{x}J_j(z,t) \tag{36}$$

There are two components of the electric field, one between the superconductors and the other parallel to the plane of the superconductors as shown in Chapter 7 for film transmission lines. For b small, the field takes the form

[19]A. Barone and G. Paternó, *Physics and applications of the ac Josephson effect*, (wiley-Interscience, New York, 1982) sec. 6.2.2.

[20]T. Van Duzer and C. W. Turner, *Principles of Superconductive Devices and Circuits* (Elsevier, New York, 1981) sec. 5.03.

$$\mathbf{E} = \hat{\mathbf{x}} E_j(z,t) - \hat{\mathbf{z}} \frac{2x}{b} E_g(z,t) \tag{37}$$

Energy flow has two components as described by the Poynting vector

$$S = \mathbf{E} \times \mathbf{H} = \hat{\mathbf{x}} \frac{2x}{b} E_g H_j + \hat{\mathbf{z}} \frac{2x}{b} E_j H_j \tag{38}$$

The principal component of energy flow is along the junction with transverse flow into the two superconductors.

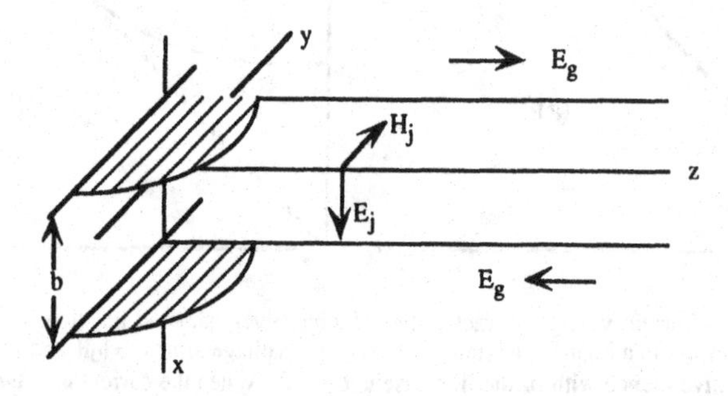

Figure 3. A transverse electromagnetic wave propagates with magnetic field in the y-direction between two superconducting surfaces parallel to the yz-plane and separated by a distance b.

10.3.1. Field equations

Faraday's law in the junction [Eq. (3-36)] gives for the relation between fields

$$\frac{\partial E_j}{\partial z} + \frac{2}{b} E_g + \mu \frac{\partial H_j}{\partial t} = 0 \tag{39}$$

Ampère's law [Eq. (3-37)] gives for b small the relation between fields and current

$$\varepsilon \frac{\partial E_j}{\partial t} + J_j + \frac{\partial H_j}{\partial z} = 0 \tag{40}$$

For an ideal London superconductor with b small, E_g is related to H_j by

$$E_g = \mu \lambda_g \frac{\partial H_j}{\partial t} \tag{41}$$

Eliminating the surface electric field between Eqs. (38) and (41) gives

$$\frac{\partial E_j}{\partial z} + \mu \left(1 + \frac{2\lambda_g}{b}\right) \frac{\partial H_j}{\partial t} = 0 \tag{42}$$

Finally, eliminating the magnetic field between Eqs. (40) and (42) gives

$$\frac{1}{\mu\left(1+2\lambda_g/b\right)}\frac{\partial^2 E_j}{\partial z^2}-\varepsilon\frac{\partial^2 E_j}{\partial t^2}=\frac{\partial J_j}{\partial t} \quad (43)$$

The electrodynamics of Josephson junctions are obtained by combining Eq. (43) with the relation between current and phase difference, Eq. (14), where the phase difference depends on the electric fields in the junction[21] through Eq. (15), which may be written as

$$\frac{1}{2\pi}\Phi_0\frac{\partial\delta}{\partial t}=bE_j \quad (44)$$

Substituting Eqs. (14) and (44) into Eq. (43) gives

$$\frac{1}{\mu\left(b+2\lambda_g\right)}\frac{\partial^2 E_j}{\partial z^2}-\frac{\varepsilon}{b}\frac{\partial^2 E_j}{\partial t^2}=2\pi\cos\delta\frac{J_0}{\Phi_0}E_j \quad (45)$$

10.3.2. Dispersion relation

We examine solutions of Eq. (45) for a weak electric field

$$E_j(z,t)=E_j e^{i(kz-\omega t)} \quad (46)$$

with δ determined by an applied bias current. We obtain for the dispersion relation

$$\omega^2=\omega_J^2\cos\delta+v^2k^2 \quad (47)$$

where the Josephson plasma frequency is

$$\omega_J=\left(2\pi\frac{bJ_0}{\varepsilon\Phi_0}\right)^{1/2} \quad (48)$$

and v is the velocity of a film transmission line as discussed in Ch. 7

$$v=\frac{1}{\sqrt{\varepsilon\mu(1+2\lambda_g/b)}} \quad (49)$$

For $b \ll \lambda_g$ the velocity is substantially reduced as shown by Swihart.[22]

At frequencies well below ω_J, fields attenuate as $k=(i/\lambda_J)\sqrt{\cos\delta}$ with

[21] D. J. Scalapino, "The theory of Josephson tunneling," Ch. 32 in *Tunneling Phenomena in Solids*, eds. E. Burstein and S. Lundqvist (Plenum, New York, 1969).

[22] J. C. Swihart, *J. Appl. Phys.* **32**, 461 (1961).

$$\lambda_J = \frac{v}{\omega_J} = \sqrt{\frac{\Phi_0}{2\pi\mu(b+2\lambda_g)J_0}} \tag{50}$$

10.3.3. Magnetic penetration

Substituting Eq. (43) into Eq. (44) and integrating once, we obtain,

$$\lambda_J{}^2 \frac{\partial^2\delta}{\partial z^2} - \frac{1}{\omega_J{}^2}\frac{\partial^2\delta}{\partial t^2} = \sin\delta \tag{51}$$

Equation (51) has an exact solution in Jacobian elliptic functions.[23,24] We now examine the solutions of Eq. (51) in the limits of weak and strong magnetic fields at frequencies well below the Josephson plasma frequency.

Weak fields

In weak fields $H_j << \lambda_J J_0$ δ is small leading to

$$\lambda_J{}^2 \frac{\partial^2\delta}{\partial z^2} - \frac{1}{\omega_J{}^2}\frac{\partial^2\delta}{\partial t^2} = \delta \tag{52}$$

which gives for the phase difference across a long junction

$$\delta(z,t) = -\frac{H_s}{J_0\lambda_J}\left(1-\frac{\omega^2}{\omega_J{}^2}\right)^{1/2} \exp\left[-\frac{z}{\lambda_J}\left(1-\frac{\omega^2}{\omega_J{}^2}\right)^{1/2} - i\omega t\right] \tag{53}$$

Junction critical field H_{c1j}

Integrating Eq. (51) at zero frequency between δ_1 at z_1 and δ_2 at z_2 as $z_2 \to \infty$ gives

$$\frac{1}{2}\lambda_J{}^2\left(\frac{\partial\delta_1}{\partial z}\right)^2 + \cos\delta_1 = \frac{1}{2}\lambda_J{}^2\left(\frac{\partial\delta_2}{\partial z}\right)^2 + \cos\delta_2 = 1+\varepsilon \tag{54}$$

Fields are screened for $\varepsilon = 0$ and penetrate for $\varepsilon > 0$. The limiting case with $\varepsilon = 0$ is

$$\delta_1 = \pi \qquad \frac{\partial\delta_1}{\partial z} = 2\pi\mu(b+2\lambda_g)\frac{H_{c1j}}{\Phi_0} = \frac{2}{\lambda_J} \tag{55}$$

with the connection between H and $\partial\delta/\partial z$ obtained from Eq. (21). Eq. (55) gives for H_{c1j}

---

[23] C. S. Owen and D. J. Scalapino, *Phys. Rev.* **164**, 538 (1967).

[24] A. Barone and G. Paternò, *Physics and applications of the Josephson effect*, (Wiley-Interscience, New York, 1982) sec. 5.2.

$$H_{c1j} = \frac{2\Phi_0}{2\pi\mu\lambda_J(b+2\lambda_g)} = 2J_0\lambda_J \tag{56}$$

Strong fields

The strong field limit is characterized by $\varepsilon >> 1$ in Eq. (54), leading to a phase difference δ across the junction that varies approximately linearly with depth

$$\delta \approx -\frac{2H}{H_{c1j}}\frac{z}{\lambda_J} = -2\pi\mu\frac{b+2\lambda_g}{\Phi_0}zH_j \tag{57}$$

where the critical field H_{c1j} is defined by Eq. (56). Substituting Eq. (57) into Eq. (51) and integrating twice gives to order $(H_{c1j}/H_s)^2$ for an unbiased junction at $\omega = 0$

$$\delta \approx -\frac{2H_s}{H_{c1j}}\frac{z}{\lambda_J} + \left(\frac{H_{c1j}}{2H_s}\right)^2 \sin\left(\frac{2H_s}{H_{c1j}}\frac{z}{\lambda_J}\right) \tag{58}$$

The quantization condition that the change in phase $\Delta\delta$ between the shorted ends of a junction of length d is $2\pi n$ leads to an intergral number of flux quanta in the junction

$$\mu H_s(b+2\lambda_g)d = n\pi\mu\lambda_J(b+2\lambda_g)H_{c1j} = n\Phi_0 \tag{59}$$

Diffraction pattern

Magnetic flux configurations within junctions across which current is transported may be calculated from the theory of Ferrell and Prange.[25] The Josephson current across a junction containing flux is

$$I' = \int_0^d dz\, J_0(z)\sin\delta(z) \tag{60}$$

For magnetic flux penetrating the junction symetrically it is useful to take $z = 0$ at the center of the junction and write from Eq. (60)

$$\delta(z) = \delta(0) + \frac{H}{H_J}\frac{z}{\lambda_J} \tag{61}$$

The surface current density is obtained by integrating through the junction

$$I' = J_0 \int_{-d/2}^{d/2} dz \sin\delta(z) = J_0 d \sin\delta(0)\frac{\sin(\pi\Phi/\Phi_0)}{(\pi\Phi/\Phi_0)} \tag{62}$$

---

[25] R. A. Ferrell and R. E. Prange, *Phys. Rev. Lett.* **10**, 479 (1963).

where Φ is the flux in the junction. The critical current is the Fourier transform of $J_0(z)$ and is the analog of the Fourier transform of the aperture function. Eq. (62) is the Fraunhofer diffraction pattern of a single slit. Note that Φ may be negligibly small for a thin film.

10.3.4. Flux flow transistor

The superconducting flux flow transistor is a single film, active device that has been made from high temperature superconducting materials.[26,27] The device is based on current-control of flux flow through Josephson junctions and can operate to 40 GHz.

The body of the device consists of a parallel array of junctions across which a bias current flows. A separate control line provides a local magnetic field that modulates the flux density in the junctions. When the body of the device is biased below the critical current I_0, no flux is admitted into the junctions. Above I_0, flux is admitted and flows through the junctions. A voltage is developed across the device proportional to the rate at which flux flows through the junctions.

10.4. Pulse Propagation

10.4.1. Solitons

Voltage pulse profiles that agree well with solitary wave or *soliton* theory[28] have been observed in Josephson transmission lines.[29] We look for a ballistic solution to Eq. (51) of the form[30]

$$\delta = \delta(z - vt) \tag{63}$$

Substituting into Eq. (51) leads to

$$\left(1 - \frac{v^2}{v_J^2}\right)\frac{\partial^2\delta}{\partial z^2} = \frac{\sin\delta}{\lambda_J^2} \tag{64}$$

Integrating this equation gives

$$z = vt + \int_{\delta(0)}^{\delta(z-vt)} \frac{\lambda d\delta}{\sqrt{2(1-\cos\delta)}} \tag{65}$$

with $\lambda / \lambda_J = \sqrt{1-(v/v_J)^2}$ as sketched in Fig. 4 and

[26] J. S. Martens, J. B. Beyer, J. E. Nordman, G. K. G. Hohenwarter and D. S. Ginley, *IEEE Trans. Magn.* **27**, 3284 (1991).

[27] J. S. Martens, D. S. Ginley, J. B. Beyer, J. E. Nordman and G. K. G. Hohenwarter, *IEEE Trans. Appl. Supercond.* **1**, (1991).

[28] G. B. Witham, *Linear and Nonlinear Waves* (Wiley-Interscience, New York, 1975).

[29] A. Matsuda and T. Kawakami, *Phys. Rev. Lett.* **51**, 694 (1983).

[30] A. Barone and G. Paternó, *Physics and Applications of the Josephson Effect* (Wiley-Interscience, New York, 1982) pp. 264-271.

$$\delta(z,t) = 4\tan^{-1}\left[\exp\left(\pm\frac{z - vt}{\lambda}\right)\right] \tag{66}$$

As λ approaches λ_J, the soliton velocity goes to zero. In the opposite limit, as v approaches v_J, λ goes to zero and the wavefront becomes more abrupt.

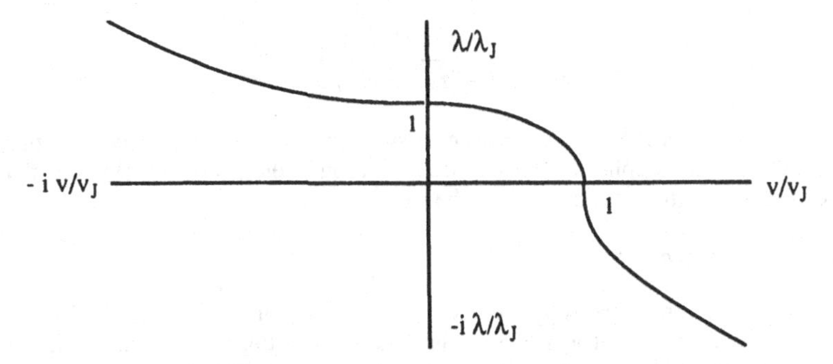

Figure 4. As λ approaches λ_J, the soliton velocity goes to zero. In the opposite limit, λ goes to zero and the wavefront becomes more abrupt with v approaching v_J.

The soliton is an electromagnetic voltage pulse with integrated value from Eq. (2)

$$\int V dt = \frac{1}{2\pi}\Phi_0 \int d\delta = \pm\Phi_0 \tag{67}$$

The flux carried by the soliton is of order $(\lambda/\lambda_J)^2\,\Phi_0$, going to zero as v approaches v_J.

Since the phase change across a soliton *must* be $\pm\,2\pi$, a single soliton can not be divided and must be either wholly transmitted or reflected. Since V is zero, and thus δ constant, at the end of a shorted Josephson transmission line, a voltage pulse must be reflected out-of- phase. At the end of an open transmission line the longitudinal current is zero, and thus the gradient of V and of δ must be zero, which means that a voltage pulse is reflected in-phase. A vortex is reflected by a shorted line as a vortex but by an open line as an antivortex.

The effect of dissipation on soliton propagation is described by what is called the perturbed sine-Gordon equation, an extension of Eq. (51)

$$\left(1 + \tau_s\frac{\partial}{\partial t}\right)\frac{\partial^2\delta}{\partial z^2} - \frac{1}{v_J{}^2}\left(\frac{\partial}{\partial t} + \frac{1}{\tau_j}\right)\frac{\partial\delta}{\partial t} = \frac{1}{\lambda_j{}^2}\left(\sin\delta - \frac{J_b}{J_0}\right) \tag{68}$$

where $v_J = \omega_J\lambda_J$ is the maximum soliton velocity. The parameter $\tau_j = \rho_j c_j$ is the charge relaxation time across the junction with ρ_j and c_j the specific resistance and capacitance per unit area across the junction. The parameter $\tau_s = G_s/B_s$ is the current relaxation time along the junction with G_s the conductance and B_s the susceptance. The quantity J_b is the junction bias current.

Substituting a solution of the form

$$\delta = \delta\left(\frac{z - vt}{\lambda}\right) \tag{69}$$

into Eq. (68), with λ and v functions of z, leads to a rate of increase in λ near the center of the soliton to first order in α and β

$$\frac{d\lambda}{dz} = \frac{1}{2}\frac{(\tau_s v/\lambda)\delta'' - (v\lambda/v_J{}^2\tau_j)\delta'}{\delta' - (z/\lambda)(\lambda v_J/\lambda_J v)^2\delta''} \tag{70}$$

Matsuda and Kawakami[27] find the dominant spreading of λ to be associated with surface conductance of the Josephson transmission line through the parameter β rather than with conductance across the junction through the parameter α.

10.4.2. Soliton resonators

Long Josephson junctions support resonant modes containing integral numbers of solitons.[31] Series-biased arrays of long junctions operating in a resonant soliton mode have possible application as local oscillators in integrated superconducting microwave and millimeter wave circuits.[32]

Operating an array of soliton resonators in coherent phase-lock offers increased power and reduced linewidth. Davidson and Pederson[33] have reported the coupling of long resonant Josephson junction soliton resonators to a coplanar twin-strip resonator. The junctions, about 200 μm long, were across and in series with one of the strips. The junctions were constructed with a $Nb–Al_2O_x–Nb$ trilayer technology and the coplanar resonator was Nb. The Q of the coplanar resonator with two junctions in series was 350 at the fundamental resonance around 30 GHz. A resonator could be operated with up to 32 such junctions in series.

---

[31] B. Dueholm, O. A. Levring, J. Mygind, N. F. Pedersen, O. H. Soerensen and M. Cirillo, *Phys. Rev. Lett.* **46**, 1299 (1981).

[32] N. F. Pedersen, *IEEE Trans. Magn.* **27**, 3328 (1991).

[33] A. Davidson and N. F. Pederson, *Appl. Phys. Lett.* **60**, 2017 (1992).

11
GRANULAR SUPERCONDUCTIVITY

11.1. Granular Superconductors

The earliest granular superconductors studied were metallic films, oxydized to form grains small compared with both the Ginzburg-Landau coherence length and magnetic penetration depth of the composite material. By controlling the oxidation a film can be brought close to the percolation limit with little metallic conduction and at the same time only moderate oxidation between grains. Below the temperature at which the grains become superconducting, Josephson tunneling may strongly couple small grains even though there had been little metallic conductivity in the normal state. The entire film then becomes superconducting. Studies of oxidized aluminum,[1,2,3] aluminum-germanium,[4] tin,[5] indium,[6] gallium[7] and NbN-Bn[8] have been carried out under these conditions. Such materials are characterized as strongly coupled granular superconductors.

---

[1] K. A. Müller, M. Pomerantz, C. M. Knoedler and D. Abraham, *Phys. Rev. Lett.* **45**, 832 (1980).

[2] E. Stocker and J. Buttet, *Solid State Comm.* **53**, 915 (1985).

[3] M. Kunchur, P. Lindenfeld, W. L. McLean and J. S. Brooks, *Phys. Rev. Lett.* **59**, 1232 (1987); M. Kunchur, Y.Z. Zhang, P. Lindenfeld, W. L. McLean and J. S. Brooks, *Phys. Rev. B* **36**, 4062 (1987).

[4] Y. Shapira and G. Deutscher, *Phys. Rev. B* **27**, 4463 (1983).

[5] B. G. Orr, H. M. Jaeger and A. M. Goldman, *Phys. Rev. B* **32**, 7586 (1985); B. G. Orr, H. M. Jaeger, A. M. Goldman and C. G. Kuper, *Phys. Rev. Lett.* **56**, 378 (1986).

[6] B. I. Belevsev, Yu. F. Komnik and A. V. Fomin, *J. Low. Temp. Phys.* **69**, 401 (1987).

[7] H. M. Jaeger, D. B. Haviland, A. M. Goldman and B. G. Orr, *Phys. Rev. B* **34**, 4920 (1986).

[8] M. Leung, U. Strom, J. C. Culbertson, J. H. Classen, S. A. Wolf and R. W. Simon, *Appl. Phys. Lett.* **50**, 1691 (1987).

Experimental investigations of tunnel-junction arrays[9,10,11,12] and related theoretical studies[13,14] have been performed in part to model the behavior of natural granular material.

With the recognition that the theory of granular superconductors is formally similar to that of a system of weakly coupled magnetic spins—a spin-glass—much of the theory that had been developed for spin-glasses has been applied to superconductors. This has included mean field theory[15,16,17] and computer simulations.[18,19,20]

Bulk granular behavior has been obtained in $Pb_{0.8}Bi_{0.2}BaO_3$ with the degree of percolation a sensitive function of heat treatment.[21] A synthetic sample of three-dimensional granular niobium has been prepared and studied by Raboutou *et al.*[22] Small metallic particles were lightly oxidized and then cold-pressed under sufficient force that they appeared to be at the percolation limit. The particles were fixed in epoxy to maintain the desired degree of contact and the ac magnetic susceptibility was studied. The penetration depth and the critical current were determined from the ac measurements. One interest in this system is that it may be a model for natural granular superconductors.

The Chrevel-phase compounds have been widely studied and are moderately granular with critical current densities sensitive to grain size.[23] Boysel *et al.*[24] have studied samples of Pb grains in a Zn matrix with intergranular coupling through the proximity effect. The intergranular ordering temperature T_g = 4 K was substantially below the superconducting transition temperature of Pb, T_c = 7.2 K.

11.2. High-Temperature Superconductors

The electrical and magnetic properties of the new high-temperature superconductors are largely dominated by intergranular processes suggesting that the grains are weakly coupled. This is certainly true of the ceramics and to some extent as well of nominal single crystals and epitaxial films.

The resistivity of the ceramics may be percolative as a result of the partial isolation of grains. The rf and microwave surface impedance of type II superconductors, the main subject of this lecture note volume, are largely determined by intergranular coupling with the surface resistance at these frequencies increased by the penetration of rf and microwave magnetic fields into the intergranular medium. Additionally, static and low frequency

---

[9] S. Chakravarty, G.-L. Ingold, S. Kivelson and A. Luther, *Phys. Rev. Lett.* **56**, 2303 (1986).

[10] R. F. Voss and R. A. Webb, *Phys. Rev. B* **25**, 3446 (1982); R. A. Webb, R. F. Voss, G. Grinstein and P. M. Horn, *Phys. Rev. Lett.* **51**, 690 (1982).

[11] B. J. van Wees, H. S. J. van der Zant and J. E. Mooij, *Phys. Rev. Lett.* **35**, 7291 (1987).

[12] B. Giovanni and L. Weiss, *Solid State Commun.* **28**, 1005 (1978).

[13] M. P. A. Fisher, *Phys. Rev. Lett.* **57**, 895 (1986).

[14] E. Simànek and R. Brown, Phys. Rev. B **34**, 3495 (1986).

[15] P. Minnhagen, *Phys. Rev. Lett.* **54**, 2351 (1985).

[16] *Percolation, Localization and Superconductivity*, eds. A. M. Goldman and S. A. Wolf (Plenum, New York, 1984.)

[17] S. John and T. C. Lubensky, *Phys. Rev. B* **34**, 4815 (1986).

[18] C. Ebner and D. Stroud, *Phys. Rev. B* **31**, 165 (1985).

[19] B. Berge, H. T. Diep, A. Ghazali and P. Lallemand, *Phys. Rev. B* **34**, 3177 (1986).

[20] I. Morgenstern, K. A. Müller and J. G. Bednorz, *Z. Phys. B*, in press.

[21] A. W. Sleight, J. L. Gillson and P. E. Bierstedt, *Solid State Commun.* **17**, 27 (1975). For recent references, see R. G. Steinmann and F. de la Cruz, *Solid State Commun.* **62**, 335 (1987).

[22] A. Raboutou, J. Rosenblatt and P. Peyral, *Phys. Rev. Lett.* **45**, 1035 (1980).

[23] C. Rossel and Ø. Fischer, *J. Phys. F* **14**, 455, 473 (1984).

[24] R. M. Boysel, A. D. Caplin, M. N. D. Dalimin and C. N. Guy, *Phys. Rev. B* **27**, 554 (1983).

magnetic flux readily penetrates the intergranular medium at fields well below H_{c1}.[25,26]

The history of the improvement of the microwave properties of classical high-field superconductors had been one of the control and elimination of grain boundaries.[27] This has been just as true of the new superconductors with the potential usefulness of these materials depending on the extent to which intergranular processes can be understood and controlled.

11.2.1. Ceramics

Soon after the discovery by Bednorz and Müller[28] of superconductivity in the compound $(La_{1-x}Ba_x)_2CuO_{4-y}$, Blazey *et al.*[29] undertook a systematic investigation of the magnetic field dependence of magnetically modulated microwave absorption in the La-Ba-Cu-O and La-Sr-Cu-O systems, using a commercial 9-GHz electron spin resonance (ESR) spectrometer. Their samples were compressed sintered pellets, prepared from stoichiometric quantities of the metallic oxalates that had been precipitated from solution with oxalic acid. These and other early studies of microwave absorption were limited to pressed polycrystalline samples and to powders prepared by regrinding sintered material.

Cohen *et al.*[30] compared the surface resistance R_s and surface reactance X_s of a platelet of $YBa_2Cu_3O_{7-\delta}$ placed within a copper stripline resonant at 5 GHz. Above the transition, R_s and X_s were each about 1.2 Ω per square. Surprisingly, the two components decreased together below T_c with X_s dropping toward zero and R_s to a residual value of 0.3 Ω per square. The observed behavior near T_c suggests the penetration of flux into a resistive medium.

11.2.2. Powders

Gould *et al.*[31] have systematically studied zero-field microwave absorption in micron-size powders of $YBa_2Cu_3O_{7-\delta}$ with diameters 2, 6, 10 and 50 µm. Measurements were made at 2, 10, 22 and 35 GHz. Although sintered ceramics show substantial microwave absorption down to 0.3 T_c or lower, the smallest particles showed intrinsic temperature-dependent absorption that was negligible below 0.8 T_c, compatible with a two-fluid model and London screening.[32] Agglomerates of particles show a broadened transition with significant absorption down to 0.7 T_c. Particles of 100 µm diameter were carefully prepared to avoid agglomeration[33] and showed a sharpened transition, which may have resulted from increased screening of microwave fields.

The application of low static magnetic fields[34,35] to $YBa_2Cu_3O_{7-\delta}$ powders increases

---

[25] J. R. Clem, *Physica C* **153-155**, 50 (1988).

[26] M. Tinkham and C. J. Lobb, *Physical Properties of the New Superconductors* (Academic Press, San Diego, 1989).

[27] H. Piel, M. Hein, N. Klein, U. Klein, A. Michalke, G. Müller and L. Ponto, *Physica C* **153-155**, 1604 (1988).

[28] J. G. Bednorz and K. A. Müller, *Z. Phys. B* **64**, 189 (1986).

[29] K. W. Blazey, K. A. Müller, J. G. Bednorz, W. Berlinger, G. Amoretti, E. Buluggiu, A. Vera and F. C. Mattacotta, *Phys. Rev. B* **36**, 7241 (1987).

[30] L. Cohen, I. R. Gray, A. Porch and J. R. Waldram, *J. Phys. F* **17**, L179 (1987).

[31] A. Gould, E. M. Jackson, K. Renouard, R. Crittenden, S. Bhagat, N. D. Spencer, L. E. Dolhert and R. F. Wormsbecher, *Physica C* **156**, 555 (1988).

[32] F. London, *Superfluids, Volume 1, Macroscopic Theory of Superconductivity* (Wiley, New York, 1950); reprinted (Dover, New York, 1961).

[33] A. Gould, S. D. Tyagi, S. M. Bhagat and M. A. Manheimer, *IEEE Trans. Mag.* **25**, 3224 (1989).

[34] A. Gould, S. M. Bhagat, M. A. Manheimer and S. Tyagi, "Field induced microwave absorption in high T_c

the low-temperature microwave absorption even of those particles that show no residual absorption in zero magnetic field. The absorption at first increases quadratically, saturating in fields of a few tens of Oe at a fraction of the normal-phase absorption. The microwave absorption is hysteretic in a cycled magnetic field with the observation of substantial "coercive" fields at 1.3 K in 10 μm diameter particles. Raising the temperature to 16 K reduces the hysteresis.[34] The field dependence of the magnetization, measured separately, is substantially different from that of the microwave absorption.

Farrell *et al.*[36] have used the anisotropic magnetic susceptibility of the paramagnetic phase of $YBa_2Cu_3O_{7-\delta}$ to produce oriented composites of fine powders 2-4 μm in diameter. Similarly prepared samples have been produced by Livingston and Hart.[37] Padamsee *et al.*[38] and Poirier et al.[39] have measured the surface resistance of oriented powders for microwave magnetic fields parallel and perpendicular to the *c*-axis. At low power levels[38] the surface resistance is an order of magnitude lower for the microwave field parallel to the *ab*-plane. As the power is increased, the surface resistance for this orientation rises more rapidly than for the microwave magnetic field parallel to the *c*-axis.

11.2.3. Thick films and coatings

Thick films of high-temperature superconductors have been prepared by plasma-spraying, screen-printing, sol-gel, spinning, settling, spreading and by electrophoresis, which utilizes the migration of charged particles through a stationary liquid under the action of an electrostatic field. Such thick-film fabrication techniques have been widely investigated because the prospect of coating superconducting materials on a variety of substrates and surfaces offers a number of promising advantages for large-scale as well as electronic device applications. Piel *et al.*[27] in addition to their development of ceramic pellets for microwave applications, have pioneered the preparation of polycrystalline layers deposited electrophoretically on silver substrates. These layers were deposited step by step from a colloidal suspension of $YBa_2Cu_3O_{7-\delta}$ powder in n-butanol onto a silver disk 12.5 cm in diameter. After each deposition step, which increased the thickness of the layer by about 10 μm, the silver disk was annealed at 930°C in an oxygen atmosphere. In this way films of about 50 μm thickness could be developed.

Chu *et al.*[40] have deposited somewhat larger particles (400 mesh) suspended in acetone onto Cu and Ag substrates. By first forming a metallic electrode by evaporation, they were able to deposit onto insulating substrates as well. Particles were deposited to achieve a coating of 200-500 μm thickness. The coatings were dried and fired in air to a fractional density of about 70%.

---

superconducting powders—evidence for a superconducting glass phase at low T," *J. Appl. Phys.* **67**, 5020 (1990).

[35]S. Tyagi, A. Gould, G. Shaw, S. M. Bhagat and M. A. Manheimer, *Phys. Lett. A* **136**, 499 (1989).

[36]D. E. Farrell, B. S. Chandrasekhar, M. R. DeGuire, M. M. Fang, V. G. Kogan, J. R. Clem and D. K. Finnemore, *Phys. Rev. B* **36**, 4025(1987).

[37]J. D. Livingston, H. R. Hart, Jr. and W. P. Wolf, *J. Appl. Phys.* **84**, 5806 (1988).

[38]H. Padamsee, J. Kirchgessner, D. Moffa, D. Rubin, Q. S. Shu, H. Hart and C. Gaddapati, "rf Surface Resistance of a Magnetically Aligned Sintered Compact of Y-Ba-Cu-O," SUNY, Bufalo, NY, 1988.

[39]M. Poirier, G. Quirion, B. Quirion, F. D'Orazio, J. P. Thiel, W. P. Halperin and K. R. Poeppelmeier, *J. Appl. Phys.* **66**, 1261 (1989).

[40]C. T. Chu and B. Dunn, *Appl. Phys. Lett.* **55**, 492 (1989).

By applying a magnetic field of 8 T perpendicular to their silver disks, Hein *et al.*[41,42] have been able to achieve a high degree of texturing with the *c*-axis of their particles orthogonal to the plane of the disk. This texturing is a consequence of the anisotropic paramagnetic susceptibility of Y-Ba-Cu-O in the normal state,[36,43] already discussed. After sintering for about 140 h in pure oxygen, the textured thick films demonstrated much lower microwave surface resistance R_S than untextured films or bulk ceramic samples as aresult of grain growth. For the best sample, measured at 22 GHz, R_s drops steeply below T_c from about 0.33 Ω to 18 mΩ at 77 K and finally to less than 3 mΩ at 4.2 K. Similarly, eddy-current measurements yield narrower transition curves and higher T_c values as a result of reduced granularity. Mooney *et al.*[44] have applied voltages up to 5×10^5 V/cm, substantially higher than normally used, and report a consequent increase in film density.

11.2.4. Single crystals

Early microwave studies of high-temperature superconductivity were limited to sintered polycrystalline samples. Although in some cases grain growth during sintering yielded individual grains up to 80μm in diameter, large enough for single crystal x-ray determination, these crystals were not large enough for the measurement of their physical properties. More recently, single-crystal platelets of $YBa_2Cu_3O_7$ with exceptional microwave properties and diameters up to 4 mm have been prepared using BaO/CuO flux methods.[45,46,47,48,49,50] These procedures appear to take advantage of a region of partial melting near 1000°C around the concentration $YBa_2Cu_3O_7$. Mixtures in this region that are rich in Cu and Ba may serve as a flux for crystal growth. A similar procedure, using CuO as a flux, had been used earlier to obtain crystals of $(La_{1-x}Sr_x)_2CuO_4$.[51,52] The *c*-axis is normal to the plane of the $YBa_2Cu_3O_7$ platelets and the aspect ratio is typically about 10. The crystals as grown may be slightly oxygen defficient. Subsequent annealing in oxygen raises the superconducting onset temperature and sharpens the transition.

Some degree of contamination results from most crucible materials. Alumina crucibles lead to aluminum-doping and a consequent reduction in transition temperature. Zirconia crucibles react somewhat with the melt, contaminating crystals that grow near the crucible wall. The use of thoria or magnesia largely avoids metallic contamination.

Wu *et al.*[50] report that their single crystals have superior microwave properties. At a frequency of 10 GHz the surface resistance R_S drops from 100 mΩ to less than 0.4 mΩ just 4 K below the superconducting transition. These crystals possess an intrinsic penetration depth measured at both microwave and rf frequencies $\lambda(0) = 1600$ Å. Very

---

[41] M. Hein, E. Mahner, G. Müller, H. Piel, L. Ponto, M. Becks, U. Klein and M. Peiniger, *Physica C* **162-164**, 111 (1989).

[42] M. Hein, G. Müller, H. Piel, L. Ponto, M. Becks, U. Klein and M. Peiniger, *J. Appl. Phys.* **66**, 5940 (1989).

[43] D. E. Farrell, C. M. Williams, S. A. Wolf, N. P. Bansal and V. G. Kogan, *Phys. Rev. Lett.* **61**, 2805 (1988).

[44] J. B. Mooney, A. Sher, M. L. Riggs, K. A. Sabo, A. Rosengreen, B. Kingsley and J. C. Terry, "Electrophoretic deposition of high-temperature superconductor thick films," MRS, Boston, 27 November–2 December 1989.

[45] C. L. Bohn, J. R. Delayen, U. Balachandran and M. T. Lanagan, *Appl. Phys. Lett.* **55**, 304 (1989).

[46] Y. Hidaka, Y. Enomoto, M. Suzuki, M. Oda and T. Murakami, *Jpn. J. Appl. Phys.* **26**, L377 (1987).

[47] K. L. Keester, R. M. Housley and D. B. Marshall, *J. Cryst. Growth* **91**, 295 (1988).

[48] N. P. Ong, Z. Z. Wang, S. Hagen, J. T. W, J. Clayhold and J. Horvath, *Physica C* **153-155**, 1072 (1988).

[49] L. F. Schneemeyer, J. V. Waszczak, T. Siegrist, R. B. van Dover, L. W. Rupp, B. Batlogg, R. J. Cava and D. W. Murphy, *Nature* **328**, 601 (1987).

[50] D. Wu, W. L. Kennedy and S. Sridhar, *Appl. Phys. Lett.* **55**, 696 (1989).

[51] Y. Hidaka, Y. Enomoto, M. Suzuki, M. Oda and T. Murakami, *Jpn. J. Appl. Phys.* **26**, L377 (1987).

[52] Y. Hidaka, Y. Enomoto, M. Suzuki, M. Oda and T. Murakami, *J. Crystal Growth* (1990).

small increases in penetration depth are induced by a magnetic field, $d\lambda(0)/dH^2 = 10^{-3}$ Å/Oe^2. Sridhar et al.[53] have analyzed these measurements and find good agreement with the BCS temperature dependence of an s-wave superconductor with a gap parameter exhibiting mean field behavior.

Tyagi *et al*,[35,54] have observed microwave absorption from single-crystal flakes of $Bi_2Sr_2CaCu_2O_8$ 2 mm × 1 mm × 0.1 mm, prepared by Wang *et al.*[54] with a small magnetic field along the *c*-axis, perpendicular to the plane of the flakes. Their results are very similar to those obtained from small particles of $YBa_2Cu_3O_{7-\delta}$. Overall, these studies suggest a more complex origin of magnetic-field induced microwave absorption than the damped flux flow identified in conventional type II superconductors.[55]

Pakulis and Chandrashekhar[56] have observed microwave absorption from thin plates of $Bi_{2.1}Sr_{1.6}CaCu_2O_x$. The crystals were placed in the electric field E of the microwave cavity and could be oriented with E either parallel or perpendicular to the *c*-axis. A static magnetic field H could also be applied either parallel or perpendicular to the *c*-axis. With E along the *c*-axis, only a weak background absorption, independent of temperature and static field H, was obtained. With E in the *ab*-plane, the absorption increased with magnetic field, doubling in a field of 1 T. Superconducting losses at 80 K were estimated to be at least three orders of magnitude larger for this orientation of E. The corresponding anisotropy in $YBa_2Cu_3O_{7-\delta}$ is only a factor of five.[57] At temperatures below 30 K, Pakulis and Chandrashekhar[58] observed losses in low magnetic fields along the *c*-axis that were double-peaked and showed considerable hysteresis with magnetic field. With the magnetic field normal to the *c*-axis there is clear evidence of hysteresis but only single, broad absorption maxima are observed. It is suggested that at temperatures below 25 K, flux parallel to the *c*-axis is strongly trapped while flux in the *ab*-planes is only weakly trapped. Yau et al.[59] have observed related modulated absorption signals that they attribute to intergranular flux-induced losses

Karim *et al.*[60] have observed modulated microwave absorption in single crystals of $YBa_2Cu_3O_{7-\delta}$ with the static magnetic field H parallel to the *c*-axis and a peak modulation field of 2 Oe. At low temperatures and also at temperatures just below T_c the observed signal indicates the absorption minimum that is expected at zero magnetic field. At intermediate temperatures between 39.5 and 52 K, the modulated signal unexpectedly reverses sign, suggesting a modulated absorption maximum at zero field.

11.2.5. Thin films

High quality thin films of $YBa_2Cu_3O_{7-\delta}$ have been produced by a number of techniques– electron beam codeposition, molecular beam epitaxy, sputtering and laser deposition. For most processes, an amorphous film is deposited with a post-annealing step at temperatures up to 900 °C necessary to form a properly oxygenated and crystalline superconducting phase. Films that are well oriented with a minimum of grain boundaries and minimal

---

[53] S. Sridhar, D.-H. Wu and W. Kennedy, *Phys. Rev. Lett.* **63**, 1873 (1989).

[54] J. Wang, G. Chen, X. Chu, Y. Yan, D. Zheng, Z. Mai, Q. Yang and Z. Zhao, *Supercond. Sci. Technol.* **1**, 27 (1988).

[55] B. Rosenblum and M. Cardona, *Phys. Rev. Lett.* **12**, 657 (1964).

[56] E. J. Pakulis and G. V. Chrandrashekhar, *Phys. Rev. B* **38**, 11,974 (1988).

[57] E. J. Pakulis, T. Osada, F. Holtzberg and D. Kaiser, *Physica C* **153-155**, 510 (1988).

[58] E. J. Pakulis and G. V. Chandrashekhar, *Phys. Rev. B* **39**, 808 (1989).

[59] W. F. Yau, A. M. Portis, E. R. Weber and N. Newman (MRS, San Francisco, 16-21 April 16-21 (1990).

[60] R. Karim, H. How, R. Seed, A. Widom, C. Vittoria, G. Balestrino and P. Paroli, *Solid State Commun.* **71**, 983 (1989).

interdiffusiuon show high critical current densities. These conditions are more readilyachieved with an *in situ* growth process in which the film is oxygenated as it is deposited. High quality laser-deposited films of $YBa_2Cu_3O_{7-\delta}$[61,62,63] have been prepared under high oxygen partial pressure with surface resistance R_s as low as 8mΩ at 86.7 GHz and temperature 77 K.[64,65]

The lower processing temperatures of laser-deposition produces films with minimal granularity and better quality surfaces than thin films processed at higher temperature. Deposition parameters—laser energy density, oxygen partial pressure, substrate temperature and deposition angle—must be optimized for each substrate material. The best films have near perfect stoichiometry and a crystalline structure comparable to or better than that of currently available bulk single crystals. A representative laser deposition process[64] consists of firing pulses of 30 ns duration at a wavelength of 248 nm once per second from a KrF excimer laser at a rotating sintered $YBa_2Cu_3O_7$ pellet, placed opposite the substrate, which is held at temperatures between 650 and 850 °C. A plume of material ablated from the target deposits on the substrate, resulting in layer by layer growth of the thin film. An ambient oxygen pressure of about 100 mTorr produces films that are fully oxygenated and do not require post-deposition annealing. Such films deposited on appropriate substrates are fully *c*-axis oriented and crystalline.

Some of the best $YBa_2Cu_3O_{7-\delta}$ films have been deposited on (001) oriented $SrTiO_3$ substrates[62,66] because of the chemical stability of this material and good lattice-match over a wide range of temperature. The high dielectric constant ($\varepsilon \approx 18{,}000$ at 4 K) and large loss ($\tan\delta \approx .02$) of $SrTiO_3$ has made this substrate unsuitable for applications where the microwave field penetrates the dielectric. Other substrates presently used are MgO and yttria-stabilized zirconia (YSZ). The lattice-match of these substrates to $YBa_2Cu_3O_7$ is not quite as good as that of $SrTiO_3$ but still close enough to permit the deposition of reasonably good films. MgO is hygroscopic, and cleaves readily and together with YSZ has a moderately high loss-tangent, which makes these materials unsuitable for microwave applications. Sapphire, which has been a standard dielectric at microwave frequencies, reacts with $YBa_2Cu_3O_7$ and requires a buffer-layer. Lanthanum-based dielectrics, such as $LaGaO_3$ or $LaAlO_3$ (with $\varepsilon \approx 16$ and $\tan\delta < 6 \times 10^{-4}$ at 10 GHz) appear to be the best choice for microwave applications. These substrates offer not only lattice parameters closely matched to the 1-2-3 compound but also have good microwave loss properties. A comparative study of several substrates[67] has assigned the lowest microwave loss to $SrLaAlO_4$ followed by $LaGaO_3$ and $LaAlO_3$. All three dielectrics have $\tan\delta < 10^{-4}$, which is sufficiently low for most microwave applications. Oates[68] has obtained $\tan\delta < 1.2 \times$

---

[61] B. Roas, L. Schultz and G. Endres, *Appl. Phys. Lett.* **53**, 1557 (1988).

[62] T. Venkatesan, X. Wu, A. Inam, C. C. Chang, M. S. Hegde and B. Dutta, *IEEE J. Quantum Electron.* **25**, (1989).

[63] T. Venkatesan, X. D. Wu, A. Inam, M. S. Hegde, E. W. Chase, C. C. Chang, P. England, D. M. Hwang, R. Krchnavek, J. B. Wachtman, W. L. McLean, R. Levi-Cetti, J. Chabala and Y. L. Wang, *Advances in Processing High-Temperature Superconducting Thin Films with Lasers* (ACS, Washington, 1988).

[64] N. Klein, G. Müller, S. Orbach, H. Piel, H. Chaloupka, B. Roas, L. Schultz, U. Klein and M. Peiniger, *Physica C* **162-164**, 1549 (1989).

[65] A. Inam, X. D. Wu, L. Nazar, M. S. Hegde, C. T. Rogers, T. Venkatesan, R. W. Simon, K. Daly, H. Padamsee, J. Kirchgessner, D. Moffat, D. Rubin, Q. S. Shu, D. Kalokitis, A. Fathy, V. Pendrick, R. Brown, B. Brycki, E. Belohoubek, L. Drabeck, G. Grüner, R. Hammond, F. Gamble, B. M. Lairson and J. C. Bravman, *Appl. Phys. Lett.* **56**, 1178 (1990).

[66] T. Venkatesan *et. al.*, *Appl. Phys. Lett.* **53**, 1431 (1988).

[67] R. Brown *et al.*, *Appl. Phys. Lett.* **57**, 1351 (1990).

[68] D. E. Oates and A. C. Anderson, "Stripline measurements of surface resistance: Relation to HTSC film properties and deposition methods," SPIE, Santa Clara, CA, 10-12 October 1989 (SPIE, Bellingham, WA, 1990).

10^{-6} at 500 MHz and 4.2 K for MgO. The measured loss of $LaGaO_3$ under the same conditions is $\tan \delta = 2 \times 10^{-6}$.

Klein *et al.*[69] have measured the surface resistance of a laser-deposited *c*-axis oriented thin film of $YBa_2Cu_3O_{7-\delta}$ with thickness 0.4 μm and $T_c = 88$ K. At a frequency 86.7 GHz the surface resistance at 77 K was less than 8 mΩ. This value is better than obtained with polycrystalline samples and approaches that expected for classical superconductors at the same reduced temperature. Inam *et al.*[66] have determined the frequency-dependence of R_s between 1 and 100 GHz for a high quality $YBa_2Cu_3O_{7-\delta}$ thin film laser-deposited on (100) $LaAlO_3$.[70, 71] The values of R_s are quadratic in frequency as expected.

11.3. Experimental Studies

11.3.1. Transport

Transport determinations of critical current have been made on ceramic $YBa_2Cu_3O_{7-\delta}$ by detecting the onset of the resistive state.[72,73,74,75,76] From Sec. 10.3.3, the expected intergranular *decoupling* current is

$$J_{dec}(H, T) = J_0(T) \frac{\sin \pi\Phi/\Phi_0}{\pi\Phi/\Phi_0} \tag{1}$$

For the granular penetration depth λ_g much larger than the junction thickness b with a_g the granular diameter, the flux within a junction is

$$\Phi = 2\mu a_g \lambda_g H \tag{2}$$

and the decoupling field is

$$H_{dec} = \Phi/2\mu a_g \lambda_g \tag{3}$$

Substituting Eqs. (2) and (3) into Eq. (1) and expanding for small fields we obtain

$$J_{dec}(H, T) \approx J_0(T) [1 - (\pi^2/3) (H/H_{dec})^2] \tag{4}$$

Measured decoupling fields and penetration depths computed from Eq. (3) are given in Table 1. Although λ_g appears to decrease in this temperature range, it does increase as expected between 76 K and T_c.

It should be noted that the intergranular critical currents determined in this way

---

[69] N. Klein, G. Müller, H. Piel, B. Roas, L. Schultz, U. Klein and M. Peiniger, *Appl. Phys. Lett.* **54**, 757 (1989).

[70] A. Inam, M. S. Hegde, X. D. Wu, T. Venkatesan, P. England, P. F. Miceli, E. W. Chase, C. C. Chang, J. M. Tarascon and J. B. Wachtman, *Appl. Phys. Lett.* **53**, 908 (1988).

[71] T. Venkatesan, X. D. Wu, B. Dutta, A. Inam, M. S. Hegde, D. M. Hwang, C. C. Chang, L. Nazar and B. J. Wilkens, *Appl. Phys. Lett.* **54**, 581 (1989).

[72] Y. Yamada, N. Fukushima, S. Nakayama, H. Yoshino and S. Murase, *Jpn. J. Appl. Phys.* **26**, L8675 (1987).

[73] U. Dai, G. Deutscher and R. Rosenbaum, *Appl. Phys. Lett.* **51**, 460 (1987).

[74] J. F. Kwak, E. L. Venturini, D. S. Ginley and W. Fu, *Proceedings of the International Workshop on Novel Mechanisms of Superconductivity*, eds. S. A. Wolf and V. Z. Kresin, (Plenum, New York, 1987).

[75] D. S. Ginley, D. L. Venturini, J. F. Kwak, R. J. Baughman, B. Morosin and J. E. Schirber, *Phys. Rev. B* **36**, 829 (1987).

[76] J. F. Kwak, E. L. Venturini and D. S. Ginley, *Physica* **148B**, 426 (1987).

presumably indicate the development of a resistive state between grains and not the onset of intergranular flux flow. The two phenomena may not be entirely distinct, however.

We might expect in the experiments of Kwak *et al.*[74-76] that so long as flux remains in intergranular junctions and does not enter grains there should be relatively little hysteresis in the decoupling current J_d. For fields larger than the lower critical field H_{c1} flux can be expected to enter grains and be trapped. Kwak *et al.* increase their external field to H and, after reducing the field to zero, measure the apparent critical current $J_c(0, T)$. They find that above a field that they identify with H_{c1}, the critical current is reduced as a result of flux trapped within grains. It is of course possible that flux is trapped within intragranular junctions and that the detected resistive state is associated with these junctions as well as with intergranular junctions. The values obtained for H_{c1} are quite close to those obtained from single crystal studies with H parallel to the ab-planes as discussed below.

Table 1. Granular decoupling [74-76]

	4 K		30 K		76K	
J_t	1380	A/cm^2	1130	A/cm^2	305	A/cm^2
H_d	10	Oe	29	Oe	35.5	Oe
λ_L	100	Å	35	Å	28	Å
H_{c1}	330	Oe	250	Oe	120	Oe
M_r	45	emu	1.4	emu		
J_c	1380	A/cm^3	1130	A/cm^3	305	A/cm^3

Carolan *et al.*[77] have measured the ac resistance at 100 Hz of ceramic $YBa_2Cu_3O_{7-\delta}$ for fields in the range from 1 Oe to 5 kOe near T_c. A temperature T(H) was found for the onset of hysteresis obeying the relationship

$$H = H_0[1 - T(H)/T(0)]^{\gamma} \tag{5}$$

with H_0 = 51 Oe and γ = 1.49. This result is in good agreement with the de Almeida-Thouless[78] mean field value $\gamma = 3/2$ for a spin-glass and close to the estimate of Müller, Takashige and Bednorz.[79] Magnetization measurements on the same material were in agreement although they showed larger errors.

Lathrop *et al.*[80] have employed photo- and e-beam lithography to fabricate micro-wide bridges across tilt boundaries in $YBa_2Cu_3O_7$ polycrystalline thin films grown on MgO substrates. The density of large angle tilt boundaries was found to depend sensitively on the treatment of the substrate. Mechanical polishing of the leads to misalignment of about 20% of the material through 45°. If the substrate is mechanically polished and then annealed prior to deposition, nearly complete alignment is obtained. If, on the other hand, the substrate is chemically polished, a wide range of tilt angles are observed.

Depending on the growth parameters, critical current densities between 10^4 and 10^6

---

[77] J. F. Carolan, W. N. Hardy, R. Krahn, J. H. Brewer, R. C. Thompson and A. C. D. Chaklader, *Solid State Commun.* **64**, 717 (1987).

[78] J. R. L. de Almeida and D. J. Thouless, *J. Phys. A* **11**, 983 (1978)

[79] K. A. Müller, M. Takashige and J. G. Bednorz, *Phys. Rev. Lett.* **58**, 1143 (1987).

[80] D. K. Lathrop, S. E. Russek, B. H. Moeckly, D. Chamberlain, L. Pesenson, R. A. Buhrman, D. H. Shin and J. Silcox, *IEEE Trans. Magn.* **27**, 3203 (1991).

A/cm^2 are obtained at 4.2 K. Those microbridges with high critical currents exhibit magnetic self-shielding[81] with the appearance of "excess current" arising from flow flow and creep.[82] Microbridges with lower critical currents show typical RSJ behavior as described in Sec. 10.2.

The application of weak dc magnetic fields normal to the plane of the film gives results that deviate strongly from Eq. (1) with maxima that do not decrease monotonically with field and minima that do not go to zero. These observations suggest that the highest critical currents may be characteristic of 5 to 10% of the width of the microbridge and are carried by segments as narrow as 10 nm.

As observed universally in the high temperature superconductors, the I_cR_n product is only a fraction of the expected value $\pi\Delta/q$ described in Sec. 10.1.4. For a range of high-angle-grain boundaries crossing a patterned microbridge, the measured critical current was found to scale with the square-root of the normal resistance. The product $I_c^2R_n$ was also found to be invariant for a given grain boundary under oxygen depletion and reoxygenation. These effects may be associated with the depression of Δ at a high-angle grain boundary[83] or with low-resistance shunting of weak links.

11.3.2. Magnetization

Kwak *et al.*[74-76] have measured the remanent magnetization M_{rem} of their samples as a function of temperature for magnetizing fields up to 50 kOe. From these measurements they compute the critical current J_c from the Bean model. For a spherical grain of uniaxial superconductor we expect a remanent magnetization density

$$M_{rem} \approx \frac{\pi a}{320} J_c \tag{6}$$

The values of J_c given in Table 1 are obtained for a granular diameter of 10 μm. The computed currents are comparable to those measured in single crystals and good quality epitaxial films. Kwak *et al.* conclude that intragranular currents are not interrupted by junctions within grains.

If flux were able to enter grains reversibly through intragranular junctions the remanent magnetization would be substantially reduced with the granular diameter $2a_g$ replaced by the distance between junctions. If, however, the flux within intragranular junctions is also subject to pinning, as the reduction in $J_c(0, T)$ may indicate, the remanent magnetization will not be reduced. Thus the two experiments taken together do not appear to exclude the possibility of intragranular junctions but only the possibility of junctions that are sufficiently weak that flux enters and leaves at very low fields without hysteresis.

11.3.3. ac susceptibility

Raboutou *et al.*[84] have measured the mean ac susceptibility[22] of $YBa_2Cu_3O_{7-\delta}$ as a function of the amplitude of the ac field over more than three orders of magnitude from a few mOe to several Oe. At the lowest amplitudes the sample is presumed to be in the Meissner phase with flux penetrating exponentially into the grains. As the ac amplitude is

---

[81] R. A. Ferrell and R. E. Prange, *Phys. Rev. Lett.* **10**, 479 (1963).

[82] J. R. Waldram, A. B. Pippard and J. Clarke, *Phil. Trans. Roy. Soc. Lond. A* **268**, 265 (1970).

[83] G. Deutscher and K. A. Müller, *Phys. Rev. Lett.* **59**, 1745 (1987).

[84] A. Raboutou, P. Peyral, J. Rosenblatt, C. Lebeau, O. Peña, A. Perrin, C. Perrin and M. Sergent, *Europhys. Lett.* **4**, 1321 (1987).

increased, the sample is expected to go into an Abrikosov phase for an increasing fraction of the ac cycle. At the highest amplitudes the sample is almost entirely in the Abrikosov phase. The transition from the Meissner to the Abrikosov phase permits a determination of the temperature dependence of the intergranular penetration length

$$\lambda(T) = (\Phi_0/2\pi)\sqrt{6a/\mu\, z\, J_t\, |\psi|^2} \tag{7}$$

where z is the number of neighbors and ψ is the order parameter of the transition. Raboutou *et al.* obtain for the temperature dependence of λ in the vicinity of the transition

$$\lambda \approx \lambda_0(1 - T/T_c)^{\beta} \tag{8}$$

with $\beta = 0.7 \pm 0.1$. This is the same value obtained for Nb grains[22] as expected for disordered systems near the percolation limit.

11.3.4. Rotational hysteresis

Giovannella *et al.*[85] have made torque measurements on ceramic $La_{1.85}Sr_{0.15}CuO_4$ and have observed two field ranges. Below a rotation field of 350 Oe the torque is small, indicating weak hysteresis. Above this field the torque becomes large. The torques are interpreted in terms of the granular structure of the ceramic with the transitional field corresponding to H_{c1}, the field at which flux begins to enter the grains.

Measurements have also been made[86] on $YBa_2Cu_3O_{7-\delta}$ with similar results except that strong torques persist to higher fields than in $La_{1.85}Sr_{0.15}CuO_4$. It is suggested that the high field torque results from twin boundaries within grains.

11.4. Theories of Granular Superconductivity

11.4.1. Mean field theory

Clem *et al.*[87] have used a Ginzburg-Landau-like free-energy functional (Sec. 4.2) that takes into account the intragranular condensation energy

$$U_g = \frac{1}{2}\mu\, H_{cg}^2\, V_g \tag{9}$$

where H_{cg} is the thermodynamic critical field of the granular material, and the intergranular coupling energy (Sec. 10.2)

$$U_J = \hbar\frac{I_0}{q} = \frac{1}{2\pi}\Phi_0 I_0 \tag{10}$$

The parameter that characterizes the material is the ratio of these two energies $\varepsilon = U_J/U_g$.

---

[85] C. Giovannella, G. Collin, P. Rouault and I. A. Campbell, *Europhys. Lett.*, **4**, 109 (1987); C. Giovannella, G. Collin and I. A. Campbell, *J. Physique*, **48**, 1835 (1987); C. Giovannella, G. Collin and I. A. Campbell, preprint.

[86] C. Giovannella, P. Rouault, I. A. Campbell and G. Collin, preprint; C. Giovannella, I. A. Campbell and G. Collin, preprint; C. Giovannella, P. Rouault, I. A. Campbell and G. Collin, preprint.

[87] J. R. Clem, B. Bumble, S. I. Rader, W. J. Gallagher and Y. C. Shih, *Phys. Rev. B* **35**, (1987).

We follow the discussion of Clem.[25]

In the limit $\varepsilon \ll 1$, the effects of granularity are very pronounced and the current-carrying capacity and magnetic behavior of the medium are both dominated by the Josephson weak links between grains. The critical current density of the medium is $J_c = I_0/a_g^2$ with too little current flow between grains to suppress the order parameter.

In the opposite limit $\varepsilon \gg 1$, current-induced suppression of the effective-medium order parameter becomes very important. Clem *et al.* find that the theory reduces to the dirty limit of the Ginzburg-Landau theory. The characteristic lengths of the theory

$$\xi_J = \frac{1}{2}\sqrt{\frac{U_J}{U_g}}a_g \qquad \lambda_J = \sqrt{\frac{U_f}{2U_J}}a_g \tag{11}$$

are both much larger than a grain diameter a_g. Here $U_f = \Phi_0^2/2\mu a_g$ is a normalizing energy. The critical fields are

$$H_{c1J} = \frac{\Phi_0}{\mu\lambda_J^2}\left(\ln\frac{\lambda_J}{\xi_J} + 0.5\right) \qquad H_{c2J} = \frac{2\Phi_0}{\mu\xi_J^2} \tag{12}$$

Flux penetration

As we have seen, the phase of the London wave function may be written as the line integral of a wavevector

$$\phi = \int \mathbf{k} \cdot d\mathbf{r} \tag{13}$$

with

$$\hbar\,\mathbf{k} = m\,\mathbf{v} + q\,\mathbf{A} \tag{14}$$

Taking $\mathbf{J} = nq\mathbf{v}$, $\Phi_0 = h/q$ and $\lambda_g^2 = m/\mu nq^2$ we obtain from Eq. (14)

$$\mathbf{k} = \nabla\,\phi = -\frac{2\pi}{\Phi_0}(\ell\mathbf{J} + \mathbf{A}) \tag{15}$$

where $\ell = \mu\lambda_g^2$ is the intragranular inductivity. We next obtain a phenomenological relation[25] between $\mathbf{J}$ and $\mathbf{k}$ that is the macroscopic equivalent of the Josephson relation $J = J_0 \sin\delta$. Taking the Josephson current proportional to the gradient of the phase, we write the local relation

$$\mathbf{J} = (\Phi_0/2\pi\mu)\,\mathbf{k} \,/\, (\lambda^2 - \lambda_g^2) \tag{16}$$

which gives for the penetration depth

$$\lambda \approx \sqrt{\lambda_g^2 + (\Phi k/2\pi\mu J_0)} \tag{17}$$

We observe from Eq. (17) that the smaller J_0, the larger is λ. In the London limit we have $k = 0$ with J_0 finite, which gives $\lambda = \lambda_g$. Substituting into Eq. (15) we obtain the mean field London relation for a granular medium

$$\mu \lambda^2 \mathbf{J} + \mathbf{A} = 0 \tag{18}$$

and we see that λ is the distance over which flux is screened.

From the Maxwell equation

$$\nabla \times \mathbf{H} = \mathbf{J} \tag{19}$$

with $\mathbf{H} = \mathbf{B}/\mu$ we obtain the differential equation

$$\nabla^2 \mathbf{A} + \mu \mathbf{J} = 0 \tag{20}$$

Finally, combining Eq. (17) and Eq. (19) we obtain the mean field equivalent of the Josephson penetration equation

$$\nabla^2 \phi = \frac{1}{\lambda^2} \phi \tag{21}$$

with λ defined by Eq. (17). Because **k** is the gradient of ϕ it also satisfies Eq. (21) as do **J**, **B** and **A**. From Eqs. (17) and (19) we have the expression for the wavevector

$$k = 2\pi\lambda B/\Phi_0 \tag{22}$$

For $k < \pi/a_g$, Eq. (22) gives $B < \Phi_0/a_g\lambda \approx 0.2$ G for $\lambda \approx 100\mu$m. This is about the same condition that we obtained above and indicates that both vortex localization and a transition to the glass-phase may take place in quite low fields.

At magnetic fields in the mOe range we expect very little change in phase across the sample and flux is screened. As **H** is increased, the phase changes by 2π or more across the sample with the nucleation of vortices. At still higher fields k approaches π/a_g and we move into the superconducting glass phase as discussed in Sec. 11.4.3.

11.4.2. Weakly coupled grains

In terms of the parameters of the preceding section, granular high-temperature superconductors are extremely weakly coupled. Taking the thermodynamic critical field $H_{cg} \approx 10^4$ Oe and $a_g \approx 1$ µm with $J_0 \approx 10^4$ A/cm^2 leads to the estimate $\varepsilon \approx 4 \times 10^{-8}$.

As the transition is approached U_g goes to zero faster than does U_J [as $(1 - t)^2$ compared with $(1 - t)$ for U_J] and we might expect that very close to the transition the grains become strongly coupled. However, since $U_J(T)$ goes to zero as $(1 - t)$ we can expect that at a temperature T_{cJ} close to T_c the coupling energy will drop below $k_B T_{cJ}$ and thermal fluctuations will destroy the phase coherence between grains. The coupling energy $U_J(0)$ between grains corresponds to a temperature 2×10^3 K. With the assumed values, Clem estimates that T_{cJ} is about 1.4 K below T_c.

At temperatures below T_{cJ}, the theory of weakly coupled grains is the theory of the penetration of flux into intergranular Josephson junctions and of the transport of current across these junctions.

Intergranular flux penetration

Clem[25] has described the penetration of magnetic fields into Josephson junctions. The current density across a rectangular Josephson junction that contains flux Φ is

$$J = J_c(0)\sin\delta\,\frac{\sin\pi\Phi/\Phi_0}{\pi\Phi/\Phi_0} \tag{23}$$

where $J_c(0)$ is the critical Josephson current in zero applied field, δ is the phase across the junction and $\Phi_0 = h/q$ is the flux quantum. This relation assumes a rectangular junction with the magnetic field parallel to one of the edges. The flux in the junction is given by

$$\Phi = \mu H\, b_{eff}\, d \tag{24}$$

where μ is the permeability of the medium, d is the length of the junction and

$$b_{eff} = b + 2\lambda_g(T) \tag{25}$$

is the effective junction width with b the physical separation between superconductors.

The transport critical current density J_c in bulk-sintered samples decreases by about two orders of magnitude in applied fields as low as 100 Oe. Magnetization measurements, on the other hand, give magnetization critical current densities of the order of $10^6 A/cm^2$ that are relatively insensitive to applied field. These measurements taken together suggest that weak links between grains are responsible for the low-field drop-off in the transport critical current. Careful measurement and analysis of critical transport current tests this hypothesis. The maximum current density across a junction is the critical current

$$J_c(H) = J_c(0)\,\frac{\sin\pi H/H_0}{\pi H/H_0} \tag{26}$$

from Eq. (23) with $\delta = \pi/2$ and $H_0 = \Phi_0/\mu_0 Ld$.

Peterson and Ekin[88] have developed a statistical model for bulk-sintered high-T_c superconductors where the current is limited by intergranular Josephson weak links. Each weak link (SNS or SIS Josephson junctions) is described by a Fraunhofer relation. Comparison with experiment establishes that the cross-sectional areas of the links are comparable to grain areas and confirms the presence of percolation conductivity in which the transport current is limited by the weakest links. In the limit of magnetic fields $H \gg H_0$ the critical current $J_c(H)$ falls off as 1/H. For magnetic fields at a substantial angle to an edge, the critical current $J_c(H)$ falls off as $1/H^2$.

Peterson and Ekin[89] have replaced the Fraunhofer relation with an Airy current-field relation

$$J(H) = J_c \sin\delta\,\frac{J_1(\pi H/H_0)}{\pi H/2H_0} \tag{27}$$

The Airy relation is better fitting for junctions of irregular cross-section and gives $J_c(H) \propto H^{-3/2}$ in better agreement with experiment. The fact that many junctions are in series is treated numerically through the introduction of a cut-off angle Θ, which is appropriate to the weakest link in a current path.

---

[88] R. L. Peterson and J. W. Ekin, *Phys. Rev. B* **37**, 9848 (1988).

[89] R. L. Peterson and J. W. Ekin, *Physica C* **157**, 325 (1989).

11.4.3. Superconducting glass phase

At the lowest fields, flux is fully screened and the sample is in a Meissner phase. At higher fields, vortices enter the sample on an Abrikosov lattice of wavevector $k \approx \pi \times (B/\Phi_0)^{1/2}$. So long as ka_g is small (where a_g is the grain diameter) the medium appears homogeneous. This is the Abrikosov phase. When with increasing magnetic field the wavevector k approaches π/a_g, the granularity of the medium becomes important, the distribution of vortices is determined by the granular structure, and the system enters the *superconducting glass phase* as shown schematically in Fig. 1.

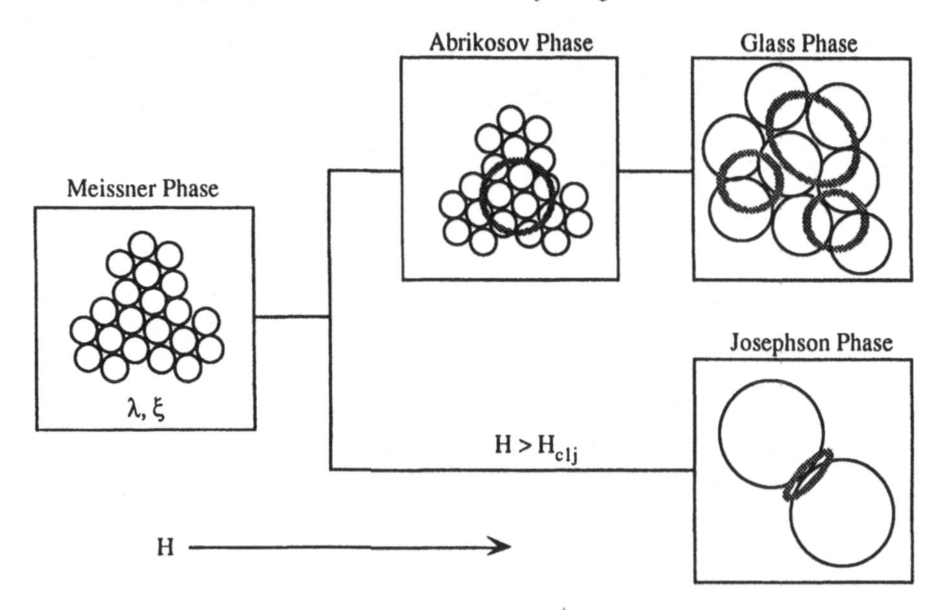

Figure 1. Schematic representation of the progression from the Meissner phase through the Abrikosov phase to the glass phase or, alternatively, directly to the Josephson phase.

How weakly must the grains be coupled for the system to progress through the sequence just described? In its range of validity the theory leads to a penetration depth λ for weakly coupled grains that can be very much longer than the intragranular screening distance λ_g. Where flux does go into junctions, the screening distance is of the order of λ_J.

For grains comparable in size to λ_g, the only way for flux to penetrate a distance $\lambda >> \lambda_g$ from the surface is between grains. The intergranular coupling must be sufficiently weak that the Josephson penetration depth λ_J is much longer than the granular diameter. A second condition on the mean field theory is that vortices not localize in intergranular junctions at fields for which k is much smaller than π/a_g which requires

$$\mu H_{c1j} < \Phi_0/a_g2 \qquad (33)$$

Feigel'man *et al.*[90] propose the careful study of critical behavior as a way of distinguishing the development of a superconductive glass phase from the Abrikosov phase. Approached from higher temperature, the glass phase should show critical slowing down as well as enhancement of the intergranular conductivity, the diamagnetic screening and the penetration depth.

[90] M. V. Feigel'man, L. B. Ioffe, A. I. Larkin and V. M. Vinokur, *Technical Report of the International Centre for Theoretical Physics, Miramare-Trieste*, July 1987. V. M. Vinokur, L. B. Ioffe, A. I. Larkin and M. V. Feigel'man, *ZhETF* **93**, 342 (1987).

12
ELECTRODYNAMICS OF INTERGRANULAR JUNCTIONS*

12.1. Introduction

The development of much of this chapter was originally motivated by an attempt to understand studies of the surface resistance of electrophoretically deposited films of $YBa_2Cu_3O_{7-\delta}$ in dc fields.[1,2,3,4] The effect of a dc magnetic field is to reduce the critical current density in the grain boundary, leading to an increase in the intergranular inductivity ℓ_j and increasing both the surface resistance R_s and the surface reactance X_s of the sample. The conventional junction model suggests also that the resistivity ρ_j in parallel with ℓ_j should increase linearly with magnetic field. Such increases are not observed and may indicate that ρ_j arises from normal-conducting bridges between grains.

This chapter continues the approach of Ch. 9 in representing the electrical properties of superconductors by lumped-element impedances. The specific impedance of granular samples has been modeled with an intragranular impedance in series with an intergranular impedance. Highly granular materials are dominated by the intergranular impedance and the intragranular impedance may commonly be neglected. For epitaxial films, on the other hand, the losses arise from defects, which are modeled as isolated grain boundaries

---

* This chapter draws on A. M. Portis and D. W. Cooke, "Effect of magnetic fields and variable power on the microwave properties of granular superconductors," in *High Temperature Superconductors*, J. J. Pouch, S. A. Alterovitz and R. R. Romanofsky, eds., Materials Science Forum (Trans Tech Publications, Aedermannsdorf, Switzerland, 1992).

[1] M. Hein, G. Müller, H. Piel, L. Ponto, M. Becks, U. Klein and M. Peiniger, *J. Appl. Phys.* **66**, 5040 (1989).

[2] M. Hein, E. Mahner, G. Müller, H. Piel, L. Ponto, M. Becks, U. Klein and M. Peiniger, *Physica* **162-164**, 111 (1989). M. Hein, E. Mahner, G. Müller, H. Piel, L. Ponto, M. Becks, U. Klein and M. Peiniger, *Physica* **162-164**, 111 (1989).

[3] M. Hein, S. Kraut, E. Mahner, G. Müller, D. Opie, H. Piel, L. Ponto, D. Wehler, M. Becks, U. Klein and M. Peiniger, "Electromagnetic properties of electrophoretic $YBa_2Cu_3O_{7-\delta}$ films," *J. Supercond.* **3**, 323 (1990).

[4] M. Hein, S. Kraut, E. Mahner, G. Müller, D. Opie, H. Piel, L. Ponto, D. Wehler, M. Becks, U. Klein and M. Peiniger, *ICMC '90 Conference on High -Temperature Superconductors: Materials Aspects*, 9-11 May 1990, Garmisch-Partenkirchen, Germany.

through which microwave fields penetrate the material.

The introduction of a sample with surface impedance Z_s into a resonator leads to a change in the complex frequency of the resonator

$$\tilde{\omega} = \omega - \frac{i}{2\tau} \tag{1}$$

where ω is the real frequency and τ is the energy relaxation time. From Ch. 8, the change in complex frequency is related to the complex surface impedance by

$$\Delta\tilde{\omega} = -\frac{i\omega}{2G}(R_s - iX_s) \tag{2}$$

with G the sample geometry factor. The surface resistance and reactance of a linear sample are then

$$R_s = \frac{G}{\omega}\Delta\frac{1}{\tau} = G\Delta\frac{1}{Q} \tag{3}$$

$$X_s = -2G\frac{\Delta\omega}{\omega} \tag{4}$$

12.2. Effect of dc Magnetic Fields

12.2.1. TBCCO films

Portis *et al.*[5] have reported a study of the effect of dc magnetic fields on the surface impedance of 15-μm-thick $Tl_mBa_2Ca_nCu_pO_x$ (TBCCO) films deposited by dc magnetron-sputtering onto 2.5-cm diameter disks of Consil 995, a Co-Ni-Ag alloy, buffered with BaF_2. Such films are promising for rf cavity applications because of their moderately low surface resistance and good heat extraction. The first films were deposited onto stock Consil 995 substrates from a target of Tl-2212 stoichiometry. Deposition from targets of Tl-2212 or Tl-1223 stoichiometry but onto rolled Consil produces textured films with substantially sharper resistive transitions.

Measurements of R_s at 18 GHz were made in a cylindrical Nb cavity, resonant in its fundamental TE_{011} mode and cooled by liquid He to 4 K. The Nb end-wall of the cavity was replaced by the superconducting film or by a stainless steel standard to allow a determination of the sample geometry factor. The surface resistance of the film was obtained from the half-width of the resonator transmission as described by Eq. (3). Simultaneous determination of R_s and the surface reactance X_s from Eq. (4) permits the determination of the variation of both the real and imaginary parts of the conductivity in an applied dc magnetic field.

The surface resistance and the cavity frequency ν near 18 GHz have been measured in a Cu cavity[5] with H perpendicular to the plane of the film and are plotted in Fig. 1. The fractional frequency shift may be related to the change in surface reactance, with the square of the wavevector written as

---

[5]A. M. Portis, D. W. Cooke, E. R. Gray, P. N. Arendt, C. L. Bohn, J. R. Delayen, C. T. Roche, M. Hein, N. Klein, G. Müller, S. Orbach and H. Piel, *Appl. Phys. Lett.* **58**, 307 (1991).

$$k^2 = \omega^2 \mu \varepsilon = \frac{i\omega\mu}{z} = -\frac{1}{\lambda^2} + \frac{2i}{\delta^2} \tag{5}$$

with the real part of k^2 written in terms of a penetration depth λ and the imaginary part in terms of a skin depth δ. The advantage of defining λ and δ in this way is that they relate to the real and imaginary parts of the film admissivity $1/z$. The surface impedance is

$$Z_s = -ikz = \frac{-i\omega\mu\lambda}{\sqrt{1 - 2i(\lambda/\delta)^2}} \tag{6}$$

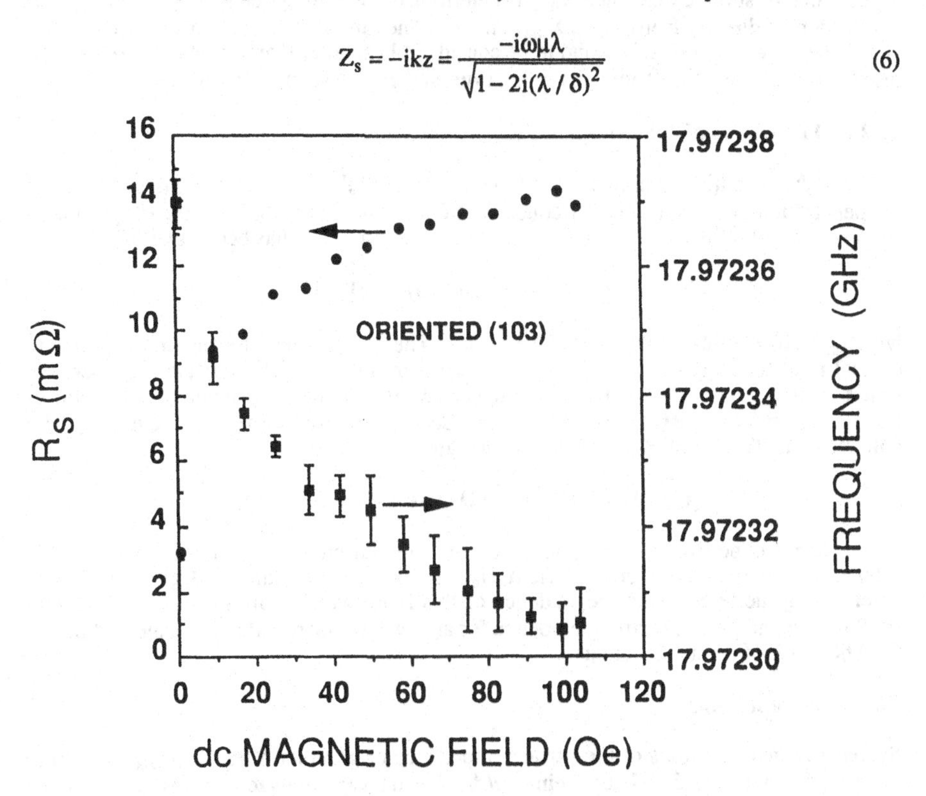

Figure 1. Sample surface resistance R_s and cavity frequency ν measured near 18 GHz for an oriented TBCCO film as a function of dc magnetic field. The observed decrease in frequency is a consequence of increased microwave penetration.

The analysis applies a two-fluid-like model to the measured surface impedance shown in Fig. 1. The films are represented by the lumped-element circuit shown in Fig. 1 of Ch. 9 with the normal-carrier inductivity neglected. The element $\ell_j = \mu\lambda^2$ represents the Josephson kinetic inductivity of defects with λ an effective penetration length. The resistivity $\rho_j = {}^1\!/_2\omega\mu\delta^2$ in shunt with ℓ_j represents junction losses. The central assumption of the analysis is that dc magnetic fields increase the Josephson inductivity ℓ_j but do not affect the shunt resistivity ρ_j. Thus the limiting classical surface resistance $R_c = {}^1\!/_2\omega\mu\delta$ is assumed to be independent of magnetic field. The normalized quantity R_s/R_c is plotted in Fig. 2 against X_s/R_c. To facilitate comparision with experiment, the derivative

dR_s/dX_s with R_c fixed is also shown as a function of normalized surface reactance. Values of $x = \lambda/\delta$ are indicated along the curve of R_s/R_c by filled circles. The resistive part of the surface impedance climbs from R_{min} toward R_c as λ increases with respect to δ, as a result of junction decoupling. The surface reactance similarly increases from X_{min} toward R_c as defects are decoupled. The minimum surface resistance R_{min} is the value of R_s measured in zero magnetic field. R_{max} is the asymptotic value of R_s as the dc field is increased. The slope dR_s/dX_s and the values of R_s establish R_c and thus the skin depth δ. Given R_c, the initial value of surface reactance X_{min} is determined. Screening lengths λ_{min} and λ_{max} are obtained most directly from X_{min}, X_{max} and R_c. The saturation surface resistance obtained experimentally at 18 GHz is found to be considerably smaller than the calculated classical resistance indicating that λ remains finite, even in the highest applied fields.

12.2.2. YBCO crystals and films

Wu *et al.*,[6] as mentioned in Sec. 11.2.4, have measured the magnetic field dependence of the penetration depth λ_{ab} at an rf frequency of 6 MHz for flux parallel to the *ab*-plane of a single crystal of $YBa_2Cu_3O_x$. An increase quadratic in the field has been obtained

$$\lambda_{ab}(T, H) = \lambda_{ab}(T, 0) + k(T)\, H^2 \tag{7}$$

with $\lambda_{ab}(0, 0) = 1400$ Å and $k(0) = 10^{-3}$ Å/Oe^2. These values lead for the thermodynamic critical field to $H_{c0}(0) = 1.02$ kOe. The coefficient k(T), which increases by over five orders of magnitude with increasing temperature, is in good agreement with Ginzburg-Landau theory[7] when BCS temperature dependences[8] are used. The increase in $\lambda_{ab}(T, H)$ with magnetic field implies a field-dependent gap

$$(1/\Delta)\, d\Delta/dH^2 = -\,(2/\lambda) d\lambda/dH^2 = -\,1.4 \times 10^{-6}/\text{Oe}^2 \tag{8}$$

In the same experiment, the appearance of a deviation from quadratic behavior is interpreted as the lower critical field H_{c1} for flux in the plane and gives a way of determining the temperature dependence of this field, which is in good agreement with BCS theory and provides strong evidence for an *s*-wave superconducting state and mean-field behavior of the gap parameter.

Electrophoretic films

Systematic investigations of the surface impedance of granular $YBa_2Cu_3O_{7-x}$ films have been carried out at 21.5 GHz by Hein *et al.*[4,9] The data are analyzed with an intergranular model based on resistively shunted Josephson junctions.[5]

The granular YBCO films that have been investigated were produced by electrophoretic deposition. Their degree of granularity could be reduced by texturing and sintering processes. The dependence of the surface impedance on sample preparation, temperature, parallel dc magnetic field and microwave power has been carefully investigated.

The application of an intergranular Josephson junction model is justified because the microwave impedance of grain boundaries dominates the grain impedance. The contribution of grain boundaries to the surface resistance R_s is expected to be proportional

---

[6] D.-H. Wu, W. L. Kennedy, C. Zahopoulos and S. Sridhar, *Appl. Phys. Lett.* **55**, 696 (1989).

[7] V. L. Ginzburg and L. D. Landau, *Soviet. Phys. JETP* **5**, 1442 (1957).

[8] J. Bardeen, L. N. Cooper and J. R. Schrieffer, *Phys. Rev.* **108**, 1175 (1957).

[9] M. Hein, H. Piel, M. Strupp, M. R. Trunin and A. M. Portis, *J. Mag. Mag. Mat.* **104-107**, 529 (1992).

to the ratio of boundary surface to grain surface and thus inversely proportional to the average grain size. Sample quality is found to increase with texturing, which promotes grain growth. The electromagnetic response of a grain boundary is that of a resistively shunted Josephson junction.[10] Its resistive part ρ_j, which is assumed constant, is in parallel with the kinetic inductance of the junction $\ell_j = \Phi_0/2\pi J_c b_{eff}$ with Φ_0 the flux

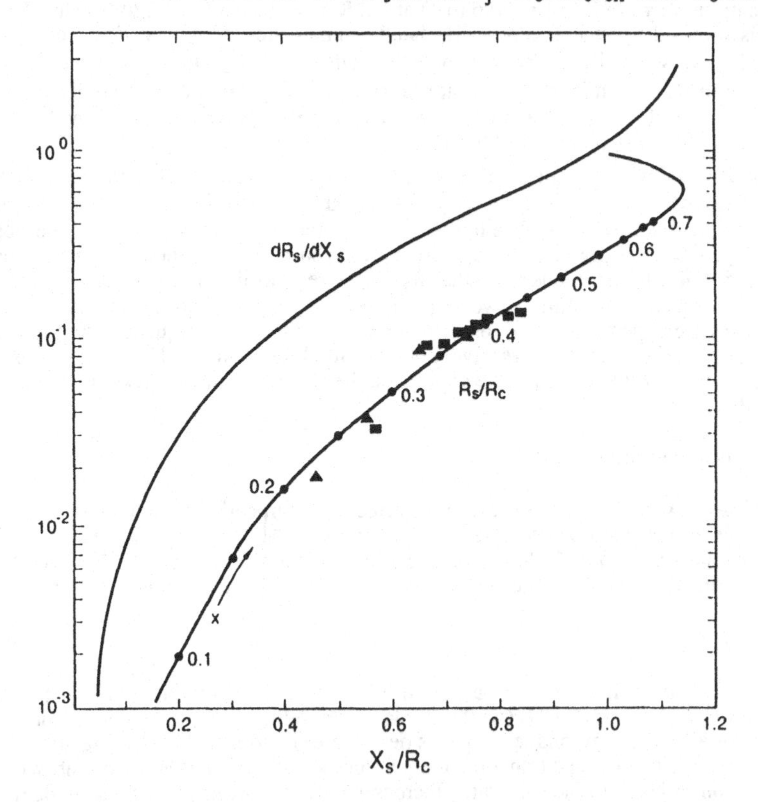

Figure 2. Plot of the normalized quantities R_s/R_c *vs.* X_s/R_c with filled circles indicating values of the parameter $x = \lambda/\delta$. The resistive part of the surface impedance climbs from R_{min} to R_{max} as the penetration depth λ increases with respect to the classical skin depth δ. The maximum surface resistance R_{max} is found to be substantially less than the classical surface resistance R_c because of residual intergranular coupling. Filled triangles are for an unoriented film and filled boxes are for an oriented film.

quantum, $J_c = J_0 \cos\delta$ the junction critical-current density and b_{eff} the magnetic width of the junction.[11] A finite limiting inductance is again required to fit the magnetic field dependence of the surface impedance.

The sensitivity of Z_s to magnetic fields arises because J_c is reduced when magnetic flux penetrates a junction. As a consequence, R_s and the surface reactance X_s both increase

---

[10] M. Gross, P. Chaudhari, M. Kawasaki and A. Gupta, *Phys. Rev. B* **42**, 10,735 (1990).

[11] F. Auracher and T. Van Duzer, *J. Appl. Phys.* **44**, 848 (1973).

with H.[12] Good agreement is obtained between the model and experiment for the reduced values R_s/R_c and X_s/R_c with $R_c = (\mu\omega\rho/2)^{1/2}$ a free parameter. For highly granular samples and at high temperature, X_s passes through a distinct maximum while R_s increases monotonically. A weak apparent dependence of ρ on H has been found and may result from frozen-in flux.[13] A typical range of values of ρ in μΩ·cm (or $\rho/\omega\ell$) for samples of increasing quality are 8 to 1 (or 0.6 to 1) at 4.2 K and 40 to 3 (or 0.2 to 0.7) at 77 K. The analysis allows the conversion of $R_s(H)$ data into an explicit field dependence of J_c. At low dc fields, the weak $R_s(H)$ dependence translates into a constant value of λ/δ. The saturation of R_s at high field levels is attributed to a finite saturation inductivity ℓ_{max}, which leads to a cutoff behavior of λ/δ at high fields. At fields above a few Oe, good agreement is obtained with the expected 1/H behavior of J_c.

The frequency exponent α for $R_s \propto \omega^\alpha$ decreases from 2 to 1/2 with increasing field. Typical values in zero dc field at 21.5 GHz are $\alpha \approx 1.8$ at 4 K and 1.3 at 77 K, again in agreement with experiment. The temperature dependence of λ/δ dominates the observed $R_s(T)$ dependence of granular samples and permits correlation of the microwave properties at 77 and at 4.2 K. It is observed that $R_s(T_{cj})$ agrees well with R_c where T_{cj} marks the onset of Josephson coupling between grains.[14,15] Finally, it is concluded that granularity determines the dependence of Z_s on dc magnetic fields and on frequency and temperature. Grain boundaries can be effectively represented by resistively shunted Josephson junctions. Improvements in sample quality require the preparation of textured large-grained material.

Microwave transmission

Golosovsky *et al.*[26] have observed the effect of a perpendicular magnetic field through thick films of YBCO, prepared by spray-pyrolysis onto MgO. The magnetic field enhances the transmission for T > 65 K but reduces the transmission for T < 65 K. These results may be indicative of interference between two modes of flux penetration.

12.2.3. BSCCO films

Remillard *et al.*[16] have investigated the effect of a dc magnetic field on the surface impedance of $Bi_2Sr_2CaCu_2O_x$ (Bi-2212) electrophoretically deposited onto Consil 995, a Co-Ni-Ag alloy. Superconductivity was detected only after melt-texturing and even then the microwave-determined transition temperature was reduced to 60 K, possibly the result of diffusion of Ni from the substrate. Increases in R_s and X_s were obtained in dc magnetic fields parallel to the film surface. The changes were identical and smaller than observed by the same authors on sputtered Tl-2212 films, saturating in fields of 50 to 75 Oe. These observations suggest that the intergranular coupling in these films is substantially resistive.

---

[12] M. Hein, *Dissertation*, University of Wuppertal (1992) unpublished.

[13] G. Müller, N. Klein, A. Brust, H. Chaloupka, M. Hein, S. Orbach, H. Piel and D. Reschke, *J. Supercond.* **3**, 235 (1990).

[14] J. Clem, *Physica C* **153-155**, 50 (1988).

[15] J. Dumas, S. Revenaz, C. J. Liu, R. Buder, C. Schlenker, S. Orbach, G. Müller, M. Hein, N. Klein and H. Piel, *J. Less-Comm. Met.* **164 & 165**, 1252 (1990).

[16] S. K. Remillard, P. N. Arendt and N. E. Elliott, *Physica C* **177**, 345 (1991).

12.3. Grain-Boundary Models

The current through a Josephson junction shunted by an impedance z_j takes the form[17]

$$J(t) = J_0 \sin \delta(t) + \frac{1}{z_j} E(t) \tag{9}$$

with the electric field across the junction

$$E(t) = \frac{1}{2\pi b_{eff}} \Phi_0 \frac{\partial \delta}{\partial t} \tag{10}$$

where $b_{eff} = b + 2\lambda_g$ is the *magnetic* thickness of the junction. Differentiating Eq. (9) with respect to the time and using Eq. (10) yields

$$\frac{\partial J}{\partial t} = \frac{1}{\ell_j} E + \frac{1}{z_j} \frac{\partial E}{\partial t} \tag{11}$$

with the Josephson inductivity

$$\ell_j = \frac{\Phi_0}{2\pi b_{eff} J_0 \cos \delta} \tag{12}$$

Note that the Josephson inductivity is independent of current only for $J << J_0$. For $J > J_0$ the phase across the junction becomes $\pi/2$ and the Josephson inductivity goes to infinity, leaving only the shunt impedance z_j.

For J a periodic function of time, z_j may be a complex frequency-dependent quantity. The equivalent lumped-element circuit is shown in Fig. 3. Note the analogy to the two-fluid model represented by Fig. 1 of Ch. 9 and to the flux-creep model represented by Fig. 4 of Ch. 9.

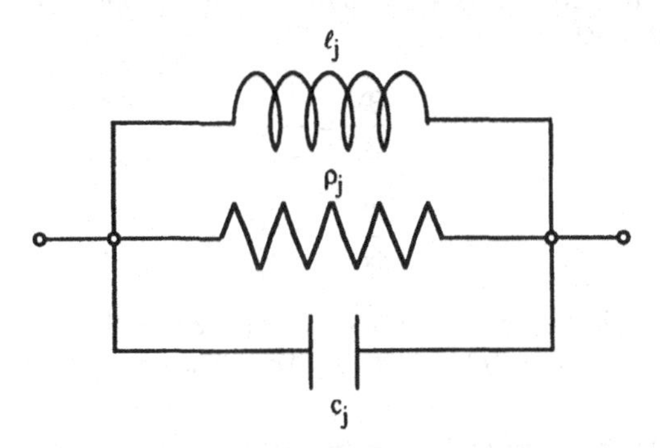

Figure 3. Equivalent lumped-element circuit of a Josephson junction. The Josephson inductivity ℓ_j is a function of current, going to infinity for $J = J_0$.

[17] J. D. Franson and J. E. Mercereau, *J. Appl. Phys.* **47**, 3261 (1976).

When the specific impedance of the sample is dominated by Josephson junctions, the sample surface impedance becomes, neglecting capacitance

$$Z_s = \sqrt{-i\omega\mu z} = -i\omega\sqrt{\frac{\mu_j \ell_j}{1 - i\omega\ell_j / \rho_j}} \tag{13}$$

At low frequency the surface impedance is approximately

$$Z_s \approx -i\omega\sqrt{\mu_j \ell_j} + \frac{\omega^2\sqrt{\mu_j \ell_j^3}}{2\rho_j} \tag{14}$$

which has of the same form as the two-fluid surface impedance.

12.3.1. Transmission-line model

Portis and Cooke[18] have applied the transmission-line model to an analysis of the surface impedance of granular superconductors. When the core of a stripline is a resistively shunted Josephson junction, the input resistance increases with the square of the frequency. A dc magnetic field increases the Josephson inductivity, leading to increased surface resistance and additional flux penetration.

Superconducting stripline

Swihart's analysis[19] of superconducting strip transmission lines provides an understanding of their slow-wave character . In this section we apply the theory to the intergranular penetration of microwave fields. For ease of representation and analysis we treat the stripline using transmission-line methods.[20] The equivalent-circuit representation of the intergranular medium is shown in Fig. 4. The series inductance per unit length $L_j = \mu_j b/a$ represents flux within the physical confines of strips of width a and separation b.

The element $2Z_g$ represents the surface impedance of the strips. The capacitance per unit area across the junction is $C_j'' = \varepsilon_j/b$. Wave propagation follows from the telegrapher's equations[21]

$$\frac{\partial V}{\partial x} = -ZI' \tag{15}$$

$$\frac{\partial I'}{\partial x} = -Y''V \tag{16}$$

where $Z = 2Z_g - i\omega L_j$ is the series impedance per square with the shunt admittance per unit area given by $Y'' = -i\omega C_j'' + 1/\rho_j b - 1/i\omega\ell_j b$. The voltage across the line is V and $I' = I/a$

---

[18] A. M. Portis and D. W. Cooke, *Supercond. Sci. Technol.* **5**, S395 (1992).

[19] J. C. Swihart, *J. Appl. Phys.* **32**, 461 (1961).

[20] T. Van Duzer and C. W. Turner, *Principles of Superconductive Devices and Circuits* (Elsevier, New York, 1981) secs 3.16 and 4.04.

[21] S. Ramo, J. R. Whinnery, and T. Van Duzer, *Fields and Waves in Communication Electronics* (Wiley, New York, 1954) sec 5.2.

is the current per unit width of strip.

For propagating voltage and current density

$$V(x,t) = Ve^{i(kx-\omega t)} \tag{17}$$

$$I'(x,t) = \frac{V}{aZ_j} e^{i(kx-\omega t)} \tag{18}$$

the characteristic impedance of the line and the propagation vector are

$$Z_j = \frac{1}{a}\sqrt{\frac{Z}{Y''}} \qquad k_j = \sqrt{-Y''Z} \tag{19}$$

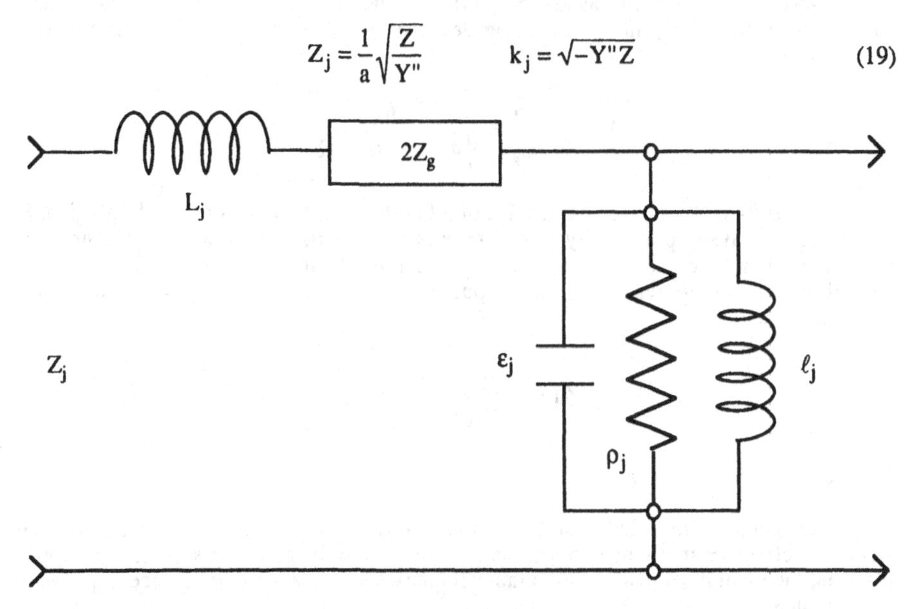

Fig. 4. Transmission line representation of an intergranular junction. The inductance L_j represents flux within the physical width b_j of the junction. The surface impedance Z_g characterizes the granular region. The specific impedance across the junction is represented by permittivity ε_j, resistivity ρ_j and inductivity ℓ_j.

A familiar example is obtained for $Z_g << \omega L_j$ and for $\omega C_j''$ greater than both $1/\rho_j b$ and $1/\omega\ell_j b$ leading to the values $Z \approx -i\omega L_j$ and $Y'' \approx -i\omega C_j''$, which gives $Z_j = (b/a)\sqrt{\mu_j/\varepsilon_j}$ and $k_j = \omega\sqrt{\mu_j\varepsilon_j}$ with phase velocity $v_j = \omega/k_j = 1/\sqrt{\mu_j\varepsilon_j}$. When the surface impedance of the bounding medium can not be neglected, the phase velocity becomes

$$v_j = \frac{1}{\sqrt{C_j''\left(L_j + \frac{2i}{\omega} Z_g\right)}} \tag{20}$$

Taking the bounding medium to be a lossless superconducting grain, we have

$$Z_g = \sqrt{-i\omega\mu_g z_g} = -\, i\omega\sqrt{\mu_g \ell_g} = -\, i\omega\mu_g\lambda_g \qquad (21)$$

where we have taken for the specific impedance $z_g = -i\omega\ell_g$ with $\ell_g = \mu_g\lambda_g^2$. The phase velocity $v_j = 1/\sqrt{C_j''(\mu_j b + 2\mu_g\lambda_g)}$ is substantially reduced by the penetration of rf fields for $\lambda_g > b$.

Josephson stripline

We next consider Josephson tunneling across the shunted junction shown in Fig. 4. If the series impedance Z is reactive as given by Eq. (21) and the shunt admittance is dominated by Josephson tunneling, the propagation vector $k = i/\lambda_j$ is imaginary with Josephson penetration depth

$$\lambda_j = \sqrt{\frac{\ell_j b}{2L_g}} = \sqrt{\frac{\Phi_0}{4\pi\mu_g\lambda_g J_0\cos\delta}} \qquad (22)$$

where $L_g = \mu_g\lambda_g$ is the surface inductance of a London superconductor. For k imaginary, the voltage decays exponentially along the junction. A dc current across the junction reduces $\cos\delta$ and increases the Josephson penetration depth λ_j. Considering both shunt inductivity and shunt resistivity, the input impedance of a junction of width a and separation b is

$$Z_j = -i\frac{\omega}{a}\sqrt{\frac{2\ell_j b L_g}{1 - i\omega\ell_j/\rho_j}} \qquad (23)$$

Sample surface impedance

A granular sample is modeled as an array of junctions on a square mesh of cell size a_g. We write the effective-medium surface impedance of a fully granular sample by simply inserting the input impedance of the grain boundaries in series with the surface impedance of the grains

$$Z_s = Z_g + Z_j \qquad (24)$$

We generalize Eqs. (21) and (22) to allow for the possibility that the impedance Z_g of an intergranular surface has both resistive and inductive components, leading to a Josephson wave vector in the grain boundaries and effective surface impedance of the composite sample

$$k_j = \sqrt{-2Z_g\left(\frac{1}{\rho_j b} - \frac{1}{i\omega\ell_j b}\right)} \qquad Z_s = Z_g + \frac{1}{a_g}\sqrt{2Z_g\left(\frac{1}{\rho_j b} - \frac{1}{i\omega\ell_j b}\right)^{-1}} \qquad (25)$$

12.3.2. Effective medium model

Hylton, Beasley *et al.*[22,23] have developed a weakly coupled grain model for granular superconductors in which a granular inductance ℓ_g/a is in series with an intergranular inductance $f(b/a_g^2)\ell_j$ where $f = 2\lambda_g/a_g$ is the weighting factor that gives the appropriate flux penetration. In shunt with the intergranular inductance is the resistance $f(b/a_g^2)\rho_j$.

The specific impedance of the medium is

$$z = -i\omega\ell_g + \frac{2\lambda_g b}{a_g^2}\frac{-i\omega\ell_j}{1 - i\omega\ell_j/\rho_j} \tag{26}$$

and the surface impedance is

$$Z_s = \sqrt{-i\omega\mu z} = Z_g\sqrt{1 - \frac{2i\omega\ell_j b/a_g^2}{Z_g(1 - i\omega\ell_j/\rho_j)}} \tag{27}$$

Comparing Eqs. (25) and (27) we see that in the limit that the grains are weakly coupled with $bz_j >> a_g^2 Z_g$, the expressions are equivalent. In this limit and assuming $\rho_j >> \omega\ell_j$, the effective surface inductance and resistance are

$$L_s \approx \frac{1}{a_g}\sqrt{2\ell_j b L_g} = \frac{2\lambda_j}{a_g}L_g \qquad\qquad R_s = \frac{\omega\ell_j}{2\rho_j}\omega L_s \tag{28}$$

dc magnetic fields

A dc magnetic field reduces the maximum Josephson current J_c to a value less than J_0[24] and by Eq. (12) increases ℓ_j as $1/J_c$. For $\rho_j << \omega\ell_j$ [the limit opposite to that of Eq. (28)], the effective impedance in Eq. (25) with $Z_g = -i\omega L_g$ becomes

$$Z_s = Z_g + \frac{1-i}{a_g}\sqrt{\omega L_g \rho_j b} \tag{29}$$

which is the classical skin depth limit. Portis *et al.*[5] for Tl-based films and Hein *et al.*[9] for electrophoretically deposited layers of $YBa_2Cu_3O_{7-x}$ and Remillard *et al.*[16] for $Bi_2Sr_2CaCu_2O_x$ have extensively studied the magnetic field dependence of the microwave surface impedance. In no case was it possible to take Z_{eff} to the limit of Eq. (29), indicating that ℓ_j can not be increased without limit. This result suggests that grains may be linked, at least in part, by superconducting bridges that are not affected by dc fields.

---

[22] T. L. Hylton, A. Kapitulnik, M. R. Beasley, J. P. Carini, L. Drabeck and G. Grüner, *Appl. Phys. Lett.* **53**, 1343 (1988).

[23] T. L. Hylton and M. R. Beasley, *Phys. Rev. B* **39**, 9042 (1989).

[24] M. Tinkham, *Introduction to Superconductivity* (McGraw-Hill, New York, 1975) sec 6-2; reprinted (Krieger, Malabar, FL, 1980).

12.4. Surface Impedance

12.4.1. Granular materials

We examine the field dependence of the surface impedance for the Josephson inductivity shunted by resistivity, approximating the two-fluid model. The surface impedance of an inductor and resistor in parallel is given by Eq. (13). Introducing the loss tangent, $\omega\ell/\rho = \tan\phi$, the surface impedance becomes

$$Z_s = -iR_c\sqrt{2\sin\phi}\; e^{i\phi/2} \tag{30}$$

with the surface resistance

$$R_s = R_c\sqrt{2\sin\phi}\;\sin\frac{\phi}{2} \tag{31}$$

and the surface reactance

$$X_s = R_c\sqrt{2\sin\phi}\;\cos\frac{\phi}{2} \tag{32}$$

and with the classical surface resistance $R_c = \sqrt{\omega\mu\rho/2}$. Portis *et al.*[5] interpret the effect of a static magnetic field on the surface impedance by assuming that the field acts to increase the Josephson inductivity without affecting the shunt impedance. This is equivalent to taking ϕ variable with ρ fixed. Fig. 5 shows R_s/R_c and X_s/R_c as functions of ϕ with the physical range up to $\pi/2$. Note that X_s goes through a maximum within this range while R_s does not.

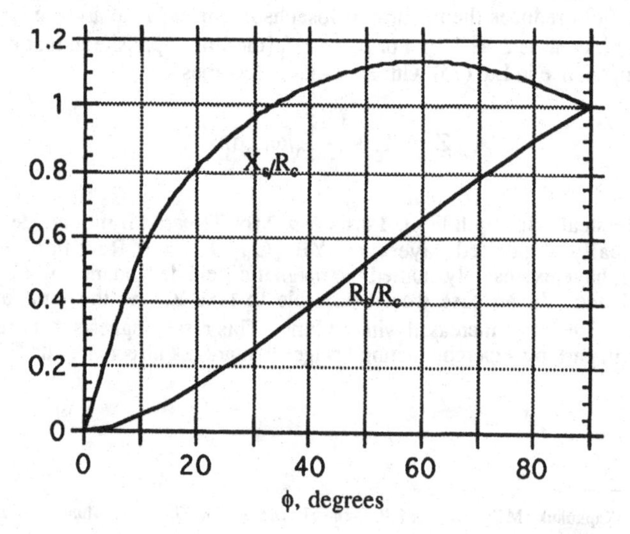

Figure 5. R_s/R_c and X_s/X_c are plotted as functions of ϕ with the physical range up to $\pi/2$. Note that X_s goes through a maximum within this range while R_s does not.

The normalized surface resistance R_s/R_c is plotted against X_s/R_c in Fig. 6. Although R_s may be determined unambigously, X_s is known only to within a constant. In order to locate the operating point on a plot of R_s/R_c *vs.* X_s/R_c we take the derivative with changing magnetic field, unambigously locating the operating point

$$\frac{dR_s}{dX_s} = \tan\tfrac{3}{2}\phi \tag{33}$$

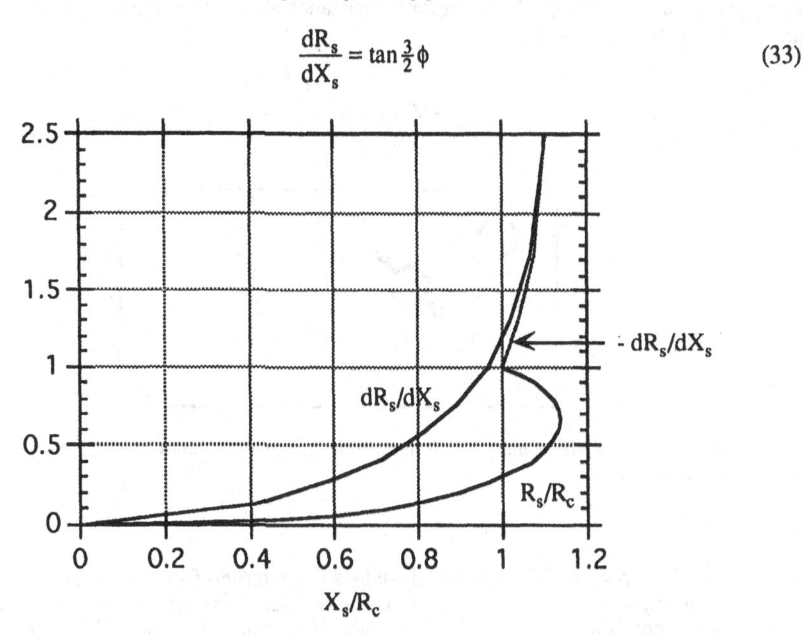

Figure 6. The normalized surface resistance R_s/R_c is plotted against X_s/R_c. Although R_s may be determined unambigously, X_s is known only to within a constant. In order to locate the operating point on a plot of R_s/R_c *vs.* X_s/R_c we use the derivative dR_s/dX_s, which may be determined experimentally.

12.4.2. Granular inclusions

The discussion of the preceding section was concerned with fully granular material, where currents are forced to flow across grain boundaries. In this section we model high quality thin films in terms of granular *inclusions* for which we apply an effective-medium surface model. We assume grains of diameter a_g separated by a distance d_g as sketched in Fig. 7.

The sample surface impedance is approximately

$$Z_s \approx (1-f)Z_g + f\frac{Z_g[Z_g + (2-f)Z_j]}{Z_g + (1-f)(2-f)Z_j} \tag{34}$$

with the fraction $f = a_g/d_g$. Note for $Z_j >> Z_g$ that the surface impedance approaches

$$Z_s \approx \left(1+\frac{f^2}{1-f}\right)Z_g \tag{35}$$

independent of Z_j. For f small, the surface impedance increases as the fractional area, which is quadratic in f

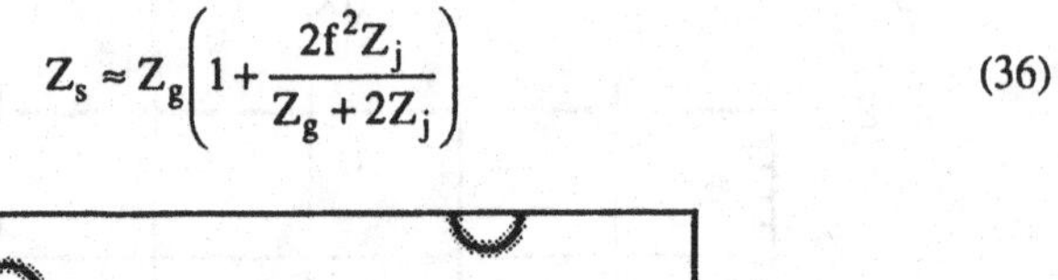

$$Z_s \approx Z_g\left(1+\frac{2f^2Z_j}{Z_g+2Z_j}\right) \tag{36}$$

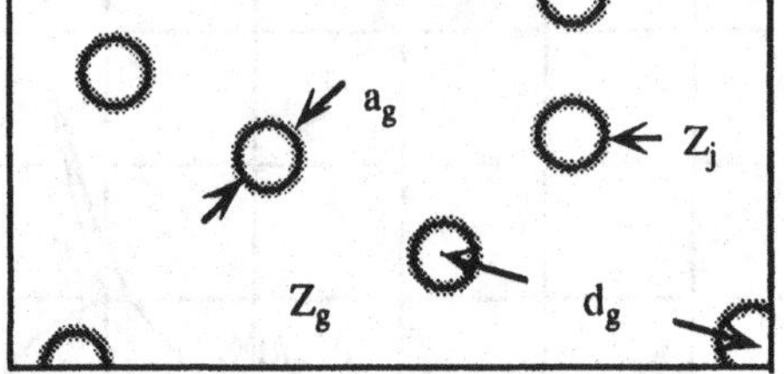

Figure 7. A weakly granular film is modeled by granular inclusions of diameter a_g and mean separation d_g.

Patterned films

Because of the crowding of current at the edges of patterned films, as discussed in Ch. 7, current is strongly constrained to flow across grain boundaries that that intersect the edges. Under these conditions the effective surface impedance is much closer to the one-dimensional result

$$Z_s \approx Z_g + \frac{d_f}{d_g} Z_j \tag{37}$$

where d_f is a characteristic distance for current-crowding at the edges. Except for the reduction factor d_f/d_g, this is the expression for a fully granular film, suggesting why patterned structures are of much lower quality than the films from which they originate.

13
MICROWAVE ABSORPTION IN TRANSIENT MAGNETIC FIELDS

13.1. Absorption in a Swept Magnetic Field

13.1.1. Introduction

Cardona *et al.*[1] first reported hysteresis in the microwave surface resistance of Pb-In and Pb-Tl alloys as well as Nb_3Sn. Their data showed a curious loop when an increasing magnetic field was reversed. Gittleman and Rosenblum[2] reported generally similar results for Pb-In alloys plated with Cu. Walton *et al.*[3,4] have also observed similar loops in the surface resistance at field-reversal as shown in Fig. 1. Stalder *et al.*[5,6] have reported the observation of absorption troughs in scanned signals in ceramic samples of $YBa_2Cu_3O_7$ as shown in Fig. 2.

13.1.2. Flux-induced loss

The hysteresis shown in Fig. 1 results from trapped magnetic flux and establishes that at least a component of the surface resistance is associated with flux density. The generally linear increase in surface resistance indicates that another component of the signal is

---

[1] M. Cardona, J. Gittleman and B. Rosenblum, "An intrinsic hysteresis in type II superconductors," *Phys. Lett.* **17**, 92 (1965).

[2] J. I. Gittleman and B. Rosenblum, "Intrinsic hysteresis in a copper plated type II superconductor," *Phys. Lett.* **20**, 453 (1966).

[3] B. L. Walton, B. Rosenblum and F. Bridges, "Nucleation of vortices in the superconducting mixed state: Nascent vortices," *Phys. Rev. Lett.* ***32***, 1047 (1974).

[4] B. L. Walton and B. Rosenblum, in *Low Temperature Physics-LT13*, Vol. 3, eds. W. J. O'Sullivan, K. D. Timmerhaus and E. F. Hammel (Plenum, New York, 1973) pp. 172-176.

[5] M. Stalder, G. Stefanicki, M. Warden, A. M. Portis and F. Waldner, "Nonlinear microwave response to oscillating fields: Critical current and field penetration in high-T_c oxides," *Physica C* **153-155**, 659 (1988).

[6] M. Warden, M. Stalder, G. Stefanicki, A. M. Portis and F. Waldner, "Nonlinear microwave response to scanning fields in high-T_c oxides," *J. Appl. Phys.* **64**, 5800 (1988).

associated with flux-flow loss.

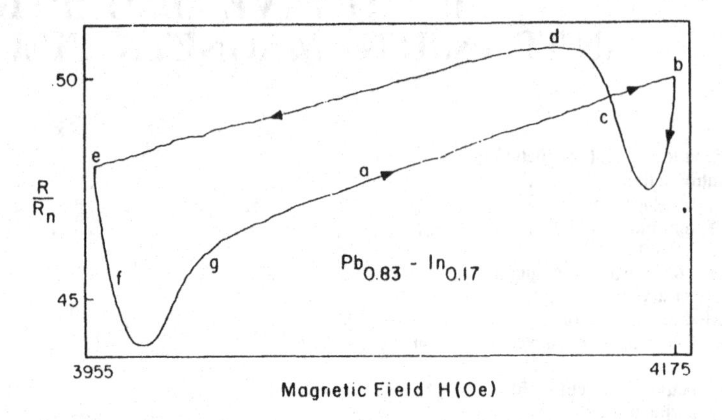

Figure 1. A recorder tracing of the hysteresis loop of the microwave absorption at $\nu = 35$ GHz and $H \approx 0.8\ H_{c2}$ and $T = 1.7$ K. The magnetic field is parallel to the surface and perpendicular to the microwave currents. Data of Walton *et al.*[3,4].

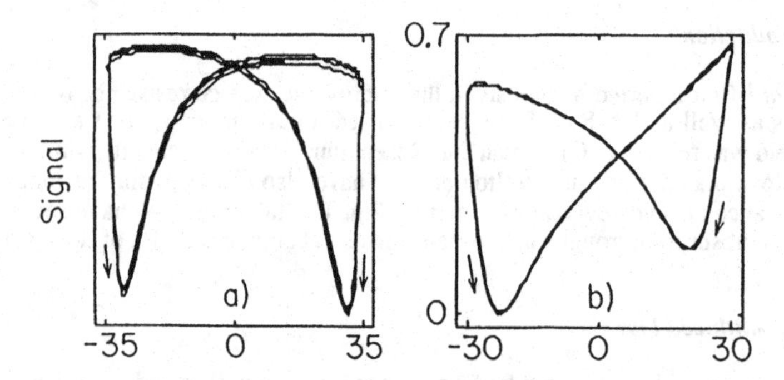

Figure 2. Measured microwave absorption signal *vs.* scanning field H in μT. (a) T = 4.2 K, (b) T = 77 K. Experimental data of Stalder *et al.* [5,6].

13.1.3. Current-induced loss

Portis *et al.*[7,8,9] found that the signals of Stalder *et al.*[5,6] could be fit with a calculation of the dc surface current density associated with gradients in the dc magnetic field that penetrates the superconductor and weighted by the local microwave energy density. Calculated curves, which are to be compared with Fig. 2, are shown in Fig. 3.

---

[7] A. M. Portis, K. W. Blazey and F. Waldner, "Critical state and flux pinning in high-T_c superconductors," *Physica C* **153-155**, 308 (1988).

[8] K. W. Blazey, A. M. Portis and J. G. Bednorz, "Microwave study of the critical state in high-T_c superconductors," *Solid State Commun.* **65**, 1153 (1988).

[9] A. M. Portis, M. Stalder, G. Stefanicki, F. Waldner and M. Warden, " Critical state model for cuprate superconductors," *J. de Phys. Colloque C* **8**, 2231 (1988).

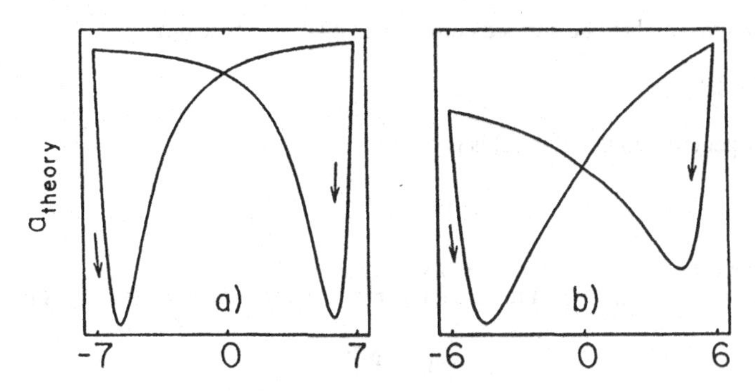

Figure 3. Weighted surface dc current *vs.* the scaled scanning field H/H'. (a) $\lambda' = \lambda_{rf}$. (b) $\lambda' = {}^1/_3\, \lambda_{rf}$.

The analysis recognizes five field ranges as sketched in Fig. 4.

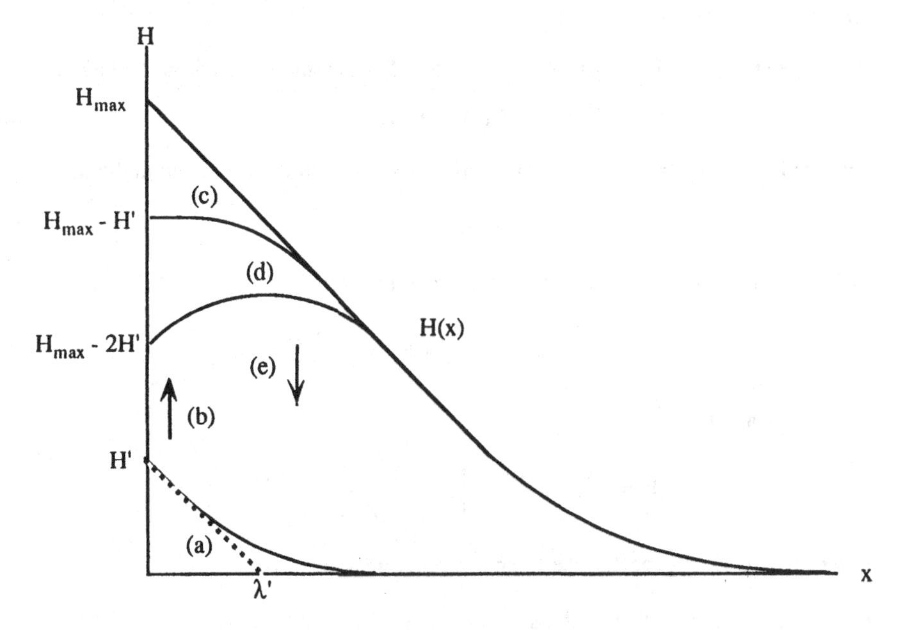

Figure 4. Sketch of the five field ranges associated with current-modulated surface resistance.

(a) The first range is for fields H less than the minimum field H' at which hysteresis develops.
(b) The second range for H > H' extends indefinitely.
(c) The third range is below the maximum field H_{max} and extends down to H_{max} - H'.
(d) The fourth range is between H_{max} - H' and H_{max} - 2H'.
(e) The fifth and final range extends indefinitely below H_{max} - 2H'.

We now obtain the weighted *dc* surface current density in these field ranges.

Range (a): $H < H'$

The initial flux penetration in range (a) is given by

$$H(x) = He^{-x/\lambda'} \tag{1}$$

where λ' is the elastic relaxation distance of vortices.
The microwave energy density and resultant mean of the absolute dc current density are

$$u(x) = u(0)e^{-2x/\lambda_{rf}} \tag{2}$$

$$\langle|J|\rangle = \frac{2}{\lambda_{rf}}\int_0^{\infty}\frac{dH(x)}{dx}e^{-2x/\lambda_{rf}}dx = \frac{H}{\lambda'+\lambda_{rf}} \tag{3}$$

Range (b): $H > H'$

In the open range (b) the magnetic field exceeds H' at the surface and drops linearly as

$$H(x) = H - J_c x \tag{4}$$

down to $H = H'$ at $x = (H - H')/J_c = x'$. For $x > x'$ the field drops exponentially as

$$H(x) = H' e^{-(x-x')/\lambda'} \tag{5}$$

The mean of the absolute dc current is the sum of two terms

$$\langle|J|\rangle = \frac{2J_c}{\lambda_{rf}}\left[\int_0^{x'} e^{-2x/\lambda_{rf}}dx + \int_{x'}^{\infty} e^{-(1/\lambda'+2/\lambda_{rf})x}dx\right] \tag{6}$$

which evaluates to

$$\langle|J|\rangle = J_c\left[1 - e^{-2x'/\lambda_{rf}}\right] + \frac{H'}{\lambda'+\lambda_{rf}/2}e^{-(1/\lambda'+2/\lambda_{rf})x'} \tag{7}$$

For $x >> x'$, $\langle|J|\rangle$ approaches the Bean critical current density J_c.

Range (c): $H_{max} - H' < H < H_{max}$

The third range immediately follows reversal of the magnetic field at a field H_{max} assumed here for simplicity to be much larger than H'. The magnetic field is

$$H(x) = H_{max} - xJ_c + (H - H_{max})e^{-x/\lambda'} \tag{8}$$

The expression for the mean of the absolute dc current is

$$\langle|J|\rangle = \frac{2J_c}{\lambda_{rf}}\int_0^\infty \left(1 - \frac{H_{max} - H}{H'} e^{-x/\lambda'}\right) e^{-2x/\lambda_{rf}} dx \tag{9}$$

This integral evaluates to

$$\langle|J|\rangle = \left[1 - \frac{H_{max} - H}{H'} \frac{1}{1 + \lambda_{rf}/2\lambda'}\right] J_c \tag{10}$$

Range (d): $H_{max} - 2H' < H < H_{max} - H'$

In range (d) the field has been reduced to the second interval of H' below H_{max}. In this region the field initially increases with x at a rate less than J_c with the current positive for $x < x_0$. The mean of the absolute current is the sum of two terms

$$\langle|J|\rangle = \frac{2J_c}{\lambda_{rf}}\int_0^{x_0}\left[e^{-(x-x_0)/\lambda'} - 1\right]e^{-2x/\lambda_{rf}}dx - \frac{2J_c}{\lambda_{rf}}\int_{x_0}^\infty\left[1 - e^{-(x-x_0)/\lambda'}\right]e^{-2x/\lambda_{rf}}dx \tag{11}$$

with $x_0 = \lambda' \ln [(H_{max} - H)/H']$. The mean of the absolute current is

$$\langle|J|\rangle = \frac{H_{max} - H}{\lambda' + \frac{1}{2}\lambda_{rf}} + \frac{2J_c e^{-2x_0/\lambda_{rf}}}{1 + 2\lambda'/\lambda_{rf}} - J_c \tag{12}$$

Range (e): $H < H_{max} - 2H'$

In this open field range, H increases linearly at a rate J_c up to $x = x'_{min}$. For $x > x'_{min}$ the current decreases and goes through zero, ultimately reaching $-J_c$. The dc current is the sum of three terms

$$\langle|J|\rangle = \frac{2J_c}{\lambda_{rf}}\int_0^{x'_{min}} e^{-2x/\lambda_{rf}}dx + \frac{2J_c}{\lambda_{rf}}\int_{x'_{min}}^{x_0}\left[e^{-(x-x_0)/\lambda'} - 1\right]e^{-2x/\lambda_{rf}}dx + \frac{2J_c}{\lambda_{rf}}\int_{x_0}^\infty\left[1 - e^{-(x-x_0)/\lambda'}\right]e^{-2x/\lambda_{rf}}dx \tag{13}$$

with $x'_{min} = (H_{max} - 2H' - H)/J_c$. The dc current in this range evaluates to

$$\langle|J|\rangle = (1 - 2e^{-2x'_{min}/\lambda_{rf}})J_c + \frac{2J_c e^{-2x_0/\lambda_{rf}}}{1 + 2\lambda'/\lambda_{rf}} + \frac{J_c e^{(x_0 - x'_{min})/\lambda'} e^{-2x'_{min}/\lambda_{rf}}}{1 + \frac{1}{2}\lambda_{rf}/\lambda'} \tag{14}$$

In Fig. 5 is sketched the mean of the absolute dc current in these five ranges under the assumption that H_{max} is large compared with H' so that the trough at the turnaround does not impinge on the initial rise. (This condition is lifted for the simulations in Fig. 3.)

After discussing field scans in which the field is sinusoidally modulated, we return to a consideration of the connection between the mean of the absolute current and the scanned surface resistance.

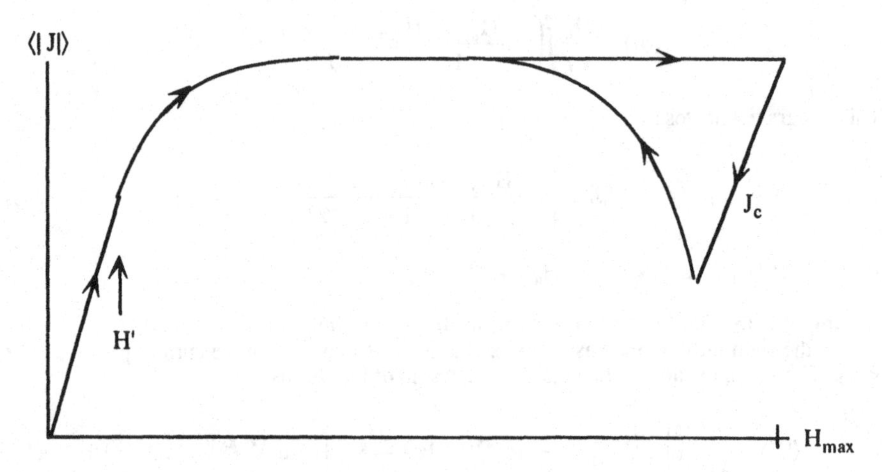

Figure 5. Sketch of the weighted mean of the absolute current in the five identified ranges of magnetic field.

13.2. Modulated Microwave Absorption

13.2.1. Materials

Metal films

Kim *et al.*[10] first studied the magnetic field-modulated microwave surface resistance of Pb and Sn films and found that for modulation fields below 0.1 Oe, the phase of the signal reversed with the direction of sweep of the magnetic field. For larger modulation fields the signal is independent of sweep direction. It was recognized that the dominant effect in the large-amplitude case is modulation of the flux density. In the small-amplitude case, on the other hand, the effect of modulation is primarily to change the boundary current. Given the swept-field behavior shown in Fig. 1, such a phase reversal is to be expected for modulation fields $H_m < H'$.

Copper oxide ceramics

Soon after the discovery by Bednorz and Müller[11] of superconductivity in $(La_{1-x}Ba_x)_2 CuO_{4-y}$, Blazey *et al.*[12] undertook a systematic investigation of the magnetic field dependence of magnetically modulated microwave absorption in the La-Ba-Cu-O and La-Sr-Cu-O systems, using a commercial 9-GHz electron-spin-resonance (ESR) spectrometer. As mentioned in Sec. 11.2.1, their samples were compressed sintered pellets, prepared from stoichiometric quantities of the metallic oxalates that had been precipitated from

[10] Y. W. Kim, A. M. de Graaf, J. T. Chen, E. J. Friedman and S. H. Kim, "Phase reversal and modulated flux motion in superconducting thin films," *Phys. Rev. B* **6**, 887 (1972).

[11] J. G. Bednorz and K. A. Müller, *Z. Phys. B* **64**, 189 (1986).

[12] K. W. Blazey, K. A. Müller, J. G. Bednorz, W. Berlinger, G. Amoretti, E. Buluggiu, A. Vera and F. C. Mattacotta, *Phys. Rev. B.* **36**, 7241 (1987).

solution with oxalic acid. These and other early studies of microwave absorption were limited to pressed and sintered polycrystalline samples and to powders prepared by regrinding sintered material.

Bhat *et al.*[13,14] were among the first to publish a study of magnetically modulated microwave absorption in the Y-Ba-Cu-O system[15] following the report of 90 K superconductivity in this material. As with the earlier systems, a strong modulated absorption appeared only below the superconducting transition and was most pronounced near zero magnetic field. Bhat *et al.*[14] also found strongly field- and temperature-dependent absorption in the radio frequency range 8 to 22 MHz. Particularly striking was their observation that ambient oxygen substantially increased the strength of the modulated absorption and their conclusion that the observed absorption must be associated with the oxygenation of Josephson junctions.

A systematic study of the effect of modulation amplitudes in the high-T_c superconducting oxides has been reported by Khachaturyan *et al.*[16] as shown in Fig. 6 with results very similar to those of Kim *et al.*[10]

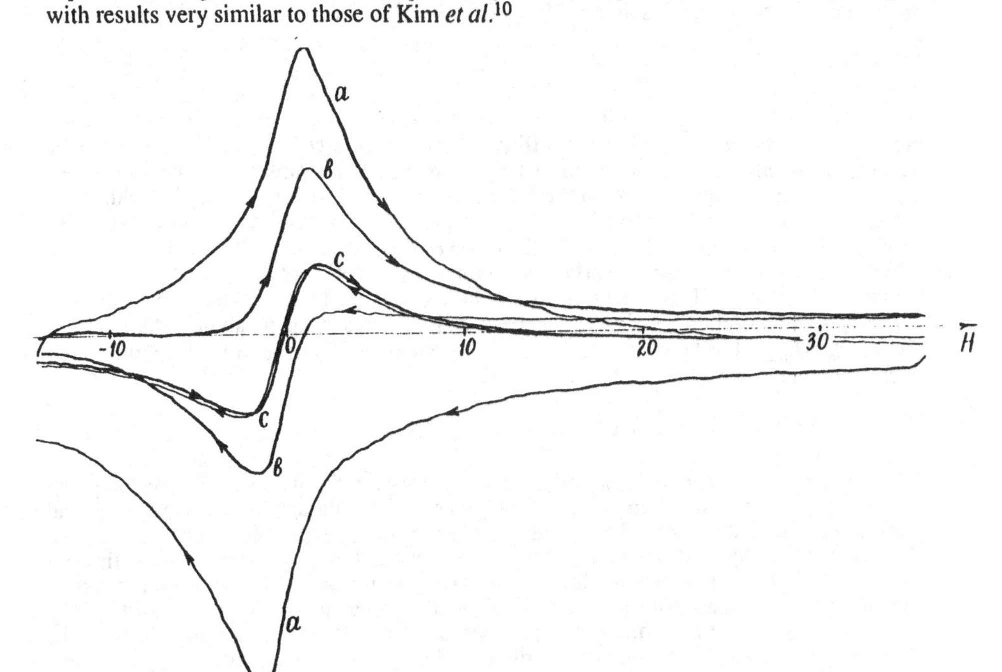

Figure 6. Absorption signal from $La_{1-x}Sr_xCuO_4$ for different modulation amplitudes at T = 15 K. (a) H_m = 0.01 Oe and receiver gain G = 2500. (b) H_m = 0.1 Oe and G = 250. (c) H_m = 1 Oe and G = 25. Incident microwave power is 20 dB below 200 mW. Measurements of Khachaturyan *et al.* [16].

---

[13] S. V. Bhat, P. Ganguly, T. V. Ramakrishnan and C. N. R. Rao, *J. Phys. C* **20**, L559 (1987).

[14] S. V. Bhat, P. Ganguly and C. N. R. Rao, *Pramana-J. Phys.* **28**, L425 (1987).

[15] M. K. Wu, J. R. Ashburn, C. J. Torng, P. H. Hor, R. L. Meng, L. Gao, Z. J. Huang, Y. Q. Wang and C. W. Chu, *Phys. Rev. B* **58**, 908 (1987).

[16] K. Khachaturyan, E. R. Weber, P. Teledor, A. M. Stacy and A. M. Portis, "Microwave observation of magnetic field penetration of high-T_c superconducting oxides," *Phys. Rev. B* **36**, 8309 (1987).

Single crystals

Karim *et al.*[17] have observed modulated microwave absorption in single crystals of $YBa_2Cu_3O_{7-\delta}$ with the static magnetic field H parallel to the *c*-axis and a peak modulation field of 2 Oe. At low temperatures and also at temperatures just below T_c the signal indicates the absorption minimum that is expected at zero magnetic field. At intermediate temperatures between 39.5 and 52 K, the modulated signal unexpectedly reverses sign, indicating an absorption maximum at zero field.

Blazey and Mangelschots[18] have observed the temperature dependence of the modulated microwave absorption in the n-type high termperature superconductor $Nd_{2-x}Ce_xCuO_4$. The absorption signal is a maximum near T_c rather than a minimum as observed in ceramic samples. Dulcic *et al.*[19] have measured the absorption in $YBa_2Cu_3O_x$ single crystals as a function of field and temperature and identify the various signals that develop as the temperature is lowered from T_c to 4.2 K.

Films

Yau *et al.*[20] have used magnetically modulated microwave absorption to study critical currents in sputtered Y-Ba-Cu-O thin films. A film sputtered onto (100) single crystal MgO is oriented with the *c*-axis normal to the plane of the substrate. The modulated absorption signal is observed with the film initially cooled in zero magnetic field. H is increased to + H_{max}, then reduced to - H_{max} and finally increased to + H_{max} again to complete the cycle. With H normal to the plane of the film and parallel to the *c*-axis, the initial modulated signal rises linearly to a maximum and then decreases slowly as the field is increased toward + H_{max}. As the field is reduced, the signal is reversible down to a field + ΔH at which it changes sign. The field ΔH increases up to about 500 Oe with increasing H_{max}. For H parallel to the substrate and in the *ab*-plane of the film, the signal is greatly reduced and exhibits little hysteresis.

13.2.2. Modulation harmonics

Golosovsky *et al.*[21] have reported microwave transmission through YBCO films prepared by spray pyrolysis with two important innovations. First, although other investigators had studied the temperature dependence of the amplitude and phase of microwaves transmitted through YBCO films, this paper was the first to examine the effect of a magnetic field on the transmission. The second innovation was the introduction of large-amplitude modulation of the magnetic field. Although, as the paper indicates, there is little in the modulation technique that can not be discerned from the response to a swept dc field, the generation of modulation harmonics is striking and provides a ready diagnostic. The earlier studies of YBCO films have been extended to BSCCO films, also prepared by spray pyrolysis, with similar results.[22]

---

[17] R. Karim, H. How, R. Seed, A. Widom, C. Vittoria, G. Balestrino and P. Paroli, *Solid State Commun.* **71**, 983 (1989).

[18] K. W. Blazey and I. Mangelschots, *Physica C* **170**, 267 (1990).

[19] A. Dulcic, R. H. Crepeau and J. H. Freed, *Phys. Rev. B* **38**, 5002 (1988).

[20] W. F. Yau, A. M. Portis, E. R. Weber and N. Newman, *MRS Spring Meeting*, San Francisco, 16-21 April 1990.

[21] M. Golosovsky, D. Davidov, C. Rettori and A. Stern, *Phys. Rev. B* **40**, 9299 (1989).

[22] M. Golosovsky, D. Davidov, E. Farber, T. Tsach and M. Schieber, *Physica A* **168**, 353 (1990).

Revenaz *et al.*[23,24,25,26] have observed modulation harmonics from granular, textured, epitaxial and ion-irradiated thin films of $YBa_2Cu_3O_7$ as well as from granular lead films. The films were cut to 1.5 × 3 mm. and were placed within the He gas-flow crystat at the center of the TE_{102} cavity of a commercial X-band EPR spectrometer. Results at zero dc field and within a few degrees of T_c are shown in Fig. 7. As seen in Fig. 7 (a) the transmitted power is cusp-like with the detected power nearly proportional to the absolute value of the modulation field, suggesting for the amplitude of the Fourier-components with n greater than zero and even

$$A_n = \frac{1}{\pi}\int_0^{2\pi} |\sin\theta| \cos n\theta \, d\theta = -\frac{4}{\pi}\frac{1}{n^2-1} \tag{15}$$

and $A_n = 0$ for n odd. We calculate $A_0 = 2/\pi$, which is just the value for the dc component of a full-wave rectified ac signal. The measured Fourier amplitudes shown in Fig 7 (b) fall off somewhat more slowly than Eq. (15) at small n but may drop as rapidly as $1/n^2$ for larger n, a consequence of the presence of a reasonably sharp cusp in the modulated signal.

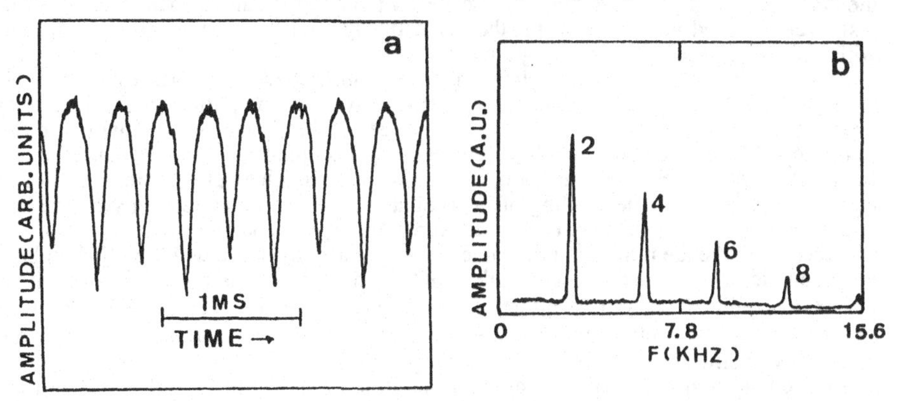

Figure 7. (a) Variation of the power reflected from the spectrometer cavity as a function of time in a sinusoidal magnetic field. (b) Simultaneously recorded Fourier spectrum. The sample is laser-ablated epitaxial $YBa_2Cu_3O_7$. The modulation frequency is 1.56 kHz, the amplitude is 5 Oe and the reduced temperature is 0.8.

Current modulation

Golosovsky *et al.*[27] have found that an ac current passed through their films also modulates the transmitted microwaves with the generation of even harmonics of the ac frequency. The addition of a dc current leads to odd harmonics as well, analogous to the effect of ac and dc magnetic fields. Sputtered BSCCO films and a laser-ablated YBCO film

---

[23] S. Revenaz, J. Dumas, R. Buder, P. L. Reydet, J. Marcus and C. Schlenker, *Physica C* **162-164**, 1585 (1989).

[24] S. Revenaz, J. Dumas, C. J. Liu, C. Schlenker, S. Orbach, C. Müller, N. Klein and H. Piel, *J. Less Comm. Met.* **164-165**, 1252 (1990).

[25] S. Revanaz, J. Dumas and A. Gerber, *Physica C* **180**, 180 (1991).

[26] S. Revenaz, J. Dumas, A. Gerber and J. Schubert in *High T_c Superconductor Thin Films*, ed. L. Correra (North-Holland, Amsterdam, 1992) pp 101-106.

[27] M. Golosovsky, Y. Naveh, D. Davidov, H. Raffy, M. Schieber and S. Chocron, *Physica C* **176**, 379 (1991).

were also examined. The laser-ablated YBCO produced no harmonics except close to T_c, where even high quality films are known to become granular. Nor were current-induced harmonics observed well below the T_c of a BSCCO film prepared by spray pyrolysis. The authors conclude that the observed modulation is a consequence of the effect of current on weak links. For those films that do yield harmonics, a dc field is found to reduce their amplitude.

13.2.3. Mechanisms of modulated absorption

The observed linear increase in surface resistance with applied field[3,28,29,30] is a strong indicator of viscous losses associated with flux flow. It is curious that these losses do not show up in magnetically modulated microwave absorption, which might be expected to yield the magnetic-field derivative of this absorption. It is because of magnetic hysteresis that the magnetically modulated absorption does not give the expected derivative signal. For example, as the magnetic field is increased above H_{c1}, the flux density within grains increases and, as observed, the microwave absorption increases. If the magnetic field is now slightly reduced, as over a modulation cycle, flux leaves grain boundaries rather than the grains, producing a signal that is associated with intergranular loss, which arises from resistively shunted junctions. Thus, the magnetically modulated ac signal is associated with grain boundaries and not with grains.

The signals obtained from modulated microwave absorption at the surface of a granular superconductor in a magnetic field depend sensitively on the amplitude of the magnetic-field modulation and on the sign of the field sweep. The signals are insensitive to the frequency of field modulation. Only at low modulation amplitudes are the signals sensitive to the rate of field sweep and then for quite low sweep rates. These effects are all shown to have their origin in the decoupling of intergranular Josephson junctions in synchronism with the modulation field, either by surface current or by magnetic flux.

Because of the identification of the microwave surface resistance with flux flow in bulk superconductors, early studies concluded that the modulated signals arose from either vortex nucleation[3,4] or from the effect of boundary current on the flux lattice.[10] Dulcic *et al.*[31] have correctly identified the current-modulated signal with the variation in the specific inductivity of critical current-biased Josephson junctions.[32]

The connection between the scanned absorption as shown in Figs. 1 and 2 and the modulated absorption as shown in Fig. 6 can readily be understood from Fig. 4 and the dynamics of the Bean critical state. For a slowly swept field

$$H(t) = v_H t + H_m \cos \omega_m t \tag{16}$$

with $v_H << \omega H_m$ and positive, the field is a maximum for $\omega t = 2n\pi$. For $H_m < H'$, the field is limited to ranges (c) and (d) in Fig. 4 and the modulated mean absolute current $\langle |J| \rangle$ is a maximum when the magnetic field is a maximum. On the other hand, for v_H negative, we can use Fig. 4 with H_{max} a negative field that corresponds to $H_{max} - H_m$ with $\langle |J| \rangle$ maximum for $\omega t = (2n + 1)\pi$. Any source of absorption that is modulated by $\langle |J| \rangle$ then leads to signals that reverse in sign with the direction of field sweep. as argued by Kim *et al.*[10]

---

[28]M. Cardona, G. Fischer and B. Rosenblum, *Phys. Rev. Lett.* **12**, 101 (1964).

[29]J. I. Gittelman and B. Rosenblum, *Phys. Rev. Lett.* **16**, 734 (1966).

[30]W. H. J. Hackett, E. Maxwell and Y. B. Kim, *Phys. Lett.* **24A**, 663 (1967).

[31]A. Dulcic, B. Ravkin and M. Pozek, *Europhys. Lett.* **10**, 593 (1989).

[32]M. Mahel, R. Hlubina and S. Benacka, "Microwave study of the critical state in granular $YBa_2Cu_3O_x$ thin films," *Physica C* **169**, 429 (1990).

On the other hand, if the modulation amplitude H_m is much larger than H', the field extends well into region (e) over a modulation cycle and $\langle|J|\rangle$ has Fourier components largely at $2\omega_m$ with very little variation at ω_m. The same argument hold for v_H negative. Thus, the signal at large modulation amplitudes can not arise from $\langle|J|\rangle$ but must be associated with modulation of magnetic flux, again as pointed out by Kim *et al.*[10]

13.3. Flux Relaxation in Swept Fields

13.3.1. Experiment

Erhart *et al.*[33,34,35] have observed the modulated microwave absorption signal in superconducting samples of $YBa_2Cu_3O_x$ as a function of the rate $v_H = dH/dt$ at which the dc field is swept. In an initial study, the reverse-sweep hysteresis signal Δ of pressed powders was compared with that of sintered samples.[26] In sintered samples a power-law relation is observed between Δ and v_H. For pressed powders, on the other hand, the signal drops rapidly at low sweep rates, suggesting that Δ might be an exponential function of v_H.

A typical absorption signal for ceramic $YBa_2Cu_3O_x$ with x = 6.946 is shown in Fig 8 (a). The measurement was performed on a Varian-E spectrometer, operating at 9 GHz with 100 kHz magnetic field modulation. The separation Δ is typically measured at 500 Oe for various sweep rates and temperatures.

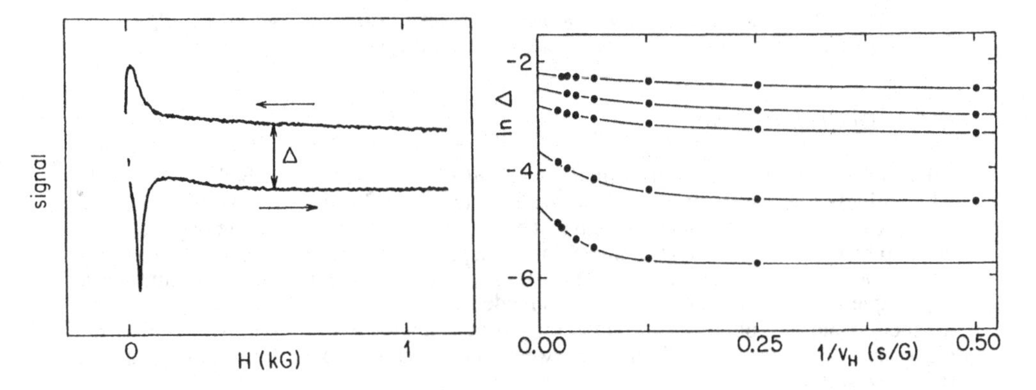

Figure 8. (a) Typical modulated microwave absorption signal obtained with ceramic $YBa_2Cu_3O_x$ at 9 GHz. The separation Δ between down and up sweep is measured as a function of field sweep. (b) Logarithm of the sweep separation Δ for $T/T_c = 0.81$, 0.85, 0.89, 0.95 and 0.99 *vs.* inverse sweep speed. The transition temperature T_c was determined from the disappearance of the low-field microwave signal. The curves are fits from the *minimal glassy model* [28].

In Fig. 8 (b) is shown a plot of the logarithm of Δ as a function of $1/v_H$. The curves are

[33] P. Erhart, J. E. Drumheller, B. Senning, S. Mini, L. Fransioli, E. Kaldis, S. Rusiecki and F. Waldner, *25'th Congress Ampere*, Stuttgart, 9-14 September 1990.

[34] P. Erhart, B. Senning, S. Mini, L. Fransioli, F. Waldner, J. E. Drumheller, A. M. Portis, E. Kaldis and S. Rusiecki, *Physica C* **185-189**, 2233 (1991).

[35] P. Erhart, J. E. Drumheller, A. M. Portis, B. Senning, S. Mini, L. Fransioli, E. Kaldis, S. Rusiecki and F. Waldner, *J. Magn. Magn Mat.* **104-107**, 487 (1992).

obtained from the *minimal glassy model* and are fitted to the data with the mean energy barrier $\langle u \rangle$ and the critical current density J_c as parameters. In Fig. 9 are plotted values of $\langle u \rangle / k_B T$ as a function of temperature and, in the inset, values of J_c. Both $\langle u \rangle$ and J_c are seen to approach zero as T approaches T_c.

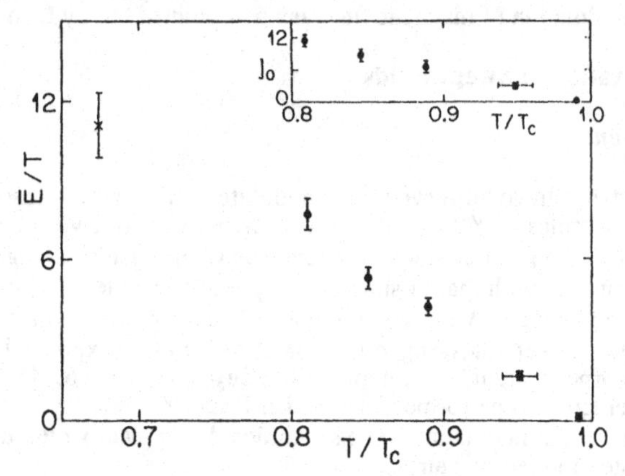

Figure 9. Values of $\langle u \rangle / k_B T$ as a function of temperature and, in the inset, values of J_c as a function of temperature. Both $\langle u \rangle$ and J_c are seen to approach zero as T approaches T_c [28].

13.3.2. Dynamical model

Erhart *et al.*[27,28] fit their results with the *minimal glassy model*, described in Sec. 6.2.1. Alternatively, the Anderson dynamical model may be used to fit the data. Although the dynamical model may be more restrictive, it has the advantage of being more physical. We develop here a phenomenological model for the dependence of boundary current on sweep-rate on the assumption that flux penetration is associated with the time decay of currents at the sample surface. We apply a dynamical model to the decay of field gradients in a swept magnetic field and obtain expressions for the dependence of Δ on v_H. To simplify the discussion, we assume initially that the sweep rate is rapid and look for deviations as the rate is reduced.

Bean critical state

Bean[36] as discussed in Sec. 6.1.2 and London and Kamper[37,38] have analyzed the penetration of magnetic flux into a type II superconductor in terms of the interaction of flux-lines (or current-vortices) with a pinning potential that arises from sample inhomogeneities. At low magnetic fields, repulsion between flux-lines can be neglected and a constant magnetic pressure drives flux into (or out of) the sample. At short times there is little relaxation of the flux distribution and we can expect the flux profile

$$B(x) = \mu(H - J_c x) \tag{17}$$

[36]C. P. Bean, *Phys. Rev. Lett.* **8**, 250 (1962).

[37]H. London, *Phys. Lett.* **6**, 162 (1963).

[38]R. A. Kamper, *Phys. Lett.* **2**, 290 (1962).

sketched in Fig. 4 for increasing and decreasing fields with the field gradient $dB/dx = -\mu J_c$ where J_c is called the Bean critical current. Elastic forces between flux lines[39] lead to smoothing of the flux profile over a characteristic distance λ' as also shown in Fig. 4. Finally, the flux gradients relax in time, leading to reduced current flow at the surface. It is this relaxation, both in time and space, that we examine here.

Rapid field sweep

Because we are concerned with the relaxation of flux-gradients rather than with relaxation of the flux itself, it is preferable to work with currents rather than with flux.

$$\mu J(x) = -dB/dx \tag{18}$$

We can always integrate the current distribution from 0 to x

$$B(x) = \mu\left[H + \int_0^x J(x)dx\right] \tag{19}$$

with H the field applied at the surface, to obtain the flux density B(x). For a rapidly increasing magnetic field, the current density

$$J^+(x) = J_c \tag{20}$$

is uniform. Application of a weak modulation field $H_m \cos \omega t$ leads to a reverse current

$$J^-(x) = J_c \pm J_m e^{-x/\lambda'} \tag{21}$$

When the sweep direction is reversed, J_c changes sign and the signal changes sign as well. The difference between the signals for positive and negative field scans is called the reverse scan hysteresis Δ and is proportional to the weighted current density

$$\Delta = R\int\left[J^+(x) - J^-(x)\right]e^{-2x/\lambda_{rf}}dx = \frac{RJ_m}{1/\lambda' + 2/\lambda_{rf}} \tag{22}$$

Slow field sweep

When the magnetic field is swept slowly, currents well within the sample have had sufficient time to relax and the current distribution may be substantially modified. Our phenomenological model is that of current laminae parallel to the surface. These laminae enter at the surface and are displaced into the medium. Moving into the medium with the laminae we write their relaxation as a convective derivative

$$dJ/dt = -J/\tau \tag{23}$$

where τ may be a function of the local current density J(x). Writing out the convective derivative and including the entry of laminae at the surface leads to the differential equation

---

[39] A. M. Campbell and J. E. Evetts, *Adv. Phys.* **21**, 199 (1972); reprinted as *Critical Currents in Superconductors* (Taylor and Francis, London, 1972).

$$\partial J/\partial t + J/\tau + v\, dJ/dx = v_H\, \delta(x) \tag{24}$$

where $v_H = dH/dt$ is the sweep rate and

$$v = - v_H/J_c \tag{25}$$

is the rate at which current laminae move into the medium. For the magnetic field swept at a steady rate, the current density J(x) is independent of time and satisfies the equation

$$v\, dJ/dx + J/\tau = 0 \tag{26}$$

Assuming a constant relaxation rate, Eq. (26) may be integrated to give for the spatial distribution of screening current

$$J = J_m e^{-x/v\tau} \tag{27}$$

which leads to the magnetic field

$$B(x) = \mu H + \mu \int_0^x J(x)dx = \mu H - \mu\, v_H \tau\left(1 - e^{-x/v\tau}\right) \tag{28}$$

The initial decay follows Eq. (15). Well within the medium the flux density approaches the value $\mu(H - v_H\tau)$. With weak field-modulation, the current alternates between $J^+(x)$ as given by Eq. (26) and a partially relaxed distribution, which we assume to be of the form

$$J^-(x) = [1 - (J_m/J_0)\exp(-x/\lambda')]J^+(x) \tag{29}$$

The difference current is

$$\Delta J(x) = J_m \exp[-(1/v\tau + 1/\lambda')] \tag{30}$$

and the reverse scan hysteresis is

$$\Delta = R \int \Delta J(x) \exp(-2x/\lambda_{rf})\, dx = RJ_m\, [(2/\lambda_{rf} + 1/\lambda') + 1/v\tau]^{-1} \tag{31}$$

Comparing this expression with Eq. (22), the relaxation is seen to add an additional term in the denominator. The reciprocal of Eq. (31) is

$$1/\Delta = (1/RJ_m)\, [(2/\lambda_{rf} + 1/\lambda') + 1/v\tau] \tag{32}$$

which suggests the usefulness of plotting $1/\Delta$ as a function of $1/v_H$,

Anderson-type relaxation

We know from the early experiments of Kim *et al.*[40] that screening currents decay

---

[40] Y. B. Kim, C. F. Hempstead and A. R. Strnad, *Phys. Rev. Lett.* **9**, 306 (1962); *Phys. Rev.* **129**, 528 (1963); *Phys. Rev.* **131**, 2486 (1963); *Rev. Mod. Phys.* **36**, 43 (1964).

logarithmically. As discussed in Sec. 6..2.1, Anderson[41,42] has developed a model of thermally activated relaxation with a barrier that is reduced by the presence of screening currents. Thus, as the current relaxes, the barriers become higher and the relaxation rates become slower. The current at long times is expected to decay with the logarithm of the time as observed experimentally. In this section we examine the short-time consequences of such a model, which should be applicable to flux gradients near the surface. We discussed in Sec. 6.2.3 a kinematic approach, the minimal glassy model, that also leads to logarithmic relaxation at long times.

Anderson has assumed current relaxation

$$dJ/dt + J/\tau = 0 \tag{33}$$

with a relaxation rate of the form

$$1/\tau = (1/\tau_0) \exp [(J - J_c)/J_1] \tag{34}$$

where J_1 is a characteristic current that establishes the current sensitivity of the barrier height. Anderson[32] found that at long times the current decreased linearly with the logarithm of the time, as observed experimentally.[40] Because our concern is with fields near the surface, we are interested in the short term rather than the long term solution. We assume that the change in J is relatively small and write a power expansion in the time

$$J(t) = J_c + J_1 t + J_2 t^2 + \cdots \tag{35}$$

Substituting into Eqs. (33) and (34) we obtain

$$J(t) = [1 - t/\tau_0 + {}^1/_2\,(1 + J_c/J_1)(t/\tau_0)^2 + \cdots]J_c \tag{36}$$

In a uniformly swept field J(x) is obtained from J(t) simply by replacing t in Eq. (36) by x/v to obtain

$$J(x) = [1 - x/v\tau_0 + {}^1/_2\,(1 + J_c/J_1)(x/v\tau_0)^2 + \cdots]J_c \tag{37}$$

The difference current with weak modulation is similar to that given by Eq. (30)

$$\Delta J(x) = J_m\, [\, [1 + (J_c/2J_1)(x/v\tau_0)^2\,]\, \exp[- (1/v\tau_0 + 1/\lambda')x] \tag{38}$$

except for reduced relaxation. Integrating Eq. (38) with the weighting factor $\exp(-2x/\lambda_{rf})$ gives for the reverse scan hysteresis as compared with Eq. (30)

$$1/\Delta \approx (2/\lambda R J_m)\, [\, [1 + (\lambda/2v\tau_0) - (J_c/J_1)(\lambda/2v\tau_0)^2\,] \tag{39}$$

where to simplify the notation we have written

$$1/\lambda = 1/\lambda_{rf} + 1/2\lambda' \tag{40}$$

We thus expect for the behavior of $1/\Delta$ as a function of $1/v$ an initial linear increase followed by reduced slope for $v > (\lambda/2\tau_0)\sqrt{J_1/J_c}$ or $v_H > (\lambda/2\tau_0)\sqrt{J_c J_1}$.

---

[41] P. W. Anderson, *Phys. Rev. Lett.* **9**, 309 (1963).

[42] P. W. Anderson and Y. B. Kim, *Rev. Mod. Phys.* **36**, 39 (1964).

Intergranular flux relaxation

The relaxation of flux within *grains* was discussed in Sec. 6.2. We first considered the Anderson dynamical model[41,42] and then examined kinetic models with special attention to Waldner's *minimal glassy model*.[34,35]

One may well ask what this all has to do with the relaxation of flux within grain *boundaries*, the evident origin of the phenomena described in this chapter. The answer is that all these models are phenomenological and with suitable modification of parameters should be applicable to grain boundaries as well as to the granular interior. An issue that remains is the source of intergranular flux-pinning. Although most authors have ascribed the pinning to variation in Josephson current, or in the extreme to microbridging of the grains, pinning of Josephson current at grain surfaces can not be ruled out. This issue is considered in Ch. 14, where power-induced nonlinearity is discussed.

14
NONLINEAR MICROWAVE ELECTRODYNAMICS*

14.1. Introduction

This chapter reviews the effect of elevated levels of microwave power on the surface impedance of granular and thick-film superconductors as well as on structures patterned from thin-film crystalline superconductors. Similar power effects have been observed in granular samples of conventional superconductors.

It is now clear that power effects originate in grain boundaries and other defects through which microwave flux penetrates the superconductor. We are able to account for a wide range of phenomena on the assumption that the dominant processes are microwave field-induced granular decoupling and hysteretic penetration of grain boundaries by microwave fields. A proposed model treats junctions as transmission lines with series impedance arising from the surface impedance of the grains and shunt admittance given by the resistively shunted Josephson susceptance of the junctions. The principal approximation made is to allow the shunt admittance to vary along the transmission line as a function of the amplitude of the rf magnetic field. As the rf field is increased, the Josephson current drops and the shunt admittance is reduced. The overall effect is to increase penetration into junctions.

An interesting observation is that whereas power effects are not seen in crystalline films, such effects are clearly apparent in patterned films. Power-effects evidently take place at the edges of patterned films where there is crowding of current and where defects have been exposed by patterning.

---

* This chapter is based in part on A. M. Portis, "Microwave power-induced flux penetration and loss in high-temperature superconductors," *J. Supercond.*, October 1992.

The study of nonlinear processes in Josephson junctions warrants further experimental study through the measurement at elevated levels of microwave power not only of the surface resistance but also of the surface reactance together with the application of dc magnetic fields that decouple the grains. Such effects should be closely related to the low-frequency magnetic properties of granular superconductors so long as the ac magnetic field is below the lower critical field H_{c1}.

14.2. Materials

14.2.1. Ceramics

Delayen *et al.*[1,2] have studied the surface resistance of thin ceramic rods of $YBa_2Cu_3O_{7-\delta}$ in coaxial TEM copper cavities operating from 150 to 1500 MHz. Because the rf fields are large at the surface of a thin central conductor, increases in R_s with rf field may be observed at moderate rf power levels. Sample-heating is substantially reduced by filling the cavity with cryogen and by pulsing the rf excitation. The rods were formed from superconducting powder that had been ball-milled in a slurry of organic binder, plasticizer, dispersant and solvent. After casting, the material was dried and then extruded through a metal die to form rods. The rods were heated to burn off the organics and finally sintered and cooled in oxygen. The rods formed half-wave resonant elements of a coaxial cavity. Above an rf field of about 0.1 Oe the surface resistance increased up to a field between 10 and 100 Oe, where the increase saturated. The field for saturation decreased with frequency approximately as $1/\sqrt{\omega}$. At low rf fields, the induced R_s increased as ω^2 but only as $\sqrt{\omega}$ at high rf fields. The frequency dependence of the surface resistance just above T_c dropped to $\sqrt{\omega}$ as expected for a normal conductor.

Bielski *et al.*[3] were among the first to determine the surface resistance R_s of ceramic disks of $YBa_2Cu_3O_{7-\delta}$ from radio frequencies well into the microwave range. Values of R_s were determined from the quality factor of the resonant structure in which the disks were placed.[4] In contrast to the results of Delayen *et al.*,[1,2] R_s was found to increase linearly with frequency. The increase in R_s with the strength of the surface rf magnetic field H_s was also found to be approximately linear. This observed linearity in field and frequency has led to the suggestion that the increase in surface resistance is associated with critical state behavior.[5]

Rezende and de Aguiar[6] have also measured the microwave surface resistance R_s of ceramic superconducting $YBa_2Cu_3O_{7-\delta}$ as a function of the surface rf field and observed a linear increase. Macêdo *et al.*[7] have reported a quadratic increase in R_s at 9.4 GHz in rf fields up to 0.3 Oe for a highly compacted ceramic sample of $YBa_2Cu_3O_{7-\delta}$ held at 80 K. This initial rf field dependence was followed by a slow linear increase in R_s.

---

[1] J. R. Delayen, K. C. Goretta, R. B. Poeppel and K. W. Shepard, *Appl. Phys. Lett.* **52**, 930 (1988).

[2] J. R. Delayen and C. L. Bohn, *Phys. Rev. B* **40**, 5151 (1989).

[3] M. Bielski, O. G. Vendik, M. M. Gaidukov, E. K. Gol'man, S. F. Karmanenko, A. B. Kozyrev, S. G. Kolesov and T. B. Samoilova, *JETP Lett.* **46**, S145 (1987).

[4] A. M. Portis, D. W. Cooke and E. R. Gray, *J. Supercond.* **3**, 297 (1990).

[5] J. Halbritter, *J. Appl. Phys.* **68**, 6315 (1990).

[6] S. M. Rezende and F. M. de Aguiar, *Phys. Rev. B* **39**, 9715 (1989).

[7] M. A. Macêdo, F. L. A. Macado and S. M. Rezende in *Proceedings of the International Conference on Transport Properties of Superconductors*, Rio de Janeiro, Spring 1990 (World Scientific, Singapore, 1991).

Delayen *et al.*[8,9] have constructed a niobium quarter-wave cavity resonant at 821 MHz for the study of disk-shaped samples of the high-temperature superconductors in surface rf fields up to 300 Oe. Measurements on a wide range of samples are shown in Fig. 1.

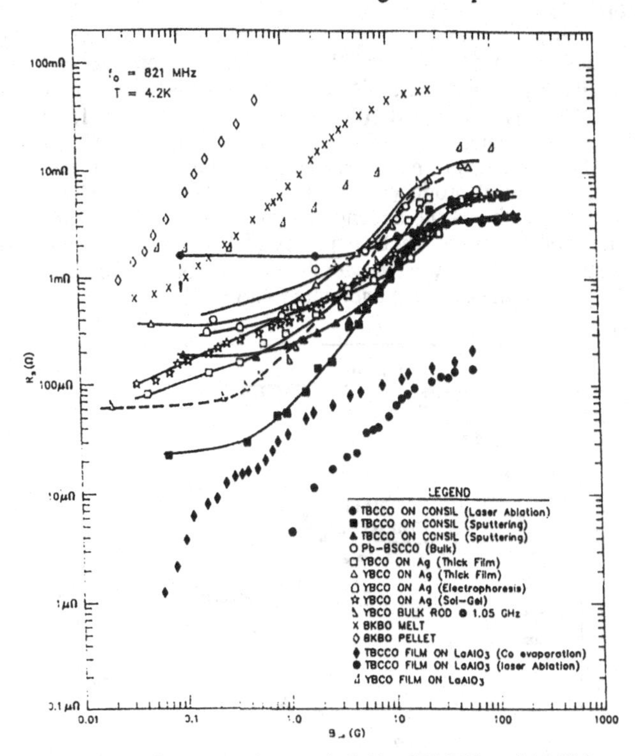

Figure 1. Surface resistance *vs.* rf magnetic field at 821 MHz and 4.2 K for a variety of samples, [9] and J. R. Delayen, private communication. The behavior of a ceramic rod [1,2] at 1.05 GHz is shown for comparison. The high apparent surface resistance of laser-ablated TBCCO on Consil is probably from the edges, which were not coated. Sample sources are given in the reference.

The behavior of all samples is qualitatively similar. Above an onset field between 0.1 and 1 Oe, the surface resistance rises at a decreasing rate and finally saturates in fields between 10 and 100 Oe. The surface resistance at saturation is 10 to 100 times the low power value but still only a few percent of the normal-state surface resistance.

Janes *et al.*[10] have reported an initial increase and then disappearance of the field-modulated absorption at elevated microwave power levels in ceramic samples of $YBa_2Cu_3O_{7-x}$, $Er_2Ba_4Cu_7O_{15-x}$ and $Tl_2Ba_2CaCu_2O_x$ at 77 K. The possibility of sample-heating to temperatures above T_c can not be excluded.

Kobayashi *et al.*[11] have used a dielectric rod resonator to measure the power

---

[8] J. R. Delayen, C. L. Bohn and C. T. Roche, *Rev. Sci. Instrum.* **61**, 2207 (1990).

[9] J. R. Delayen, C. L. Bohn and C. T. Roche, *J. Supercond.* **3**, 243 (1990).

[10] R. Janes, R. S. Liu, P. P. Edwards and J. L. Tallon, *Physica C* **167**, 520 (1990).

[11] Y. Kobayashi, T. Imai and H. Kayano, *IEEE Trans. Microwave Theory Tech.* **39**, 1530 (1991).

dependence of the surface impedance of a $YBa_2Cu_3O_7$ ceramic disk. Surface fields up to 10^4 A/m (126 Oe) have been produced with a four-fold increase in R_s at 11 K.

14.2.2. Granular Films

An increase of R_s at 4.0 K for 100 μm-thick TBCCO films has been reported as a function of rf power at 18 GHz by Cooke *et al.*[12,13] The surface resistance of an unoriented film was found to saturate at values around 100 mΩ in rf fields of about 15 Oe. The surface resistance of the better oriented films increases more slowly with insufficient power to reach saturation. The unoriented films had been deposited onto stock Consil 995, a Co-Ni-Ag alloy. Rolling the Consil produces partial texturing of the substrate and results in oriented films. Figure 2 shows the dependence of R_S on H_S for a number of TBCCO films, both oriented and unoriented, at a frequency of 18 GHz and a temperature of 4 K. The surface resistance at the lowest power levels was below 10 mΩ and rose with increasing H_S, reaching saturation in the unoriented film.

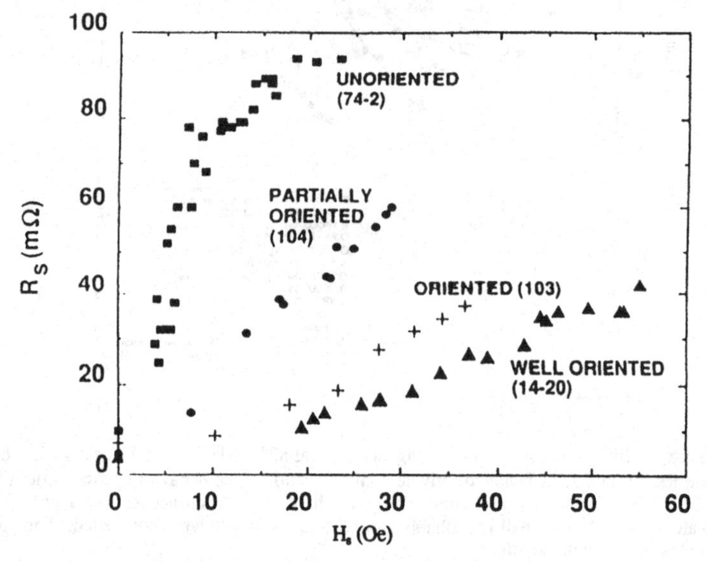

Figure 2. Surface-field dependence of R_s of TBCCO films for varying degrees of c-axis texturing measured at 18 GHz and 4.0 K [12,13].

Measurement of the rf surface resistance at 821 MHz has been made at 4.2 K in a pulsed quarter-wave coaxial Nb cavity, resonant in its fundamental transverse electromagnetic (TEM) mode.[8,9] The sample, a Consil disk coated on both sides with a partially oriented TBCCO superconducting film, was placed in a recession in the base of the cavity at the position of maximum rf magnetic field. The specimen was replaced in turn by superconducting niobium and stainless steel disks to determine the sample geometry factor. The surface resistance of the superconducting film was determined from the initial decay

[12] D. W. Cooke, P. N. Arendt, E. R. Gray, A. Meyer, D. R. Brown, N. E. Elliott, G. A. Reeves and A. M. Portis, *IEEE Trans. Magn.* **27**, 880 (1991).

[13] D. W. Cooke, P. N. Arendt, E. R. Gray, and A. M. Portis, *IEEE Trans. Microwave Theory Tech.* **39**, 1539 (1991).

time of the cavity.[14] The surface resistance was found to saturate at 5 mΩ in an rf field around 30 Oe as shown in Fig. 3. The ratio of the saturation surface resistance at the two frequencies for similar samples 100/5 = 20 is close to the ratio of the frequencies 18/0.82 = 22. The saturation fields at the two frequencies are close and do not show the reduction with increasing frequency that is observed in ceramic rods.[1,2]

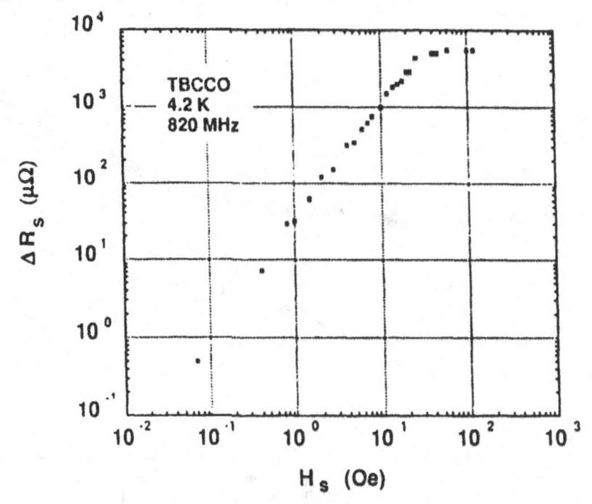

Figure 3. Increase of the surface resistance of a partially oriented TBCCO film vs. rf field at 820 MHz and 4.2 K. The surface resistance R_s at the lowest power level is 22μΩ. R_s rises quadratically at first with H_s, saturating in fields above 35 Oe [8,9].

Hein *et al.*[15,16] have developed thick electrophoretically deposited films for use in resonators at high levels of microwave power. Deposition in an 8 T magnetic field orients the particles in suspension and leads to texturing and increased grain growth on sintering. The dependence of the surface resistance on rf magnetic field was measured in a niobium host cavity at 21.5 GHz up to 1000 A/m (12.6 Oe).[17] Results for two thick films, one highly textured and one untextured, are shown in Fig. 4. For comparison are shown two thin films. The surface resistance of the highly textured film, represented by the filled circles increases as the square-root of the microwave field. The surface resistance of the untextured film, represented by the open circles, is independent of field up to about 50 A/m (0.6 Oe) and then increases linearly with field. The film represented by inverted filled triangles is an electron-beam coevaporated film of high stoichiometry and shows no power dependence to within the precision of measurement at 21.5 GHz.[18] The film represented by upright filled triangles is a Siemens CVD film with intermediate properties.

The relation between the increase in R_s and the increase in X_s in microwave fields up to 1.2 mT is represented by the filled circles in Fig. 5. The filled squares indicate the corresponding increases in dc magnetic fields.[19] The increase in R_s with dc magnetic fields

---

[14] J. R. Delayen and C. L. Bohn, *Phys. Rev. B* **40**, 5151 (1989).

[15] M. Hein, G. Müller, H. Piel and L. Ponto, *J. Appl. Phys.* **66**, 5940 (1989).

[16] M. Hein, S. Kraut, E. Mahner, G. Müller, D. Opie, H. Piel, L. Ponto, D. Wehler, M. Becks, U. Klein and M. Peiniger, *J. Supercond.* **3**, 323 (1990).

[17] M. Hein, Dissertation, University of Wuppertal, WUB-DIS 92-2, May 1992.

[18] M. Hein, S. Hensen, G. Müller, S. Orbach, H. Piel, M. Strupp, N. G. Chew, J. A. Edwards, S. W. Goodyear, J. S. Satchel and R. G. Humphreys in *High T_c Superconducting Thin Films*, ed. L. Carrera (North-Holland, Amsterdam, 1992) pp 95-100.

[19] M. Hein, H. Piel, M. Strupp, M. R. Trunin and A. M. Portis, *J. Magn. Magn. Mat.* **104-107**, 529 (1992).

is known to arise from granular decoupling.[20] Note that a peak rf field produces the same increase in X_s as does a dc field and thus must produce the same granular decoupling. R_s is seen to increase faster in an rf field than in a dc field, suggesting the presence of additonal loss mechanisms in rf fields. Hein[17] has found that the rate of increase of R_s with microwave field correlates with the square of R_s for films of varying degreees of texturing.

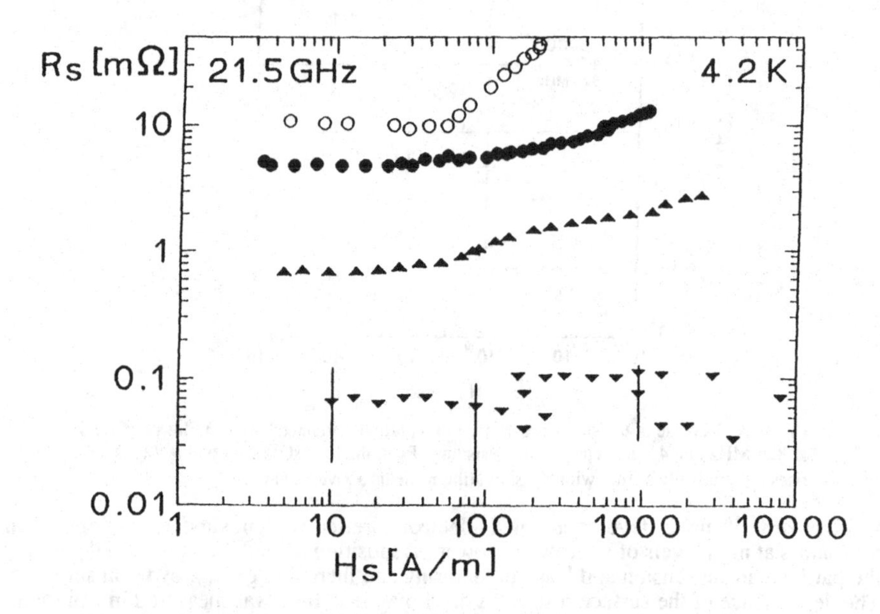

Figure 4. Dependence on rf field H_s of the surface resistance of electrophoretically deposited $YBa_2Cu_3O_7$ thick films at 4.2 K and 21.5 GHz. Results for two thick films, one highly textured (filled circles) and one untextured (open circles), are shown. For comparison are shown two thin films, an electron-beam coevaporated film (filled inverted triangles) and a CVD film (filled upright triangles) [17].

Alford *et al.*[21,22,23,24] have prepared thick films of $YBa_2Cu_3O_7$ by melt-processing and have obtained surface resistances an order of magnitude lower than for ceramics or untextured thick films. The rate of increase in surface resistance with frequency is however lower than the quadratic dependence that is observed for high quality epitaxial films and suggests that the junctions in these thick films are strongly damped.

Bonin and Safa [25] have compared the results of Delayen *et al.*[1,2] with early studies of

---

[20] A. M. Portis, D. W. Cooke, E. R. Gray, P. N. Arendt, C. L. Bohn, J. R. Delayen, C. T. Roche, M. Hein, N. Klein, G. Müller, S. Orbach and H. Piel, *Appl. Phys. Lett.* **58**, 307 (1991).

[21] T. W. Button, N. McN. Alford, F. Wellhofer, T. C. Shields, J. S. Abell and M. Day, *IEEE Trans. Magn.* **27**, 1434 (1991).

[22] P. A. Smith, L. E. Davis, N. McN. Alford and T. W. Button, *Electron. Lett.* **26**, 1476 (1990)

[23] N. McN. Alford, T. W. Button, M. J. Adams, S. Hedges, B. Nicholson and W. A. Phillips, *Nature* **349**, 680 (1991).

[24] N. McN. Alford, T. W. Button and D. Opie, *Supercond. Sci. Technol.* **4**, 433 (1991).

[25] B. Bonin and H. Safa, *Supercond. Sci. Technol.* **4**, 257 (1991).

sputtered niobium,[26] sputtered NbN[27] and sputtered NbTiN[28] at high levels of rf power. These authors point out that the power dependence obtained in granular films of classical superconductors is very similar to that reported for the high temperature superconductors.

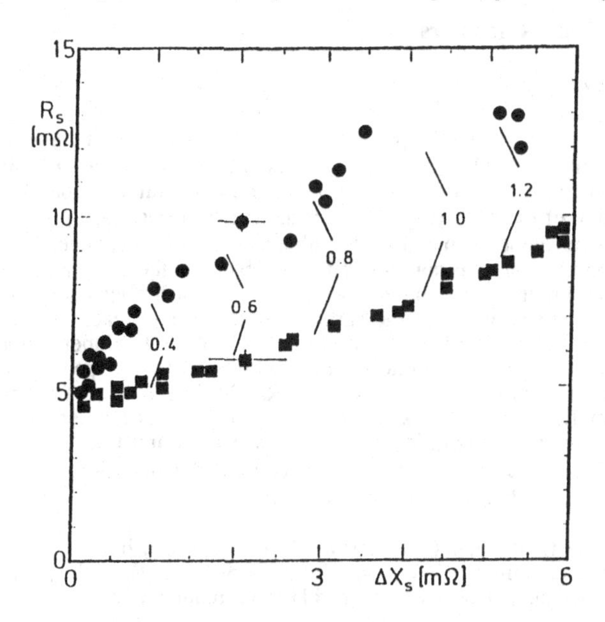

Figure 5. Surface resistance R_s plotted against the increase in X_s in microwave fields up to 1.2 mT (filled circles.) For comparison, the filled squares indicate corresponding increases in variable dc magnetic fields also up to 1.2 mT [17].

14.2.3. Crystalline Films

Humphreys *et al.*[29] have reviewed the growth and characterization of Y-Ba-Cu-O thin films deposited from the vapor. Chew *et al.*[30] have grown thin epitaxial films of $YBa_2Cu_3O_7$ *in situ* by evaporation onto (001) MgO substrates. The power dependence of the microwave surface resistance of films optimized for their microwave properties has been studied by Hein *et al.*[18] at 21.5 GHz in a niobium host cavity with little indication of an increase in R_s up to surface fields of 120 Oe.

Shiren *et al.*[31] have compared the pulsed-power response at 16.5 GHz of laser-deposited $YBa_2Cu_2O_7$ thin films and Nb films deposited onto oxydized Si wafers.

---

[26] *Groupe d'Etudes Cavités Supraconductrices* (GEC Saclay) unpublished.

[27] R. T. Kampwirth and K. E. Gray, *IEEE Trans. Magn.* **17**, 565 (1981).

[28] P. Bosland, F. Guemas and M. Juillard, *ICMAS 90*, Grenoble, 17-19 October 1990. Proceedings edited by A. Niku-Lari.

[29] R. G. Humphreys, J. S. Satchell, N. G. Chew, J. A. Edwards, S. W. Goodyear, S. E. Blenkinsop, O. D. Dosser and A. G. Cullis, *Supercond. Sci. Technol.* **3**, 38 (1990).

[30] N. G. Chew, S. W. Goodyear, J. A. Edwards, J. S. Satchell, S. E. Blenkinsop and R. G. Humphreys, *Appl. Phys. Lett.* **57**, 2016 (1990).

[31] N. S. Shiren, R. B. Laibowitz, T. G. Kazyaka and R. H. Koch, *Phys. Rev. B* **43**, 10 478 (1991).

Additional loss was obtained from $YBa_2Cu_2O_7$ at surface fields above 1000 A/m (12.6 Oe). The evaporated Nb films, which were presumably highly granular, showed increased surface resistance at fields as low as 100 A/m (1.26 Oe).

14.3. Patterned Film Resonators

14.3.1. Stripline resonators

Oates *et al.*[32,33] have examined the properties of an off-axis *in situ* magnetron-sputtered $YBa_2Cu_3O_7$ thin-film[34] stripline resonator in peak rf fields up to 300 Oe at the edges for modes ranging in frequency from 1.5 to 20 GHz and temperatures from 4 to 90 K. An explicit calculation of the magnetic field over the entire surface of a patterned thin film stripline has been developed,[35] making it possible to extract the dependence of R_s on local surface field. The calculation is complicated by the fact that the penetration depth may be comparable to the film thickness and is as well a function of surface magnetic field. Over the entire temperature range, both the surface resistance and reactance were found to increase quadratically with field at low power levels. At low temperatures the surface resistance R_s shows only a moderate increase with power up to peak values of the rf magnetic field around 250 Oe. The increase in R_s with increased rf field at 67.3 K and frequencies from 1.5 to 7.5 GHz is shown in Fig. 6. At this temperature, R_s is again found to at first increase quadratically with surface field and then more rapidly. The penetration depth λ shows a weaker dependence on rf field than does R_s.

Oates *et al.* distinguish three ranges in surface rf field:

(i) In the low field region ($H_{rf} < 10$ Oe) R_s increases very little.

(ii) In the intermediate field region ($10 < H_{rf} < 50$ Oe at 77 K) the surface resistance increases quadratically with rf field H_{rf} and frequency f as

$$R_s(H_{rf}) = a(f, T) + b(f, T)H_{rf}^2$$

with both a(f, T) and b(f, T) increasing quadratically with frequency.

(iii) At high fields $H_{rf} > 50$-300 Oe, depending on temperature and frequency, R_s increases faster than the initial quadratic increase. In this field region R_s increases linearly with frequency, indicating hysteretic processes.

Poorer films, on the other hand, show a linear increase in R_s with surface rf field, saturating at moderate rf fields. Rezende and de Aguiar[5] and Macêdo *et al.*[6] had obtained similar results with compacted ceramics.

14.3.2. Microstrip resonators

Wilker *et al.*[36] have fabricated microstrip resonators from films of $YBa_2Cu_3O_7$ and $Tl_2Ba_2CaCu_2O_8$. Both films, and especially the latter, have low surface resistance and low

---

[32] D. E. Oates, A. C. Anderson, D. M. Sheen and S. M. Ali, *IEEE Trans. Microwave Theory Tech.* **39**, 1522 (1991).

[33] D. E. Oates, P. Nguyen, G. Dresselhaus, M. S. Dresselhaus, C. W. Lam and S, M, Ali, preprint.

[34] A. C. Westerheim, L. S. Yu-Jahnes and A. C. Anderson, *IEEE Trans. Magn.* **27**, 1001 (1991).

[35] D. M. Sheen, S. M. Ali, D. E. Oates, R. S. Withers and J. A. Kong, *IEEE Trans. Appl. Superconduct.* **1**, 108 (1991).

[36] C. Wilker, Z.-Y. Shen, P. Pang, D. W. Face, W. L. Holstein, A. L. Matthews and D. B. Laubacher, *IEEE Trans. Microwave Theory Tech.* **39**, 1462 (1991).

power dependence. A resonator patterned from a film of $Tl_2Ba_2CaCu_2O_8$ had a loaded Q of 8000 at 5 GHz and 80 K and could handle 120 W of peak power (average H_s of 12 Oe) with a 25% reduction in Q.

Kozyrev *et al.*[37] have reported increased attenuation of pulsed microwave fields on a $YBa_2Cu_3O_{7-x}$ coplanar transmission line for a peak power of the order of one watt. Very similar results have been obtained for a NbN line.

Namordi *et al.*[38] have reported the power dependence of a $YBa_2Cu_3O_7$ coplanar line resonant at 4.75 GHz. The surface resistance of the line was found to degrade at rf field levels for which the peak rf current density was of the order of the dc critical current.

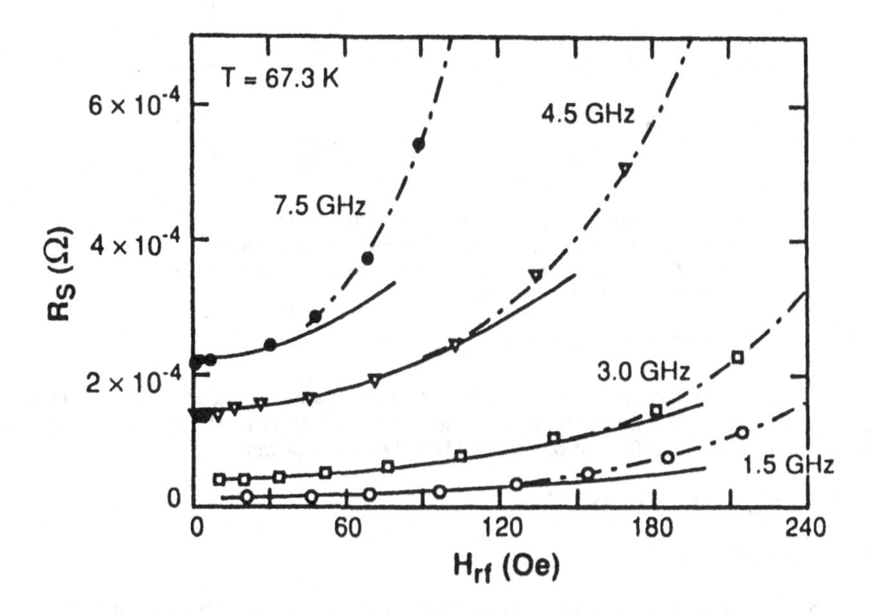

Figure 6. Surface resistance *vs.* rf magnetic field with frequency as a parameter. The solid lines fit the initial parabolic increase. At higher fields R_s increases more rapidly than the extrapolated quadratic increase [33].

14.3.3. Coplanar resonators

Porch *et al.*[39] have fabricated linear and meanderline coplanar resonators by wet-etching films of YBCO coevaporated onto MgO substrates. The geometry of coplanar devices makes them very sensitive to power effects, which arise at film edges, and provides a sensitive way of studying the effects of patterning. Figure 7 shows the surface resistance R_s and the surface reactance X_s of a linear coplanar resonator as a function of the rms rf field H_s. The actual rf field at the edges is estimated to be between one and two orders of magnitude larger than the rms value, depending on gap size.

---

[37] A. B. Kozyrev, T. B. Samoilova, O. I. Soldatenkov and O. G. Vendik, *Solid State Commun.* **77**, 441 (1991).

[38] M. R. Namordi, A. Mogro-Campero, L. G. Turner and D. W. Hogue, *IEEE Trans. Microwave Theory Tech.* **39**, 1468 (1991).

[39] A. Porch, M. J. Lancaster, H. C. H. Cheung, A. M. Portis, R. G. Humphreys and N. G. Chew, *CMMP-91, Condensed Matter and Materials Physics Conference*, Birmingham, 17-19 November 1991.

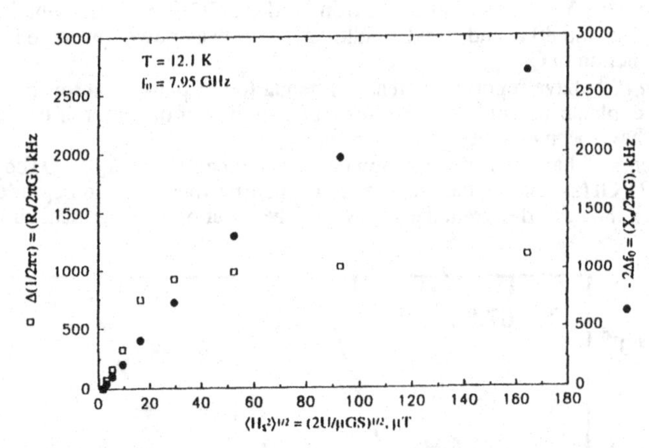

Figure 7. Surface resistance R_s plotted as an increase in bandwidth and surface reactance X_s plotted as twice the frequency reduction of a coplanar resonator as a function of rms rf field. The surface reactance appears to increase as the square root of the field with no indication of saturation while R_s saturates [39].

14.3.4. Parallel plate resonators

Taber[40] has adapted to the study of superconductors a parallel plate resonator of the type used to determine the loss tangent of dielectrics. Gallop *et al.*[41] have used a similar resonator to study the effects of both dc fields and elevated power.

14.4. Intergranular Transmission

14.4.1 Introduction

Our picture is that the increases in surface resistance and reactance observed at high power are associated with granular decoupling. In order to obtain the sample surface impedance we treat intergranular junctions as linearized transmission lines with series impedance $2Z_g$. The effect of intergranular penetration is to add to the surface impedance the input impedance of intergranular Josephson junctions.

We expect three regimes of rf power for long intergranular junctions:

(i) Just above the junction critical field H_{c1j}, both the surface resistance R_s and the surface reactance X_s are expected to increase quadratically with H_s as microwave Josephson vortices penetrate the junction. The initial increase in R_s is expected to be quadratic in frequency while the increase in X_s is linear in frequency.

(ii) At intermediate power levels, the critical Josephson current increases as H_s, increasing the input reactance of the junction as $\sqrt{H_s}$. At microwave fields $H_s > H_{c1j}$, flux is expected to further penetrate Josephson junctions. The presence of

[40] R. C. Taber, *Rev. Sci. Instrum.* **61**, 2200 (1990).

[41] J. C. Gallop, A. M. Portis, W. J. Radcliffe and C. D. Langham, *CMMP-91, Condensed Matter and Materials Physics Conference*, Birmingham, 17-19 December 1991.

microwave flux in the junction reduces the Josephson current, further increasing the flux penetration. In this regime the Josephson inductivity increases approximately as H in a dc field[19,20] and can be expected to behave similarly in an alternating field.

(iii) As the Josephson inductivity increases further, the penetration of microwave flux into junctions becomes hysteretic, contributing a component to the junction input impedance that is linear in frequency and, from experiment, independent of microwave field.

14.4.2 Transmission line model

The transmission line model shown in Fig. 12-4 takes Z_g for the surface impedance of the grains and

$$Y_j'' = \frac{1}{bz_j} = -\frac{i\omega\varepsilon_j}{b} + \frac{1}{b\rho_j} - \frac{1}{i\omega b\ell_j} \tag{1}$$

for the admittance of a Josephson junction of thickness b with ℓ_j the Josephson inductivity, ρ_j the shunt resistivity and ε_j the shunt permittivity.[42] The effect of microwave fields is expected to be much like that of dc fields,[20] increasing ℓ_j as H_s with little effect on ρ_j.

Telegrapher's equations

The so-called telegrapher's equations

$$\frac{dV}{dz} = -2I'Z_g \qquad \frac{dI'}{dz} = -VY_j'' \tag{2}$$

with V the voltage across the junction and I' the sheet current density of the grains are analogous to the Maxwell equations of the junction with appropriate boundary conditions.

We first combine these equations with Z_g and Y_j'' independent of power to obtain for the wavevector in the junction

$$k = \sqrt{-2Y_j''Z_g} \tag{3}$$

and for the input impedance of a junction of width a

$$Z_{in} = \frac{V}{I'} = \frac{1}{a}\sqrt{\frac{2Z_g}{Y_j''}} \tag{4}$$

14.4.3 Analysis

For a long junction in which Z_g and Y_j'' are field-dependent, the transmission-line equations should be integrated. We write for the gradient of the magnetic field H = I' in the junction

[42] A. M. Portis and D. W. Cooke, *Supercond. Sci. Technol.* **5**, S395 (1992).

$$\frac{dH}{dz} = -VY_j'' \tag{5}$$

and on elimination of V obtain the nonlinear wave equation

$$\frac{d^2H}{dz^2} - \frac{1}{Y_j''}\frac{dY_j''}{dH}\left(\frac{dH}{dz}\right)^2 - 2Y_j''Z_gH = 0 \tag{6}$$

with the usual expression for the junction input impedance

$$Z_{in} = -\frac{b}{a}z_j\left(\frac{1}{H}\frac{dH}{dz}\right)_s \tag{7}$$

So long as Y_j'' and Z_g do not vary too rapidly with H, Eq. (4) and the integration of Eq. (6) differ at most by a factor of the order of unity, allowing us to use the simpler Eq. (4).

Portis and Cooke[43] have developed a model of intergranular flux penetration using a transmission-line model. At sufficiently low fields, flux decays exponentially along a Josephson transmission line with penetration depth[44]

$$\lambda_j = \sqrt{\frac{\Phi_0}{4\pi J_0 L_g}} \tag{8}$$

where $L_g = \mu\lambda_g$ is the surface inductance of lossless grains that bound the junction. Above a dynamical critical field

$$H_{c1j} = \Phi_0/2\pi\mu\lambda_j\lambda_g = 2J_0\lambda_j \tag{9}$$

penetrating flux is modulated by sinusoidal Josephson currents with period $\Phi_0/2\mu\lambda_g H$.

Resonator frequency and bandwidth

Changes in the frequency and bandwidth of a resonant transmission line may be directly related to changes in the surface resistance and reactance of the component elements. The wavevector on a transmission line[45] is

$$k = \sqrt{-Y''Z} \tag{10}$$

where $Y'' = -i\omega c_j$ is the specific admittance, with c_j the shunt capacitance per unit area and

$$Z = 2R_s - 2iX_s - i\omega L \tag{11}$$

is the zeries impedance with R_s and $X_s = \omega L_s$, respectively, the surface resistance and reactance of the transmission lines and L the inductance of the gap. The resonance

43. A. M. Portis and D. W. Cooke, *Supercond. Sci. Technol.* **5**, S395 (1992).
44. T. Van Duzer and C. W. Turner, *Principles of Superconductive Devices and Circuits* (Elsevier North-Holland, New York, 1981).
45. S. Ramo, J. R. Whinnery and T. Van Duzer, *Fields and Waves in Communication Electronics* (Wiley, New York, 1984) 2nd ed.

condition for a transmission line of length d is $k = n\pi/d$. Solving for the resonant frequency f and the full bandwidth at half-power f_B leads to

$$f = \frac{1}{2\pi}\frac{1}{\sqrt{(2L_s + L)c_j}}\frac{n\pi}{d} \qquad f_B = \frac{1}{2\pi}\frac{2R_s}{2L_s + L} \tag{12}$$

Changes in surface resistance and reactance are related to changes in f and f_B by the relations

$$-2\Delta f = \frac{1}{2\pi}\frac{2\Delta X_s}{2L_s + L} \qquad \Delta f_B = \frac{1}{2\pi}\frac{2\Delta R_s}{2L_s + L} \tag{13}$$

where the second expression assumes $\Delta f/f << \Delta f_B/f_B$. Thus, the ratio $-\Delta f_B/2\Delta f$ is simply $\Delta R_s/\Delta X_s$.

Josephson transmission is described by the sine-Gordon equation[46]

$$\frac{\partial^2\delta}{\partial z^2} - \frac{1}{v_j^2}\left(\frac{\partial^2\delta}{\partial t^2} + \frac{1}{\tau}\frac{\partial\delta}{\partial t}\right) = \frac{1}{\lambda_j^2}\sin\delta \tag{14}$$

with $1/\lambda_j^2 = 4\pi J_0 L_g/\Phi_0$, $v_j^2 = 1/2L_g c_j$ and $\tau = \rho_j c_j$. Here, c_j and ρ_j are the specific capacitance and resistance across the transmission line and, as noted above, $2L_g$ is the series inductance. The electromagnetic fields in the junction are related to the phase δ across the junction by

$$V = \frac{\Phi_0}{2\pi}\frac{\partial\delta}{\partial t} \qquad H = -\frac{\Phi_0}{4\pi L_g}\frac{\partial\delta}{\partial z} \tag{15}$$

The input impedance of a Josephson transmission line is

$$Z_{in} = \frac{V}{I} = -\frac{2L_g}{a}\left(\frac{\partial\delta/\partial t}{\partial\delta/\partial z}\right)_s \tag{16}$$

At low rf fields and for $\omega << \omega_j = v_j/\lambda_j$, the propagation vector from Eq. (14) is

$$k_j = \frac{i}{\lambda_j}\sqrt{1 - \frac{i\omega}{\omega_j^2\tau}} \tag{17}$$

and the input impedance takes the form

$$Z_{in} = \frac{2\omega L_g}{k_j a} = \frac{1}{a}\sqrt{\frac{-2i\omega L_g}{1/\rho_j - 1/i\omega\ell_j}} \tag{18}$$

with the specific Josephson inductance $\ell_j = \ell_0 = \Phi_0/2\pi J_0$.

---

46. A. Barone and G. Paternò, *Physics and Applications of the Josephson Effect* (Wiley-Interscience, New York, 1982) sec. 10.1.

dc magnetic fields

Experiments in dc magnetic fields[19,20] indicate that for fields $H > H_{c1j}$, the Josephson inductance appears to increase linearly with H

$$\ell_j(H) \approx \frac{H}{H_c}\ell_j(0) \tag{19}$$

If not otherwise limited as the magnetic field is increased, we expect the input impedance as given by Eq. (19) to approach the limiting expression

$$Z_{in} = \frac{1-i}{a}\sqrt{\omega L_g b \rho_j} \tag{20}$$

Stochastic model

The observed linear increase in $\ell_j(H)$ at high fields may be traced to inhomogeneities in the Josephson current density J_0 from Fourier components of the same wavevector k as the phase δ across the junction. To analyze this problem, we use the language of scattering theory and obtain the junction response function from a scattering intensity that is the Fourier transform of the autocorrelation function of the Josephson current density J_0.[47]

The magnetic field modulates the Josephson current with wavevector

$$k = \frac{2\pi}{d} = \frac{4\pi L_g H}{\Phi_0} \tag{21}$$

because a static magnetic field takes δ through 2π in a length $d = \Phi_0/2\lambda_g\mu H = \Phi_0/2L_gH$. Screening requires a non-zero scattering intensity at wavevector k

$$I(k) = \frac{1}{\lambda_j}\int J^2(\zeta)e^{ik\zeta}d\zeta \tag{22}$$

where λ_j is the distance that microwave fields penetrate the junction and $J^2(\zeta)$ is the Josephson current autocorrelation function

$$J^2(\zeta) = \frac{1}{\lambda_j}\int J^*(z+\zeta)J(z)dz \tag{23}$$

We assume for the Josephson current autocorrelation function

$$J^2(\zeta) = J_0{}^2 + (\Delta J)^2 e^{-\zeta/\lambda_c} \tag{24}$$

where J_0 is the mean Josephson current density and $(\Delta J)^2$ is the mean-square deviation from J_0. We obtain for the scattering intensity for wavevectors $k >> 1/\lambda_j$

47. A. Barone and G. Paternò, *Physics and Applications of the Josephson Effect* (Wiley-Interscience, New York, 1982) sec. 4.4.2.

$$I(k) = \frac{2\lambda_c}{\lambda_j} \frac{(\Delta J)^2}{1+(k\lambda_c)^2} \tag{25}$$

The response function is the Josephson inductance, which increases as

$$\ell(k) = \frac{J_0}{\sqrt{I(k)}} \ell_0 = \sqrt{\frac{\lambda_j}{2\lambda_c}} \frac{J_0}{\Delta J} \sqrt{1+(k\lambda_c)^2}\, \ell_0 \tag{26}$$

which is linear in H at high fields with $k\lambda_c = H/H_c$ as required.

Note that Eq. (26) does not approach ℓ_0 as k goes to zero. This is because we have overlooked the initial reduction in ℓ_j as H increases beyond H_{c1j}. In this range with $k << 1/\lambda_c$ we have for the scattering intensity

$$I(k) = \frac{(\Delta J)^2}{1+(k\lambda_j)^2} \tag{27}$$

with $k\lambda_j = 2H/H_{c1j}$. The initial increase in the Josephson inductivity is

$$\ell(k) = \ell_0 \sqrt{1+(k\lambda_j)^2} \tag{28}$$

We finally combine the initial increase in $\ell(k)$ with the linear increase at high fields by adding the scattering intensities given by Eqs. (25) and (27)

$$I(k) = \frac{J_0^2}{1+(k\lambda_j)^2} + \frac{2\lambda_c}{\lambda_j} \frac{(\Delta J)^2}{1+(k\lambda_c)^2} \tag{29}$$

with the same definition of the response function

$$\ell(k) = \sqrt{\frac{I(0)}{I(k)}}\, \ell_0 \tag{30}$$

In effect, we approximate Eq. (14) with the equation for the averaged phase difference

$$\sqrt{\frac{I(0)}{I(k)}} \frac{\partial^2}{\partial z^2} \langle\delta\rangle - \frac{1}{v_j^2}\left(\frac{\partial^2}{\partial t^2} + \frac{1}{\tau}\frac{\partial}{\partial t}\right)\langle\delta\rangle = \frac{1}{\lambda_j^2}\langle\delta\rangle \tag{31}$$

leading to Eq. (16) for the junction input impedance.

rf magnetic fields

Experiments with electrophoretically deposited films at elevated power as shown in Fig. 5 indicate that the specific Josephson inductivity is also increased in an elevated microwave field and by about the same amount as in the corresponding dc magnetic field. The effect of microwave fields is expected to be much like that of dc fields,[20] increasing ℓ_j as H_s with little effect on ρ_j.

Patterned thin films

Patterned films are particularly sensitive to grain boundaries that intersect edges. For a long junction in which L_g and ℓ_j are field-dependent, the transmission-line equations should, strictly speaking, be integrated. We find, however, that so long as ℓ_j does not vary too rapidly with H, Eq. (16) evaluated at the surface differs at most by a numerical factor of the order of unity, allowing us to use this form.

For a lightly damped junction with $\rho_j >> \omega\ell_j$, the input impedance from Eq. (16) for variable ℓ_j is approximately

$$Z_{in} \approx -\frac{i\omega}{a}\sqrt{\ell_j L_g} + \frac{\omega}{2a}\sqrt{\ell_j L_g}\frac{\omega\ell_j}{\rho_j} \tag{32}$$

With increasing rf (or dc) fields, ℓ_j increases as H_s. This gives from Eq. (32), X_{in} increasing as $\omega\sqrt{H_s}$ and R_{in} increasing as $\omega^2 H_s^{3/2}$.

Hysteretic flux penetration

The plateau that is characteristic of power-induced loss increases linearly with ω and is a strong indicator of hysteretic loss. This is because the reactance of linear systems should vary as an odd power of ω and their resistance should vary with an even power of ω. Nonlinear processes, like hysteretic loss, are an exception to this rule. Magnetic flux penetrates the junction to a depth given by Eq. (8). For microwave fields $H_s > H_{c1j}$, with H_{c1j} given by Eq. (9), flux penetrates the junction, reducing the effective Josephson current J_0 and as a consequence increasing the penetration depth λ_j. When the field reverses, flux leaves the junction and the Josephson penetration depth contracts. Because of pinning, either in the junction as a result of inhomogeneity in J_0 or in the grains through current-pinning, there will be remanent flux and work per cycle

$$W = \oint H_s d\Phi_s \approx 2\mu H_s^2 \lambda_g \delta\lambda_j \tag{33}$$

with associated surface resistance

$$R_s = \frac{\omega}{\pi a}\frac{W}{H_s^2} = \frac{2\omega\mu}{\pi a}\lambda_g \delta\lambda_j \tag{34}$$

The fact that R_s is independent of H_s above threshold indicates that $\delta\lambda_j$ must also be independent of H_s over this field range. This absence of dependence on rf field may be a consequence of the structural inhomogeneity that we have had to assume in order to account for the dependence of the flux penetration on magnetic field.

14.4.4. Comment

dc magnetic fields

Enough work has been done on the surface resistance and reactance of granular superconductors in dc magnetic fields[19,20,48] that we can say that at least the effect of a

48. S.K. Remillard, P. N. Arendt and N. E. Elliott, *Physica C* **177**, 345 (1991).

static magnetic field is reasonably well understood. With the application of a static magnetic field above threshold, the surface reactance at first increases quadratically with the field, then as the square-root of the field and finally reaches a saturation value. The physical interpretation of this behavior is that the observed threshold field is H_{c1j}, above which Josephson vortices enter grain boundaries. The penetration of flux into grain boundaries increases the Josephson inductivity $\ell_j = \Phi_0/2\pi J_0$ at first quadratically and then linearly, as established for junctions of finite length. At fields sufficiently large that ℓ_j exceeds ρ_j/ω, a further increase in field has no effect and X_s saturates. The observed surface resistance R_s is consistent with this model and ρ_j independent of magnetic field, saturating at those fields at which X_s saturates. All samples show some residual inductivity ℓ_r, possibly from microbridges, in parallel with ρ_j.

For good quality samples in moderate magnetic fields, the surface reactance increases linearly with frequency and the surface resistance increases quadratically. As the magnetic field increases, both X_s and R_s increase more slowly with frequency, approaching $\sqrt{\omega}$ to a degree that depends on $\omega\ell_r/\rho_j$.

rf magnetic fields

The power dependence of R_s shown in Fig. 1 is qualitatively similar to the variation of R_s on a dc magnetic field. From the studies of Hein *et al.* [19] shown in Fig. 5, we know that the initial increase of X_s in an rf magnetic field is virtually the same as in a dc magnetic field. Whether this similarity continues to saturation is not known. This behavior suggests that the observed increase in X_s is a static magnetic field effect, even at microwave frequencies. As Fig. 5 also shows, R_s initially increases more rapidly in an rf field than in a dc magnetic field. Portis *et al.*[20] have found near saturation that R_s increases linearly with the frequency of the rf field, suggesting hysteretic penetration of magnetic flux. What is not clear is whether the hysteresis originates from vortex pinning in the grains or from structural inhomogeneity of J_0 in the junction.

Hysteresis in the grains can be represented by a granular surface resistance R_g proportional to ωH. For $R_g < \omega L_g$ the junction input resistance R_{in} should also be linear in frequency and field.

Chin *et al.*[49] have observed saturation of the surface resistance of a NbN thin-film stripline resonator at rf fields above 75 Oe for frequencies above 6 GHz. At lower frequencies, R_s continues to increase linearly with H_s with no evidence of saturation. These observations may be suggestive of a granular mechanism at least at high rf fields with granular vortex nucleation below 6 GHz.

The alternative is structural inhomogeneity in the junction. A small resistance ρ_s proportional to ωH in series with ℓ_j similarly gives R_{in} proportional to frequency and field. The saturation mechanism would have to be different from that at dc because of the observation that R_s is linear with frequency.

Porch *et al.*[39] have measured the field dependence of ΔX_s and ΔR_s for coevaporated films of $YBa_2Cu_3O_7$ patterned into coplanar geometry. At low power levels both ΔX_s and ΔR_s increase as H_s^2 while at high power levels ΔX_s increases as $\sqrt{H_s}$ while ΔR_s saturates as shown in Fig. 7. The observed increase in ΔX_s is presumably the result of magnetic decoupling. Assuming that ΔR_s arises from hysteretic loss, the constancy of $\delta\lambda_j$ is probably a consequence of structural inhomogeneity.

---

[49] C. C. Chin, D. E. Oates, G. Dresselhaus and M. S. Dresselhaus, *Phys. Rev. B* **45**, 4788 (1991).

15
MICROWAVE ABSORPTION AND QUANTUM INTERFERENCE*

15.1. Periodic Absorption

15.1.1. Introduction

Narrow periodic absorption lines have been observed at microwave frequencies in thin single crystals of micro-twinned $YBa_2Cu_3O_{7-\delta}$[1,2,3] up to 1 mm on a side and 50 to 100 μm thick. In Fig. 1 is shown a recording of the absorption at magnetic fields below 10 Oe in the plane of the crystal and at moderate microwave power levels. The same pattern of lines is obtained for both increasing and decreasing fields with very little hysteresis.

---

* This chapter is based in part on K. W. Blazey, A. M. Portis and F. H. Holtzberg, "Fluxon nucleation by microwave currents in Josephson junctions," *Physica C* **157**, 16 (1989).

[1] K. W. Blazey, A. M. Portis, K. A. Müller, J. G. Bednorz and F. Holtzberg, "Spin-glass phase microwave study in high-T_c superconductors," *Physica C* **153-155**, 56 (1988).

[2] K. W. Blazey, A. M. Portis, K. A. Müller and F. H. Holtzberg, "Macroscopic flux quantization and microwave excitation in single crystal $YBa_2Cu_3O_{7-\delta}$," *Europhys. Lett.* **6**, 457 (1988).

[3] K. W. Blazey, A. M. Portis and F. H. Holtzberg, "Fluxon nucleation by microwave currents in Josephson junctions," *Physica C* **157**, 16 (1989).

As the crystals are rotated, the periodic lines move in field so that the component of the magnetic field H cos ϕ along a [110] direction remains constant. Similar lines have been obtained with the external field normal to the crystal plane.[4,5]

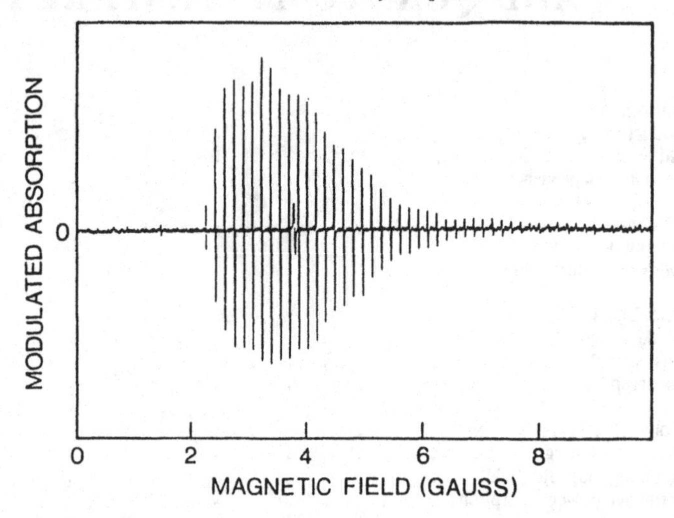

Figure 1. Modulated microwave absorption of a $YBa_2Cu_3O_{7-\delta}$ single crystal at 4.4 K and fields up to 10 Oe. The static magnetic field is in the ab-plane and nearly parallel to a [110] direction. The microwave field is normal to the ab-plane. Incident microwave power is 40 dB below 200 mW. From K. W. Blazey *et al.* [2].

Each line in the series broadens into a band with a width that increases as the square-root of the microwave power and also with the temperature. Different series of absorption lines appear at successive threshold microwave powers and temperatures. These new series also widen with the square-root of the microwave power as well as with temperature. Dulcic *et al.*[6] have reported sinilar effects with microwave power. Bugai *et al.*[7] have have observed the unmodulated absorption in single crystals of $RBa_2Cu_3O_{7-\delta}$ with R = Y, Gd or Dy and have found corresponding changes in the absorption spectrum. Foukis *et al.*[8] have attributed observed fine structure of the derivate absorption to flux slippage of single phase

[4]H. Vichery, F. Beuneu and P. Lejay, "Microwave absorption in a single crystal of YBaCuO at low magnetic field," *Physica C* **159**, 823 (1989).

[5]H. Vichery, F. Rullier-Albenque, F. Beuneu and P. Lejay, "Microwave absorption in a single crystal of $YBa_2Cu_3O_7$," *Physica C* **162-164**, 1583 (1989).

[6]A. Dulcic, R. H. Crepeau and J. H. Freed, "Discrete microwave absorption lines in $YBa_2Cu_3O_{7-\delta}$ single crystals," *Physica C* **160**, 223 (1989).

[7]A. A. Bugai, A. A. Bush, I. M. Zaritskii, A. A. Konchits, N. I. Kashirina and S. P. Kolesnik, "Macroscopic quantum microwave interference in single crystals of high-T_c superconductors," *JETP Lett.* **48**, 228 (1988).

[8]V. Foukis, O. Dobbert, K.-P. Dinse, M. Lehnig, T. Wolf and W. Goldacker, "Fine structure and hysteresis in the low-field microwave absorption of $YBa_2Cu_3O_{7-x}$ superconductors," *Physica C* **156**, 467 (1988).

domains, which function as weakly coupled rf SQUIDS,[9,10,11] operating in an inductive mode.[12,13] Fourier analysis leads to a maximum loop area of about 230 μm^2.

Narrow periodic lines have also been observed from niobium filins[3] with strong and reproducible signals in series observed up to fields of 50 Oe from a *single* irregular Nb particle of about 100 μm diameter. The best spectra were obtained from niobium filings embedded in epoxy when the epoxy contained only two or three irregular particles. While single particles of irregular shape gave uniform line spectra, single particles of regular shape gave no spectra, leading to the conclusion that flux nucleation takes place in oxidized regions[14] of irregular particles. Similar absorption spectra have been observed from a artificial superconducting rings obtained by pressing together oxidized lead particles.[15,16]

The periodic absorption lines are believed to arise from microwave-current induced nucleation of vortices within natural rf SQUID structures.[4,5,9] This process requires a critical current density $J_c(T)$ at which vortices are first nucleated near field-values at which magnetic energy-levels cross. At still higher microwave current densities, hysteretic loss leads to bands in field over which the additional absorption is observed. At much higher current densities the absorption signals are observed to attenuate, possibly the result of saturation.

15.1.2. Field orientation

Blazey *et al.*[1,2,3] found , as the $YBa_2Cu_3O_{7-\delta}$ single-crystals were rotated about the c-axis, that the periodic lines moved in field according to the relation

$$H_0 \cos \phi = \pm (p + 1/2)\, \Delta H \qquad (1)$$

with ϕ the angle from a [110] direction and $\Delta H = \Phi_0/S$ the minimum field interval, where $\Phi_0 = h/2e$ is the quantum of flux and $S = (w + 2\lambda_L)t$ is a geometrical area for the interception of flux with t the thickness of the crystal, w the width of the flux ring and λ_L the London penetration depth. This behavior indicates that flux is nucleated within domain boundaries at the regularly spaced level-crossings of vortex states.[17,18] These boundaries were observed with a polarizing microscope as striations on the face of the crystal and were found to be parallel to a single [110] direction and separated by about 1 μm. Rotation of the magnetic field out of the plane of the crystal led to additional series of lines so that it was not possible to determine the extent to which vortices were restricted to the plane of the crystal.

---

[9]T. Van Duzer and C. W. Turner, *Principles of Superconductive Devices and Circuits* (Elsevier, New York, 1981) pp. 221-224.

[10]A. H. Silver and J. E. Zimmerman, "Josephson weak-link devices," in *Applied Superconductivity*, ed. V. L. Newhouse (Academic, New York, 1975) vol. 1.

[11]J. Clarke, "SQUID concepts and systems," in eds. H. Weinstock and M. Nisenoff, *Superconducting Electronics, NATO ASI Series* (Springer-Verlag, Berlin, 1989), vol. F 59.

[12]P. K. Hansma, "Observability of Josephson pair-quasiparticle interference in superconducting interferometers," *Phys. Rev. B* **12**, 1707 (1975).

[13]S. N. Erné, H.-D. Hahlbohm and H. Lübbig, "Theory of rf-biased superconducting quantum interference device for nonhysteretic regime," *J. Appl. Phys.* **47**, 5440 (1976).

[14]J. Halbritter, *Appl. Phys. A*. **43**, 1 (1987).

[15]J. E. Drumheller, Z. Trybula and J. Stankowski, "Flux nucleation in Josephson junctions formed by touching lead pieces," *Phys. Rev. B* **41**, 4743 (1990).

[16]Z. Trybula, J. E. Drumheller and J. Stankowski, "Microwave absorption in superconductor loops formed by touching lead pieces,"*J. Appl. Phys.* **67**, 5041 (1990).

15.1.3. Temperature dependence

Blazey[17] has measured the temperature dependence of the cross-section $S = \Phi_0/\Delta H$ as shown in Fig. 2 (a). Figure 2 (b) shows S plotted against $[2(1 - t)]^{-1/2}$ and also against $(1 - t^4)^{-1/2}$ with $t = T/T_c$. The first form is expected from BCS theory near T_c while the second form follows from the two-fluid model.[18]

Surprisingly, the BCS result is fitted over the entire temperature range while the empirical two-fluid expresion fits only near T_c. Assuming a ring of height equal to the crystal thickness of 50 μm leads to a width of the ring w = 4.5 μm and a London penetration depth $\lambda_L(0) = 0.42$ μm which is about three times the expected $\lambda_{ab} = 0.14$ μm.

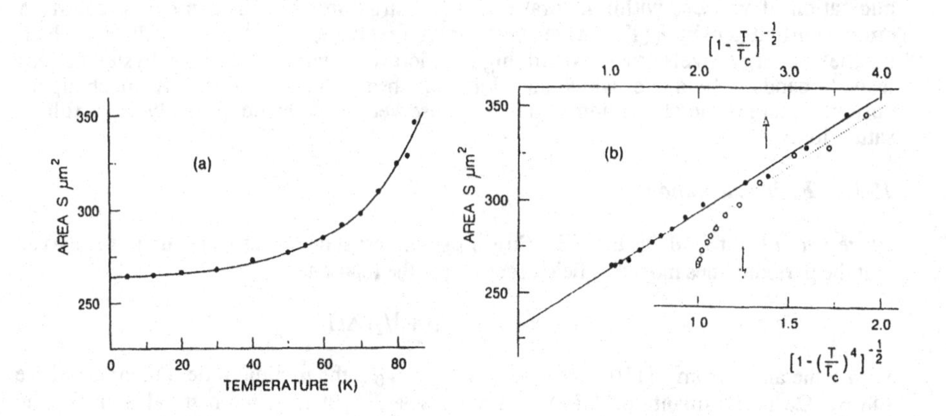

Figure 2. (a) Temperature variation of the Josephson junction cross-section $S = \Phi_0/\Delta H_0$ responsible for the absorption spectrum of Fig. 1. (b) The same temperature dependence assuming two different variations of the London penetration depth contribution to the cross section S. From K. W. Blazey [17].

15.2. rf SQUID Behavior

15.2.1. Absorption threshold

Most but not all periodic series observed in $YBa_2Cu_3O_{7-\delta}$ appear above a temperature-dependent threshold in microwave power. Some periodic series appear above a critical dc magnetic field without any microwave-power threshold. As the microwave power is increased, the series extend to lower fields.

New series of absorption lines appear at successive thresholds in microwave power and temperature. The lines also broaden into absorption bands with increasing microwave power as shown in Fig. 3 (a) where a plot of the band-width $2\delta H$ as a function of the square root of the threshold power of a weak junction is shown for several temperatures. The observed dependence of δH on microwave power can be fit by the expression

$$2\delta H/\Delta H = (P/P_0)^{1/2} - (P_c/P_0)^{1/2} \tag{2}$$

---

[17] K. W. Blazey, "Low field microwave absorption of granular superconductors," *Physica Scripta* **T29**, 92 (1989).
[18] M. Tinkham, *Introduction to Superconductivity* (McGraw-Hill, New York 1975).

where $P_c(T)$ is the threshold microwave power and $P_0(T)$ characterizes the temperature-independent rate at which $\delta H/\Delta H$ increases above threshold. The response of a junction in which vortices nucleate at even lower power is shown in Fig. 3 (b). A limitation of such studies is that with a multiplicity of domain-boundaries in the crystal, nucleation does not occur in a single junction over the whole superconducting temperature-range but switches from one junction to another as the temperature is raised.

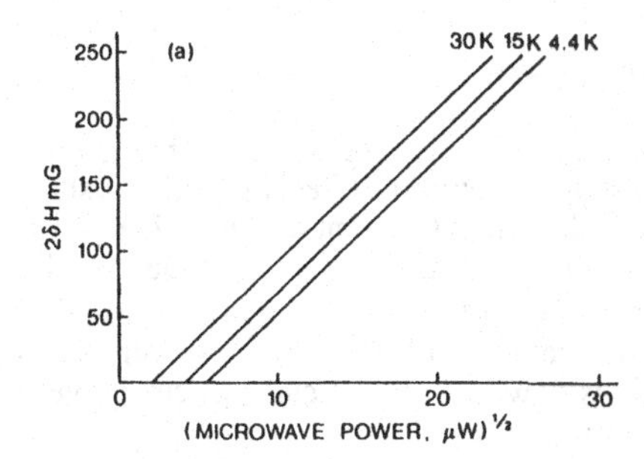

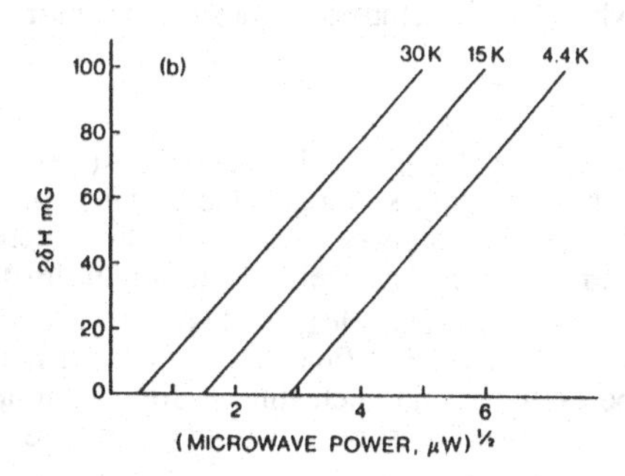

Figure 3. (a) Plot of the microwave-induced absorption band-width $2\delta H$ of a weak domain-boundary in single-crystal $YBa_2Cu_3O_{7-\delta}$ as a function of the square-root of the microwave power for several temperatures. The straight lines are Eq. (4) in the text with P_c and P_0 fitted to the data. (b) Plot of the microwave-induced absorption band-width $2\delta H$ of a weaker domain boundary in the same single crystal of $YBa_2Cu_3O_{7-\delta}$. The powers P_c and P_0 are each reduced by about a factor of 4. From K. W. Blazey *et al.* [3].

A similar plot for an irregular niobium particle is shown in Fig. 4 (a) at several temperatures. A log-log plot of P_c determined from Fig. 4 (a) as a function of $(1-T/T_c)$ is shown in Fig. 4 (b). As there appeared to be relatively few junctions in the niobium sample, it is certain that the same junction was followed in temperature. The temperature dependence shown in Fig. 2 (b) of the threshold power P_c is compatible with an approach to zero as $(1-T/T_c)^{3/2}$.

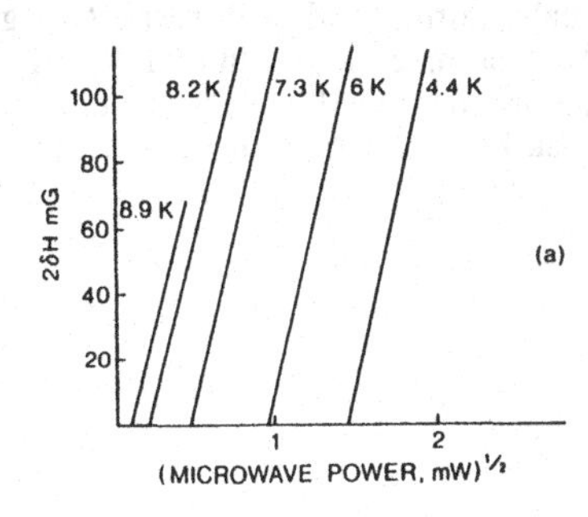

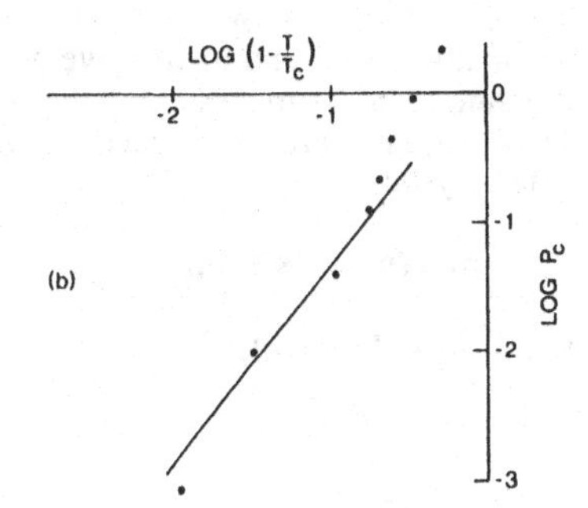

Figure 4. (a) Plot of the microwave-induced band-widening $2\delta H$ of an irregular niobium particle as a function of the square-root of the microwave power at several temperatures. The straight lines are Eq. (4) in the text with P_c and P_0 fitted to the data. (b) Log-log plot of P_c as a function of $(1-T/T_c)$. The critical power P_c is compatible with $(1-T/T_c)^{3/2}$. From K. W. Blazey *et al.* [3].

This observed behavior suggests that the power threshold is associated with the nucleation of vortices by critical currents that screen the microwave magnetic field from the interior of the superconductor. The Ginzburg-Landau critical current[19] takes the form

$$J_c(T) \approx (2/3)^{3/2} [H_c(T)/\lambda_L(T)] \tag{3}$$

where $\lambda_L(T)$ is the London penetration depth and $H_c(T)$ is the thermodynamic critical field, which has the temperature dependence near T_c

$$H_c(T) \approx 1.73\, H_c(0)(1 - T/T_c) \tag{4}$$

The critical current density of a superconducting bridge is the same as for a bulk superconductor so long as the bridge is not too short. Proximity effects stabilize a short bridge and suppress vortex nucleation, increasing the critical current density. The lower and upper critical fields of niobium are $H_{c1} \approx 2000$ Oe and $H_{c2} \approx 3000$ Oe and the thermodynamic critical field is $H_c = (\sqrt{2}\,\kappa/\ln \kappa)\, H_{c1} \approx 11{,}000$ Oe. Since the observed critical microwave fields are in the mOe range, nucleation in bulk or in bridges appears to be excluded and nucleation in Josephson junctions,[20] where critical currents can be very much lower, must be the origin of the observed absorption.

15.2.2. Subthreshold dispersion

Yau *et al.*[21] have studied the dispersion signal associated with the observed periodic absorption lines and confirm that natural rf SQUIDs structures are the origin of the observed lines.

It was found that when the microwave power is reduced below threshold, the dispersion signals do not go to zero as do the absorption signals . The magnitude of the flux changes below threshold are the double integral of $d\chi'/dH$ with H over a single line times the area S of the SQUID loop. The signals are observed to be proportional to the square-root of the incident microwave power, indicating that the mechanism causing these signals is independent of H_{rf} as expected in the inductive mode of an rf SQUID.[10,11] This observation is consistent with the rf-SQUID model and indicates that transistions between adjacent vortex states are the origin of the below-threshold signals. From the absolute strength of the below-threshold dispersion signals, normalized with respect to gain, modulation field, and microwave power, Yau *et al.* estimate the length of a SQUID-like structure to be 3 μm, assuming that H_{rf} fully penetrates the SQUID. This value indicates that flux quanta are constrained by rather short weak-links on the top and bottom surfaces of the crystal.

15.3. Active Weak Links

15.3.1. Field dependence

---

[19] R. P. Huebener, *Magnetic flux structures in superconductors*, Volume 6 in Springer Series in Solid-State Sciences, ed. Peter Fulde (Springer-Verlag, Berlin, 1979) sec. 3.3.

[20] J. R. Clem, "Granular and superconducting-glass properties of the high-temperature superconductors," *Physica C* **153-155**, 50 (1988).

[21] W. Yau, A. M. Portis, E. R. Weber, Z. Z. Wang and N. P. Ong, *Physica C* **162-164**, 1047 (1989).

All long periodic series exhibit a threshold $P_c(H)$ that varies with increasing field, going nearly to zero at one or more values of the applied field H. For these series, the field regions over which they are observed broadens with increasing microwave power. The variation in $P_c(H)$ with applied field is shown in Fig. 5 (a) for a Nb junction and in Fig. 5 (b) for $YBa_2Cu_3O_{7-\delta}$. The origin of this behavior is that flux is nucleated in a junction that extends only a limited distance δ. The critical microwave current density is then

$$J_c = \frac{1}{\delta} J_0 \left| \int_0^{\delta} dx \cos\frac{2\pi x}{\lambda_B} \right| = \frac{\lambda_B}{2\pi\delta} J_0 \left| \sin\frac{2\pi\delta}{\lambda_B} \right| \tag{5}$$

where $\lambda_B = \Phi_0/B(2\lambda_L + b)$ is the period of the Josephson current in a junction of physical thickness b.

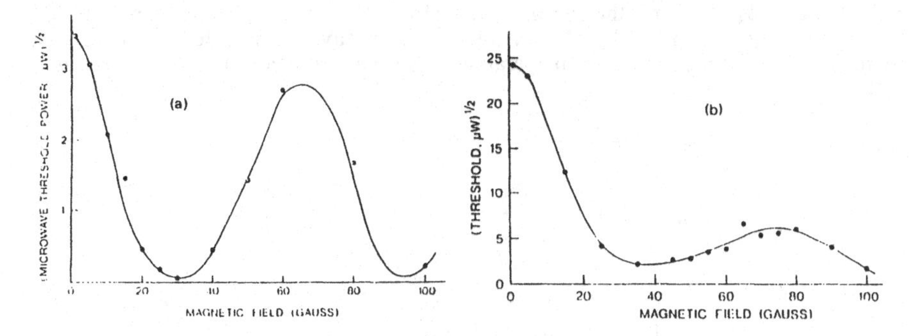

Figure 5. Variation of the square-root of the microwave threshold power $P_c(H)$ in natural rf SQUIDs at 4.3 K. The threshold power is seen to go through minima that are characteristic of the nucleating junction. (a) Niobium. From K. W. Blazey *et al.* [3]. (b) $YBa_2Cu_3O_{7-\delta}$. From K. W. Blazey [12].

For 2δ an integral multiple n of λ_B or $B = n\Phi_0/2\delta(2\lambda_L + b)$ the critical current density J_c goes to zero and the microweave threshold power $P_c(H)$ vanishes. As the flux within a niobium particle of radius a equals the total intercepted flux, $2Ba(2\lambda_L + b) = HS$, the threshold vanishes at fields $H = n(a/\delta)\Delta H$. Taking $a \approx 50$ μm, the data shown in Fig. 5 (a) for a series with $\Delta H = 95$ mOe gives for the length of the active junction $2\Delta\delta \approx 0.16$ μm.

Similar behavior has been observed in a $YBa_2Cu_3O_{7-\delta}$ junction as indicated in Fig. 1 where the line series does not extend down to zero field at the microwave power level of the measurement. The threshold power plotted in Fig. 5 (b) leads to a period of about 80 Oe with $\Delta H = 0.08$ Oe. This indicates that the active length of the junction is about 10^{-3} of the thickness t of the crystal.

15.3.2. Transient fields

Kessler *et al.*[22] have observed the dependence on field sweep of the phase of the field-modulated periodic absorption lines from natural rf SQUIDS in single crystals. The

[22] C. Kessler, B. Nebendahl, A. Dulcic, Th. Wolf and M. Mehring, *Physica C* **192**, 79 (1992).

authors conclude from the observed phase reversal that microwave absorption must occur at the crystal surface. Vortices are driven into or out of the sample depending on the sign of the surface current which in turn depends on the flux profile at the surface.

15.3.3. Flux creep

Hoffmeister *et al.*[23] have observed flux creep from SQUID structures is single crystals of $YBa_2Cu_3O_{7-x}$ by monitoring pulses in microwave absorption as flux quanta pass through the active junction. The decay flux is found to decay logarithmically

$$\frac{1}{\Phi}\frac{d\Phi}{dt} \approx -S\ln t \tag{6}$$

with $d\Phi/dt = r\,\Phi_0$ where r is the vortex counting rate, which is seen from Fig. 6 to follow Eq. (6) over at least four decades in time. The decay rate r is found to increase approximately with the cube of the preparation field H and with temperature approximately as $T^{2.7}$.

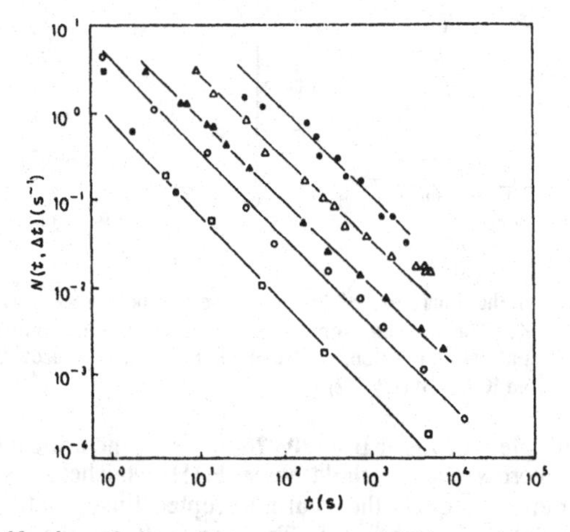

Figure 6. Number of absorption peaks r per time interval measured at various temperatures with a preparation field H = 20 Oe. Open and closed squares: T = 5 K and S = -1.05 ± 0.1. Open circles: T = 10 K and S = -1.02 ± 0.1. Closed triangles: T = 20 K and S = -0.97 ±0.1. Open triangles: T = 35 K and S = -0.96 ±0.1. Closed circles T = 40 K and S = -1.00 ±0.2 [26].

15.4. Weakly Connected Rings

15.4.1. Ground state energy

Vichery *et al.*[4,5] associate the perodic absorption observed in single crystals of $YBa_2Cu_3O_{7-\delta}$ with the existence of superconducting loops containing weak links. They show further that the power absorption is well described by the model of Silver and

---

[23] D. Hoffmeister, O. Dobbert, K.-P. Dinse, W. Goldacker and T. Wolfe, "Flux creep in single crystals of the high-temperature superconductor $YBa_2Cu_3O_{7-x}$," *Europhys. Lett.* **8**, 369 (1989).

Zimmerman,[24] which they generalize to N small identical junctions in series with a ring of surface area S. The energy of this ring in a magnetic field is the sum of the magnetic energy stored in the ring and the coupling energy of the junctions is, closely following Vichery *et al.*[4,5]

$$E = \tfrac{1}{2}LI^2 - NL_0I_0^2\cos\delta + K \qquad (7)$$

where K is a constant and δ is the phase difference of the order parameter across a Josephson junction. The current I and the phase difference δ are taken to be related by the Josephson equation

$$I = I_0 \sin \delta$$

The relationship between I, δ and the applied external magnetic flus $\Phi_{ext} = BS$ results from two equations. The first equation is

$$\Phi = \Phi_{ext} + LI \qquad (9)$$

where Φ is the magnetic flux through the loop and L is the inductance of the ring. As functions of the phase φ of the order parameter must be single-valued, the integral around the ring must lead to

$$\int ds \text{ grad } \phi = 2\pi p \qquad (10)$$

with p an integer. The Ginzburg-Landau equation leads to

$$N\delta = 2\pi p - 2\pi\, \Phi/\Phi_0 \qquad (11)$$

and Eq. (8) becomes

$$I = I_0 \sin\frac{2\pi}{N}\left(p - \frac{\Phi}{\Phi_0}\right) \qquad (12)$$

From Eqs. (7) and (11) the energy E_p of the p'th state is

$$E_p(\Phi) = \tfrac{1}{2}LI_0^2\sin^2\frac{2\pi}{N}\left(p - \frac{\Phi}{\Phi_0}\right) - NL_0I_0^2\cos\frac{2\pi}{N}\left(p - \frac{\Phi}{\Phi_0}\right) + K \qquad (13)$$

The energies of the p, p + 1 and p + 2 states are represented in Fig. 7. To obtain such a diagram, the condition $\Phi_0/s < {}^1/_4N\Phi_0 + LI_0 < \Phi_0$ must be established. The reasoning is the same for ${}^1/_4N\Phi_0 + LI_0 > \Phi_0$. Two points are important:

(i) For $\Phi_{ext} = p\Phi_0$ the current I is zero. Since there is no magnetic energy and the Josephson coupling is maximum, the energy $E_p(p\Phi_0)$ must be a minimum.
(ii) When Φ_{ext} deviates from the value $p\Phi_0$, $E_p(\Phi)$ increases and the absolute value of I increases to I_0. The current I reaches $\pm I_0$ for $\Phi_{ext} = p\Phi_0 \pm ({}^1/_4 N\Phi_0 + LI_0)$. The zero of energy is defined for $I = \pm I_0$.

---

[24] A. H. Silver and J. E. Zimmerman, "Quantum states and transitions in weakly connected superconducting rings," *Phys. Rev.* **157**, 317 (1967).

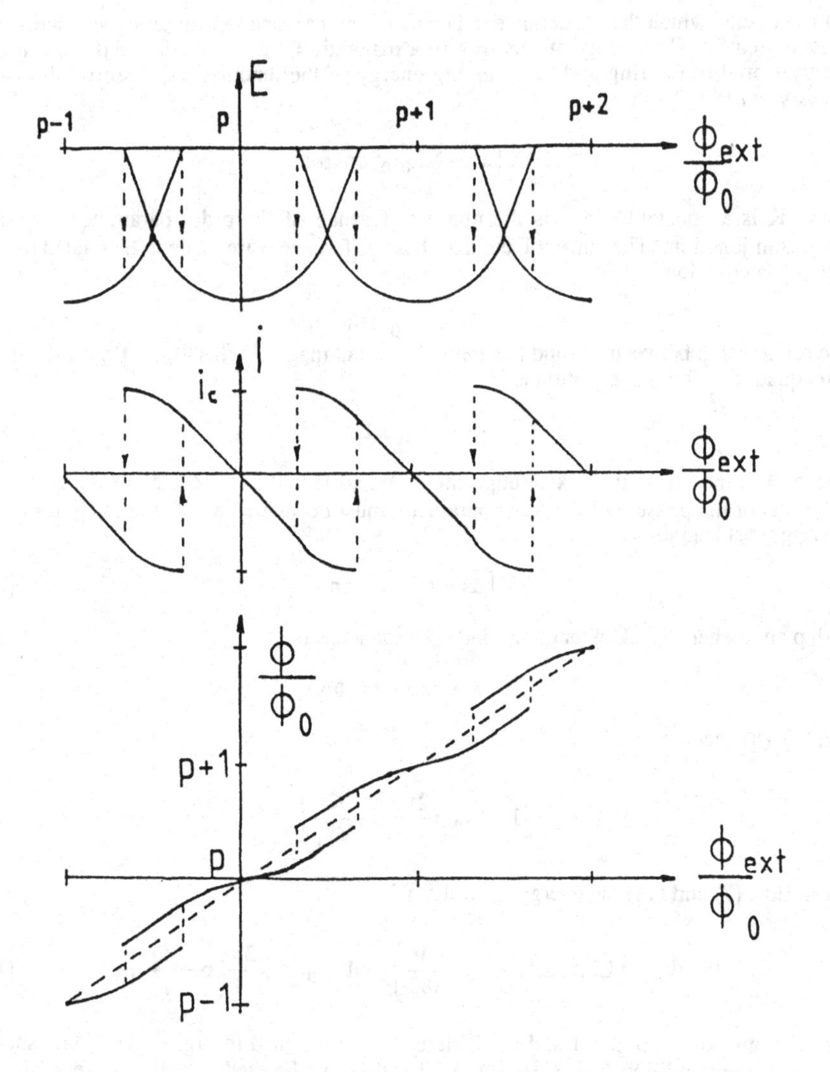

Figure 7. Energy E, circulating current I and magnetic flux Φ as functions of the external flux Φ_{ext}. The different curves have been schematically drawn for arbitrary values of L, I_0 and N. From H. Vichery *et al.* [3].

From Eqs. (9) and (11) it is easy to show that $I = I(\Phi_{ext})$ and $\Phi = \Phi(\Phi_{ext})$ are monotonic functions, which are plotted schematically in Fig. 7. It is assumed that either N or LI_0 is large enough to permit overlap of $E_p(\Phi)$ and $E_{p+1}(\Phi)$

$$\tfrac{1}{4}N\,\Phi_0 + LI_0 \geq \tfrac{1}{2}\,\Phi_0 \tag{14}$$

as shown in Fig. 7. For $LI_0 > \Phi_0$, Eq. (13) is equivalent to $N \geq 2$ and this condition is readily achieved in practice.

The key to the Silver-Zimmerman model[27] is the assumption that the transitions between p-states occur when I equals I_0. At this condition the coupling energy cancels and the system is equivalent to an open ring. The magnetic energy $\frac{1}{2} LI_0^2$ can be partly dissipated and the system condenses into an adjacent state as indicated by the dashed line in Fig. 7.

15.4.2. Oscillating magnetic field

In the discussion of the effect of an oscillating magnetic field, Vichery *et al.*[4,5] follow the work of Silver and Zimmerman.[27] The variation of the current I with magnetic field is continuous except for I = Ic where p changes by one unit, corresponding to the entry into the loop or exit from the loop of one flux quantum. This jump in magnetic field leads to a sharp peak in the emf induced in the loop. A magnetic field, which is the sum of two terms $H = H_0 + H_{rf} \sin \omega t$ where H_0 varies slowly in time. The oscillating field H_{rf} is taken parallel to the the static field H_0 in contrast with the usual geometry in an ESR spectrometer. (In actual experiment the only requirement is that rf currents flow through the junctions.) This problem had been carefully examined by Silver and Zimmerman[27] and Vichery *et al.*[4,5] present only qualitative arguments to explain the physics.

The relevant parameter is the magnetic flux overlap Ω between states p and p + 1

$$\Omega = 2LI_0 + \tfrac{1}{2} N \Phi_0 - \Phi_0 \tag{15}$$

Two regimes may be distinguished:

(i) Below absorption threshold —$2H_{rf} < \Omega/S$. In this case there is only one integer p with $I < I_c$ over the entire microwave period. As a result there is no sharp peak in the variation of the emf with time. If H_0 is increased slowly there is a single sharp peak.

(ii) Above threshold—$2H_{rf} > \Omega/S$. There are values of H_0 for which it is not possible to obtain a single value of p at which the inequality $I < I_x$ is satisfied over the microwave period. For these field values p oscillates between two values with the periodicity ν of the microwave magnetic field and the microwave loss is hysteretic. The width in applied field of the hysteretic loss is $2\delta H = 2H_{rf} - \Omega/S$ which is equivalent to Eq. (2).

15.4.3. Microwave power absorption

As Vichery *et al.*[4,5] point out, an important feature of the Silver-Zimmerman model[27] is that the time-constant characterizing the relaxation of the system toward equilibrium is very short, typically 10^{-12} sec. This means that even at X-band frequencies, no transient phenomena are to be expected. Below threshold there is no hysteretic loss and no microwave power is absorbed. Above threshold, the two jumps in p over an rf cycle lead to hysteretic absorption at a rate

$$\Omega = 2LI_0 + \tfrac{1}{2}N\Phi_0 - \Phi_0 \tag{16}$$

The interesting point is that, as observed, P is independent of H_0 so long as the threshold condition is satisfied and there are only two absorption rates, 0 and P.

15.5. Josephson Systems

15.5.1. Quantum interference devices

Clarke[25,26] has reviewed the use of dc and rf SQUIDS in the detection of extremely small magnetic fields. The dc SQUID consists of two Josephson junctions in a superconducting loop of inductance L as shown in Fig. 8 (a). The junctions are shunted by resistance R to eliminate hysteresis, which requires for the McCumber-Stewart parameter[27,28]

$$\beta_c = 2\pi(I_0R)/[\Phi_0/RC_j] < 1 \qquad (17)$$

where I_0 is the junction critical current and C_j is the junction capacitance. The change in flux within the loop is developed by a change in input current flowing through an inductor L_{in} that is coupled to the SQUID loop with mutual inductance M and coupling coefficient k $= M/\sqrt{L_{in}L}$.

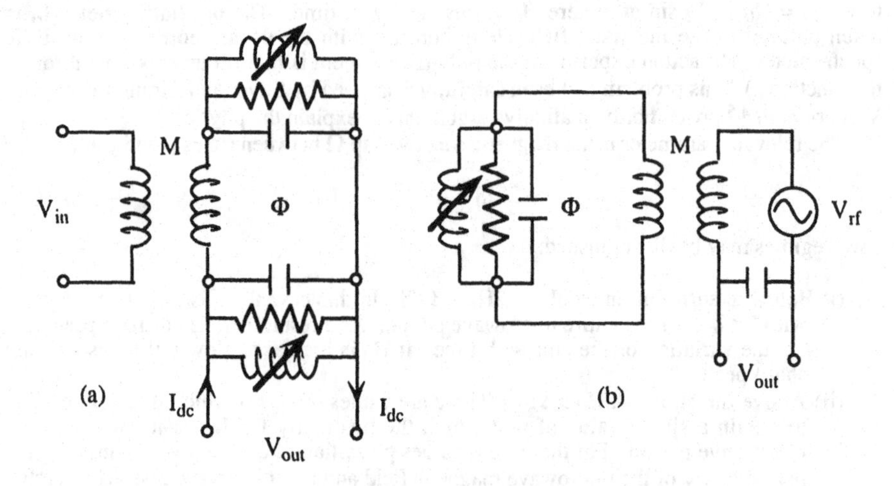

Figure 8. (a) dc SQUID. A current dc I is injected across the loop and the dc voltage V is measured as a function of the flux Φ that penetrates the loop. (b) rf SQUID. An rf current flows into an input tank circuit and the rf voltage across the tank is measured as a function of the dc flux Φ in the loop.

The SQUID may be regarded as a flux to voltage transducer with a transfer function

$$V_\Phi = dV/d\Phi \qquad (18)$$

which for optimum design is of order R/L. The dc SQUID is normally operated in a flux-locked loop where any change in applied flux is compensated to hold the flux at a quarter-integral number of quanta.

[25] J. Clarke, *IEEE Trans. Electron Devices* **27**, 1896 (1980).
[26] J. Clarke, *Physics Today*, March 1986, pp. 36-44.
[27] D. E. McCumber, *J. Appl. Phys.* **39**, 3113 (1968).
[28] W. C. Stewart, *Appl. Phys. Lett.* **12**, 277 (1968).

The rf SQUID consists of a single Josephson junction in a superconducting loop of inductance L as shown in Fig. 8 (b). The loop inductance L is usually adjusted for

$$\beta_L = 2\pi I_0 L/\Phi_0 \tag{19}$$

of the order of one, where $\Phi_0/2\pi I_0$ is the junction inductance. The voltage V_{rf} across the inductor is detected and fed back to flux-lock the SQUID at a half-integral number of flux quanta. The transfer function of the rf SQUID is

$$V_\Phi = dV_{rf}/d\Phi = \omega L_{Tf}/M \tag{20}$$

dc SQUID magnetometer

Miklich *et al.*[29] have constructed a $YBa_2Cu_3O_{7-x}$ thin-film magnetometer consisting of a dc SQUID with bi-epitaxial Josephson junctions fabricated on one chip and a flux transformer with multiturn input coil fabricated on a second chip. The magnetometer operates in a flux-locked loop with the flux transformer increasing the magnetic field sensitivity by around 80.

With the magnetometer immersed in liquid N_2, the low-frequency magnetic field noise is 0.6 $pT/\sqrt{Hz}$ at 10 Hz and 0.09 $pT/\sqrt{Hz}$ at 1 kHz, varying approximately as $1/\sqrt{\Delta f}$ over this interval. The magnetometer thus has a magnetic field sensitivity $\Delta B/\sqrt{\Delta f}$ down to 100 Hz of 0.2 $pT/\sqrt{Hz}$. Below 100 Hz, 1/f noise becomes dominant and the sensitivity drops as $1/\sqrt{f}$ with further reduction in frequency.

rf SQUID magnetometer

Groups at TRW[30] and at KFA Jülich[31,32] have fabricated step-edge junctions from $YBa_2Cu_3O_7$ and have constructed rf SQUIDS. By increasing the drive frequency from 20 MHz to about 150 MHz, Zhang *et al.*[33] have obtained a reduction in equivalent flux noise $\Delta\Phi/\sqrt{\Delta f}$ by a factor between 2 and 4 to 8-9 $\times 10^{-5}\,\Phi_0/Hz^{-1/2}$. This noise level leads to a field sensitivity above a frequency f =0.3 Hz of $\Delta B/\sqrt{\Delta f}$ = 0.9 $pT\cdot Hz^{-1/2}$ where Δf is the bandpass. For frequencies below 0.3 Hz the 1/f noise dominates and the sensitivity drops.

By going to drive frequencies in the GHz range, it should in principle be possible to further increase the sensitivity of an rf SQUID by as much as the square-root of frequency.[34,35] Daly *et al.*[36] have succeeded in placing an rf SQUID within a 10 GHz TE_{011} cylindrical cavity. With sufficient microwave power it has been possible to drive ten

---

[29] A. H. Miklich, J. J. Kingston, F. C. Wellstood, J. Clarke, M. S. Colclough, K. Char and G. Zaharchuk, *Appl. Phys. Lett.* **59**, 988 (1991); *Nature* **352**, 483 (1991).

[30] K. P. Daly, W. D. Dozier, J. F. Burch, S. B. Coons, R. Hu, C. E. Platt and R. W. Simon, *Appl. Phys. Lett.* **58**, 543 (1991).

[31] G. Cui, Y. Zhang, K. Hermann, Ch. Buchal, J. Schubert, W. Zander, A. I. Braginski and C. Heiden, *Supercond. Sci. Technol.* **4**, S130 (1991).

[32] K. Hermann, Y. Zhang, H.-M. Mück, J. Schubert, W. Zander and A. I. Braginski, *Supercond. Sci. Technol.* **4**, 583 (1991).

[33] Y. Zhang, H.-M. Mück, K. Herrmann, J. Schubert, W. Zander, A. I. Braginski and C. Heiden, *Appl. Phys. Lett.* **60**, 645 (1992).

[34] L. D. Jackel and R. A. Buhrman, *J. Low Temp. Phys.* **19**, 201 (1975).

[35] R. A. Buhrman and L. D. Jackel, *IEEE Trans. Magn.* **13**, 879 (1977).

[36] K. P. Daly, J. Burch, S. Coons and R. Hu, *IEEE Trans. Magn.* **27**, 3066 (1991).

quanta in or out of the loop per half-period, a switching frequency of 200 GHz and a switching time that must be faster than 5 ps.

Zhang *et al.*[37] have placed a double step-edge junction at the current-maximum of a 3 GHz microstrip meanderline, patterned from an epitaxial YBCO film. The sensitivity of this device was found to be little better than for the rf SQUID operating at 150 MHz. The transfer function was worse by a factor of 2, the energy resolution was better by a factor of 2 and the flux noise was comparable. The problem is evidently the coupling coefficient k between the SQUID and the meanderline. One would like k^2Q of the order of unity while it was found to be only 10^{-3}. A possible problem may be that the hysteretic loss from driving the SQUID is the dominant loss of the resonator. Under such circumstances the Q no longer increases with frequency.

Digital SQUIDs

Rylov[38] has proposed that high-performance counter-type analog–digital converters utilizing RSFQ logic (see Sec. 15.5.3) can be designed so as to preserve the high sensitivity of the analog versions without their limited slew rate and dynamic range.

15.5.2. Josephson switching

A Josephson switching device utilizes a Josephson junction connected to a voltage V_0 through a resistor R as shown in Fig. 9 (a). As discussed in Sec. 10.2, an underdamped junction exhibits the I-V curve shown in Fig. 9 (b). The device is biased by a dc current $I_b = V_0/R$ slightly below the Josephson critical current I_0 with the junction initially in state-s. A current I_{in} drives the junction above I_0, inducing switching to state-n, connected to state-s by the load-line shown in Fig. 9 (b). This process requires only a few ps.

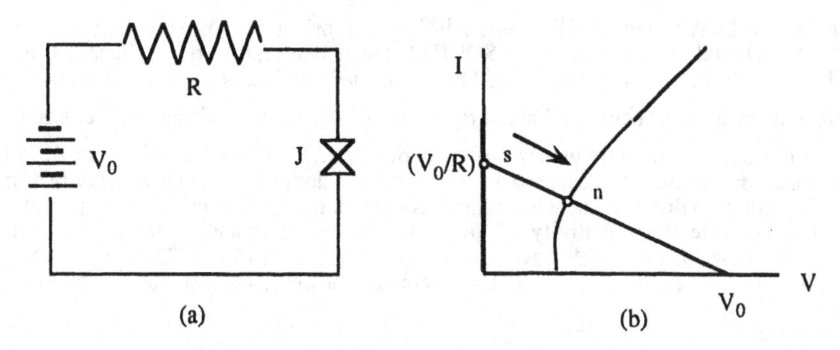

Figure 9. (a) A Josephson switching device utilizes a Josephson tunnel junction connected to a voltage V_0 through a resistor R. (b) An underdamped junction exhibits a hysteretic I-V curve. A current pulse drives the junction above I_0, inducing switching from state-s to state-n along the load-line The device may returned to state-s only by reducing the current to zero.

To reset the junction it is necessary to reduce the bias current I_b to zero and the junction is said to be *latched* in state-1. *Unlatching* the junction is accomplished in practice by

[37]Y. Zhang, H.-M. Mück, M. Bode, K. Herrmann, J. Schubert, W. Zander, A. I. Braginski and C. Heiden, *Appl. Phys. Lett.* **60**, 2303 (1992).

[38]S. V. Rylov, *IEEE Trans. Magn.* **27**, (1991). *1990 Applied Superconductivity Conference* Snowmass, CO.

utilizing an rf voltage source rather that the dc source shown in Fig. 9 (a).[39] If the rf period is much shorter than the RC-time of the junction, the device remains in its resistive state but with reversed current and voltage. Such a process, called *punch-through*, limits the frequency of the rf source and the clock speed of systems that use such devices to periods longer than the natural switching time of the junction.

15.5.3. Rapid single flux quanta

The recognition that a 2π-phase slip can be localized between two of a system of coupled Josephson junctions has led to what is called rapid-single-flux-quantum (RSFQ) logic but should more properly be called soliton logic.[40] Elementary cells have been proposed that can generate, pass, memorize and reproduce such 2π-phase slips and their associated voltage pulses of area

$$\int V dt = \frac{\hbar}{q} \int d\delta = \frac{h}{q} = \Phi_0 \tag{21}$$

Perhaps the easiest way to understand the physics of these devices is to return to the discussion of Sec. 10.4.1, which was concerned with soliton propagation on a Josephson transmission line, and to Sec. 10.4.2, which briefly reviewed soliton resonators, which are segments of a Josephson transmission line. By going from the continuous line, in which a soliton moves ballistically, to a discrete line of suitably coupled junctions, we may understand the essential physics of soliton propagation on lumped transmission lines.

Shown in Fig. 10 is a line of Josephson junctions J, coupled by inductors L. The junctions are suitably biased as well as overdamped.

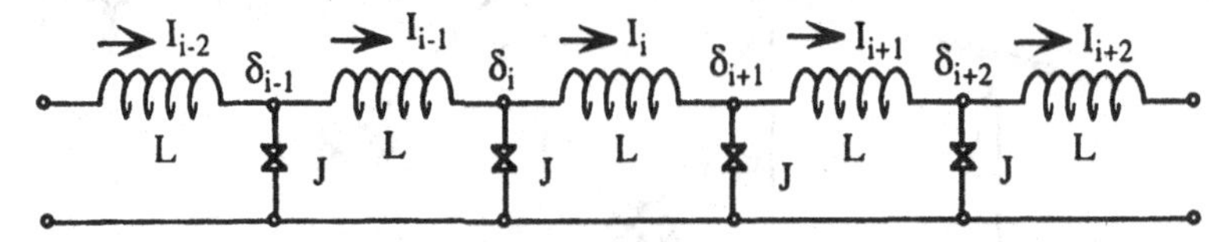

Figure 10. A line of Josephson junctions J, coupled by inductors L. The junctions are suitably biased as well as overdamped.

The equations relating voltage and current on the line are

$$I_i = I_{i-1} - C\, dV_i/dt - J_0 \sin \delta_i \tag{22}$$

$$V_i - V_{i-1} = L\, dI_{i-1}/dt \tag{23}$$

We also have the Josephson voltage-phase relation

$$(\Phi_0/2\pi)\, \partial\delta_i/\partial t = V_i \tag{24}$$

Eliminating the voltage and current gives the difference–differential equation for the phase differences across the junctions

[39]H. Hayakawa, *Physics Today* (March 1986) pp 46-52.

[40]K. K. Likharev and V. K. Semenov, *IEEE Trans. Appl. Superconduct.* 1, 3 (1991).

$$LC\, d^2\delta_i/dt^2 + (\delta_{i+1} + \delta_{i-1} - 2\delta_i) = (L/L_0) \sin \delta_i \tag{25}$$

with $L_0 = \Phi_0/2\pi I_0$. We need to find the magnitude of L such that the propagating phase-slips, which we presume to be solitons, may be localized between a single pair of Josephson junctions. That is, if the junctions are separated by a distance a, we require $\lambda \approx$ a. Taking Eq. (25) to the continuous limit with δ small leads to $v = 1/\sqrt{L'C'}$ as expected and $\lambda = a\sqrt{L_0/L}$. Thus for $\lambda \approx a$ we require $L \approx L_0$. Under these conditions we can expect soliton propagation with δ changing by 2π between adjacent junctions. Just as for solitons on a long Josephson junction, a phase change of 2π is associated with the passage of an integrated voltage pulse Φ_0. If the soliton is strongly localized, the actual amount of flux is substantially smaller than the quantum of flux Φ_0.

15.5.4. Torsional model

Scott[41] has developed the mechanical model sketched in Fig. 11 that illustrates the solutions of the sine-Gordon equation. The model torsionally connects pendula of mass m and length a to neighboring pendula. If θ_i is the angular displacement of the i'th pendulum, the kinetic energy of the system is

$$T = \sum_i \tfrac{1}{2} m a^2 \dot{\theta}^2 \tag{26}$$

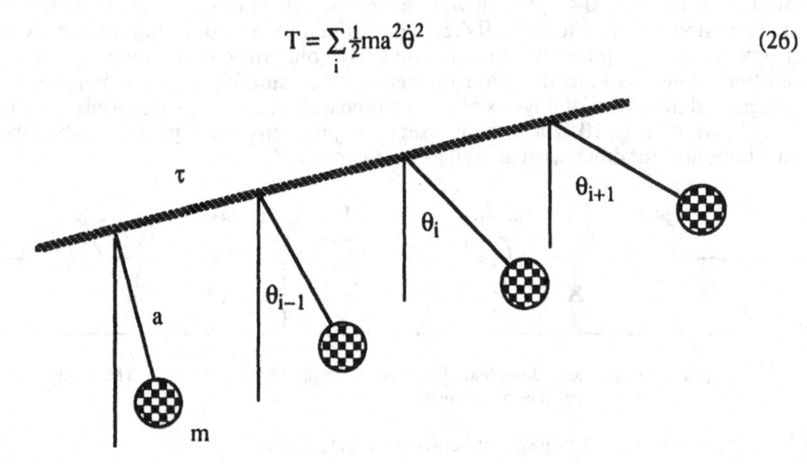

Figure 11. Mechanical model that illustrates the solutions of the sine-Gordon equation. The model torsionally connects pendula of mass m and length a to neighboring pendula.

The potential energy of the system is the sum of gravitational and torsional terms

$$U = \sum_i mga(1-\cos\theta) + \sum_i \tfrac{1}{2}\tau(\theta_{i+1}-\theta_i)^2 \tag{27}$$

with τ the torsional coupling constant. The equation of motion of the i'th pendulum is

$$ma^2\ddot{\theta}_i + \tau(\theta_{i+1}+\theta_{i-1}-2\theta_i) = mga\sin\theta_i \tag{28}$$

[41] A. C. Scott, *Am. J. Phys.* 37, 52 (1969).

which is the mechanical analog of Eq. (25).

15.5.5. Hamiltonian mechanics of Josephson systems

We end these notes with brief mention of the Hamiltonian mechanics of Josephson systems–just for fun as Richard Feynman liked to say. The kinetic energy of the system of coupled Josephson junctions that we have been considering is

$$T = \sum_i \tfrac{1}{2} C \left(\frac{\Phi_0}{2\pi} \dot{\delta}_i \right)^2 \tag{29}$$

and the potential energy is

$$U = \sum_i \frac{1}{2L} \left[\left(\frac{\Phi_0}{2\pi} \right) (\delta_{i+1} - \delta_i) \right]^2 + \sum_i \frac{\Phi_0 I_0}{2\pi} (1 - \cos\delta_i) \tag{30}$$

Lagrange's equation with the generalized coordinates $q_i = (\Phi_0/2\pi)\delta_i$ give Eq. (25). The conjugate momenta are found to be the charges on the junction capacities

$$p_i = \frac{\partial \mathcal{L}}{\partial \dot{q}_i} = C \frac{\Phi_0}{2\pi} \dot{\delta}_i = CV_i \tag{31}$$

and the Hamiltonian is

$$\mathcal{H} = \sum_i \frac{p_i^2}{2C} + U \tag{32}$$

It is interesting to note that whereas in the analysis of usual circuits, the coordinates are charge and the momenta are flux, in Josephson systems these conjugate quantities are interchanged. As we have seen, capacitance is mass and inductance is compliance.

From Sec. 10.1.4, the Josephson current may be obtained from

$$I = \frac{2\pi}{\Phi_0} \frac{\partial U}{\partial \delta} = \frac{\partial U}{\partial q} \tag{33}$$

and thus plays the role of a (negative) Newtonian force.

SUBJECT INDEX